Hans-Georg Elias

An Introduction to Plastics

Distribution:

VCH, P.O. Box 10 11 61, D-69469 Weinheim, Federal Republic of Germany

Switzerland: VCH, P.O. Box, CH-4020 Basel, Switzerland

United Kingdom and Ireland: VCH, 8 Wellington Court, Cambridge CB1 1HZ, United Kingdom

USA and Canada: VCH, 220 East 23rd Street, New York, NY 10010-4606, USA

Japan: VCH, Eikow Building, 10-9 Hongo 1-chome, Bunkyo-ku, Tokyo 113, Japan

ISBN 3-527-28578-4 (VCH Weinheim) ISBN 1-56081-784-4 (VCH, New York)

Hans-Georg Elias

An Introduction to Plastics

Weinheim · New York · Basel · Cambridge · Tokyo

Prof. Dr. Hans-Georg Elias
Michigan Molecular Institute
1910 West St. Andrews Road
Midland, MI 48640-2696
USA

1st edition 1993

Published jointly by
VCH Verlagsgesellschaft mbH, Weinheim (Federal Republic of Germany)
VCH Publishers, Inc., New York, NY (USA)

Editorial Directors: Philomena Ryan-Bugler, Louise Elsam
Production Manager: Peter J. Biel
Library of Congress Card No.: 93-4712

A CIP catalogue record for this book is available from the British Library

Die Deutsche Bibliothek – CIP-Einheitsaufnahme

Elias, Hans-Georg:
An introduction to plastics / Hans-Georg Elias. – 1. ed. –
Weinheim; New York; Basel; Cambridge; Tokyo: VCH, 1993
ISBN 3-527-28578-4 (Weinheim ...)
ISBN 1-56081-784-4 (New York)

Printed on acid-free and low chlorine paper

Printing: Strauss Offsetdruck, D-69493 Hirschberg
Bookbinding: Industrie- und Verlagsbuchbinderei Heppenheim GmbH, D-64646 Heppenheim

Printed in the Federal Republic of Germany

Preface

Plastics are ubiquitous in daily life. Major industries depend on them; neither the electronics industry nor the modern food packaging system can exist without plastics. Plastics are also used in building/construction, household goods, appliances, transportation, toys, furniture, and agriculture. Their rapid growth was due to inexpensive raw materials, easy processing, and a wide range of useful and often essential properties. It depended on serendipity and on intuition, scientific insights, and technical skills of entrepreneurs, scientists, and engineers. The scientific and technical lingo of this tribe of shakers, movers, and thinkers is bewildering; even tribal members may not understand fully the meaning of some technical terms as used by certain clans such as theoretical physicists, chemists, engineers, and practitioners.

This book thus aims to introduce, define, and clarify the most important scientific and technical terms in the field of plastics. It makes heavy use of recommendations by ISO (International Standardization Organization) and IUPAC (International Union of Pure and Applied Chemistry) but it also mentions conventional practices and the recommendations of national standardization bodies, especially ASTM (American Society for the Testing of Materials) and DIN (Das ist Norm; German standards). Physical properties are expressed in SI units (Système International d'Unités). Although "metric" systems such as the SI system are recognized by law in the United States since 1896, most U.S. practitioners still use the old Anglo-Saxon system of units; the Appendix (Chapter 15) thus provides these readers with conversion tables.

The book furthermore aims to provide an overview over all aspects of plastics: raw materials, manufacture, structure, processing, properties, and applications. Whenever possible within our present knowledge and the size of this book, phenomena, processes, and properties are explained by basic chemical and physical aspects of polymer science and engineering. It is hoped that problems of oversimplification of structure-property relationships are avoided by giving actual properties of polymer grades instead of "general" properties of a certain plastic; this procedure may indicate to the novice the wide range of property variations within one type of plastic. Whenever possible, property data were taken from the CAMPUS® data bases of various manufacturers. These data bases provide a "*C*omputer *A*ided *M*aterials *P*reselection by *U*niform *S*tandards": properties of plastics from various sources can be compared because test specimens and test procedures follow the same standards (mostly ISO).

The book is based on an article "Plastics. General Survey" in Vol. **A 20** (1992) 543-661 of the 5th edition of Ullmann's Encyclopedia of Industrial Chemistry but it is by no means identical with that essay. It has been considerably expanded from 119 to 330 pages, it now contains two new chapters on processing and types of plastics, and it was completely rewritten; no less than ca. 2/3 of the figures and most of the tables are new.

Midland, Summer 1993 Hans-G. Elias

List of Plastics Abbreviations and Acronyms

Conventional names of plastics are often abbreviated in the technical literature. The recommended abbreviations and acronyms vary: the same codes are sometimes used by ISO, IUPAC, ASTM, BS, DIN, etc., for different plastics; conversely, the same plastic may have different codes (see Chapter 15). For the convenience of the reader, an alphabetical list of abbreviations and acronyms, as used in this book, is provided.

ABS High-impact polymer from acrylonitrile, butadiene, and styrene
ACR Rubber from an acryl ester and a cross-linkable monomer
AES High-impact polymer from acrylonitrile, a rubber from (ethylene + propylene + nonconjugated diene), and styrene
ASA High-impact polymer from acrylonitrile, styrene, and an acrylate rubber

BR Butadiene rubber

CA Cellulose-2 1/2-acetate
CAB Cellulose acetobutyrate
CAP Cellulose acetopropionate
CF Cresol-formaldehyde resin
CMC Carboxy methyl cellulose
CN Cellulose nitrate
CP Cellulose propionate
CS Casein, cross-linked with formaldehyde
CTA Cellulose triacetate

EAA Copolymer from ethylene and acrylic acid
EC Ethyl cellulose
ECTFE Copolymer from ethylene and chlorotrifluoroethylene
EEA Copolymer from ethylene and acryl ester
EP Epoxy resin
EPDM Rubber from ethylene, propylene, and a nonconjugated diene
ETFE Copolymer from ethylene and tetrafluoroethylene
EVAC Copolymer from ethylene and vinyl acetate
EVOH Copolymer from the saponification of EVAC

FEP Copolymer from hexafluoropropylene and tetrafluoroethylene

HDPE High density poly(ethylene), also PE-HD

IIR Butyl rubber from isobutylene and (2-4) % isoprene

LDPE Low density poly(ethylene), also PE-LD

LLDPE Linear low density poly(ethylene)

MBS High-impact polymer from methyl methacrylate, butadiene, and styrene
MC Methyl cellulose
MF Melamine-formaldehyde resin

NBR Acrylonitrile-butadiene rubber
NR Natural rubber

PA Polyamide
PAI Polyamide-imide
PAN Poly(acrylonitrile) (PAN® is a registered trademark for fibers!)
PAR Polyarylate
PB Poly(butene-1)
PBI Poly(benzimidazole)
PBT Poly(butylene terephthalate)
PC Polycarbonate
PCTFE Poly(chlorotrifluoroethylene)
PDAP Poly(diallyl phthalate)
PDMS Poly(dimethylsiloxane)
PE Poly(ethylene)
PE-C Poly(ethylene), chlorinated
PE-HD Poly(ethylene), high density; also HDPE
PE-LD Poly(ethylene), low density; also LDPE
PE-LLD Poly(ethylene), linear low density; also LLDPE
PE-X Poly(ethylene), cross-linked
PEI Polyether-imide
PEOX Poly(ethylene oxide)
PESU Polyether sulfone (generic for all types)
PET Poly(ethylene terephthalate)
PF Phenol-formaldehyde resin
PFA Copolymer of tetrafluoroethylene and perfluoromethyl vinyl ether
PHB Poly(β-D-hydroxybutyrate)
PI Polyimide
PIB Poly(isobutylene)
PMMA Poly(methyl methacrylate)
PMP Poly(4-methylpentene-1)
POB Poly(1,4-oxybenzoate)
POM Poly(oxymethylene)
PP Poly(propylene)
PPE Poly(1,4-phenylene ether) = poly(oxy-1,4-phenylene)
PPS Poly(1,4-phenylene sulfide)
PS Poly(styrene)
PTFE Poly(tetrafluoroethylene)

PUR	Polyurethane
PVAC	Poly(vinyl acetate)
PVAL	Poly(vinyl alcohol)
PVC	Poly(vinyl chloride)
PVDC	Poly(vinylidene chloride)
PVF	Poly(vinyl fluoride)
PVK	Poly(N-vinyl carbazole)
SAN	Copolymer from styrene and acrylonitrile
SB	Thermoplastic from styrene and butadiene
SBR	Rubber from styrene and butadiene
UF	Urea-formaldehyde resin
UP	Unsaturated polyester

List of Symbols for Physical Quantities

A	Area
a	Persistence length
b	Bond length
C	Heat capacity
C_N	Characteristic ratio
c	Concentration (mass per volume)
c_p	Specific heat capacity at constant pressure
D	Diffusion coefficient
De	Deborah number
d	Diameter
d_f	Fractal dimension
E	Young's modulus
F	Force
f	Friction coefficient
f	Functionality
G	Gibbs energy
G	Shear modulus

Symbol	Quantity
H	Enthalpy
h	Planck constant
I	Intensity
K	General constant, equilibrium constant
K	Bulk modulus
k	Rate constant
k_B	Boltzmann constant
L	Length; L_K = Kuhn length
M	Molar mass; M_r = relative molecular mass ("molecular weight")
m	Mass
N	Number; N_A = Avogadro number
n	Amount of substance ("mole number")
n	Refractive index
P	Permeation coefficient
p	Pressure
R	Gas constant
R	Rate
R	Electric resistance
r	Copolymerization parameter
r	End-to-end distance
S	Entropy
S	Solubility coefficient
s	Radius of gyration
T	Temperature (both in kelvin and degree Celsius)
t	Time
V	Volume
v	Specific volume
w	Mass fraction (weight fraction)
X	Degree of polymerization
x	Amount-of-substance fraction ("mole fraction")
z	z-Average

α	Linear expansion coefficient, expansion factor
β	Cubic expansion coefficient
γ	Surface tension
γ	Shear strain; $\dot{\gamma}$ = shear rate
δ	Solubility parameter
ε	Strain, elongation; $\dot{\varepsilon}$ = extensional rate
η	Viscosity; η_r = relative viscosity; $[\eta]$ = intrinsic viscosity
Θ	Theta temperature
θ	Torsion angle
ϑ	Angle
λ	Strain ratio (draw ratio)
λ	Heat conductivity
μ	Chemical potential
μ	Poisson ratio
ν	Frequency
Π	Osmotic pressure
ρ	Density
ρ	Resistivity
σ	Stress
σ	Hindrance parameter
τ	Bond angle
ϕ	Volume fraction
χ	Interaction parameter in the theory of solutions

Contents

1. Introduction

Plastics are commercially used materials that are based on polymers or prepolymers. The name "plastics" refers to their easy processability and shaping (Greek: *plastein* = to form, to shape). "Polymers" are a group of organic, semiorganic, or inorganic chemical substances with high molar masses. Prepolymers have lower molar masses than polymers; they become polymers on processing.

"Plastics" and "polymers" are not synonyms. Polymers or prepolymers are raw materials for plastics; they become plastics only after physical compounding and, in some cases, after chemical hardening. Polymers are not only used as plastics but as fibers, binders in paints, or in other applications. Other polymers may be utilized as fibers, elastomers, thickeners, ion exchange resins, etc., but not as plastics.

1.1. Polymers

1.1.1. Fundamental Terms

Polymers are chemical substances composed of polymer molecules. The term "polymer molecule" refers to a molecule composed of many units (Greek: *poly* = many, *meros* = parts). Polymer molecules may thus consist of many atoms, usually a thousand or more, thereby having high molar masses ("molecular weights"). Benzene C_6H_6 can, for example, be "polymerized" from three acetylene molecules, C_2H_2, and was originally called a polymer. What is now called a polymer consists of molecules with hundreds and thousands of such units; it was therefore termed a "high polymer" in the older literature.

The term "polymer" carries with it the connotation of polymer molecules composed of many equal *mers*, such as the ethylene units $–CH_2CH_2–$ in poly(ethylene), $R'[CH_2CH_2]_NR''$. The number N of mers in a polymer molecule is called the degree of polymerization, X, of that molecule. Polymer molecules are called polymer homologues if they differ only in N (or X) or the amount n of mers.

There are, however, many polymer molecules (especially biopolymer molecules) with different types of mers per molecule, such as the proteins $H[NH–CHR–CO]_nOH$ with up to 16 different substituents R in irregular sequence. A less constraining and more general term for "polymer molecule" is thus "macromolecule" [Greek: *makros* = large; Latin: *molecula* = small mass (diminutive of *moles* = mass, bulk)]. No sharp dividing line with respect to the number of units per molecule exists, however, between macromolecules and low molar mass compounds (micromolecules).

Macromolecules (polymer molecules) are the simplest individual chemical constituents of a polymer. A polymer molecule may exist as an individual entity (in linear and branched polymers) or may be thought of as the primary molecule before the chemical cross-linking to polymers of "infinite" molar mass.

Macromolecules exist in nature; examples are nucleic acids, proteins, polysaccharides, polyprenes, and lignins. Some of these naturally occuring polymers are used by man as materials without further chemical transformation, e.g., cellulose for paper and cardboard. The chemical transformation of natural polymers with retainment of their chain structures leads to semisynthetic materials, for example, cellulose acetates from cellulose. Chains of other natural polymers are cross-linked before commercial use. Examples are the hardening of casein (a protein) by formaldehyde to galalith (plastic) or the vulcanization of cis-1,4-poly(isoprene) (natural rubber) to an elastomer.

Most polymers are, however, synthesized from molecules with low molar masses, the so-called monomers. These processes are called *polymerizations*. Examples are the preparations of poly(ethylene) from ethylene, poly(vinyl chloride) from vinyl chloride, nylon 6 from ε-caprolactam, or nylon 6.6 from adipic acid and hexamethylenediamine. Some industrial polymers result from the chemical conversion of other polymers, for example, cellulose acetate from the esterification of cellulose and poly(vinyl alcohol) from the saponification of poly(vinyl acetate).

1.1.2. Nomenclature

The nomenclature of individual polymers and plastics is as confusing as their classification according to properties. Various nomenclature systems are used simultaneously, often by the same author. Abbreviations and acronyms abound, sometimes with different meanings for the same letter combinations and other times without explanation. In addition, about 25 000 trade names are used worldwide for plastics, fibers, and elastomers. Furthermore, a polymer of a certain company may come in many different grades depending on the processibility and application, sometimes up to one hundred per polymer type. Some of these grades may even bear different trade names for various applications. On the other hand, the same trade name is occasionally used for plastics based on very different polymer types. An example is Bexloy® of Du Pont, which may be an amorphous polyamide (Bexloy C), a blend of a polycarbonate and a poly(ethylene terephthalate) (Bexloy J), an ionomer (Bexloy W), or a thermoplastic elastomer (Bexloy V). The following nomenclature systems are commonly used for polymers:

Trivial names are used for long known natural polymers. Examples are: cellulose, the polymeric sugar ("ose") of the plant cell; casein, the most important protein of milk and cheese (Latin: *caseus* = cheese); nucleic acids, the acids found in the cell nucleus; catalase, a catalyzing enzyme.

Poly(monomer) names indicate polymerized monomers. The polymerization of ethylene thus leads to poly(ethylene), that of styrene to poly(styrene), that of vinyl chloride to poly(vinyl chloride), and that of a lactam to a polylactam. This *poly(monomer)* nomenclature has the disadvantage that the constitution of monomeric units of the polymer molecules is not identical with the constitution of the monomers themselves. For example, the polymerization of ethylene, $CH_2=CH_2$, leads to polymers with the idealized structure ~$[CH_2-CH_2]_n$~, a saturated compound and thus a poly(alkane) and not an unsaturated "ene" compound as the name poly(ethylene) may suggest. The polymerization of lactams (cyclic amides) to polylactams does not give macromolecules with intact lactam rings in the polymer chains but open-chain polyamides ~$[NH-CO-R]_n$~, etc.

The poly(monomer) nomenclature is also ambiguous if a monomer can lead to more than one characteristic monomer unit in a polymer molecule. An example is acrolein (acrylaldehyde), $CH_2=CH(CHO)$, which polymerizes via the ethylenic double bond to carbon chain units $-CH_2-CH(CHO)-$ with aldehyde substitutents, or via the carbon-oxygen double bond to carbon-oxygen chain units $-O-CH(CH=CH_2)-$ with ethylene side groups, via both to give six-membered rings in polymer chains, or to "mixed" polymers with two or more types of these units in the chain.

For trade purposes, certain polymer names may denote not only homopolymers of one type of monomer molecules but also copolymers of two or more types of monomers, contrary to what the "chemical" names imply. For example, the copolymers of ethylene with up to 10 % of butene-1, hexene-1, or octene-1 are known as linear low density polyethylenes (LLDPEs) although they are "branched" in the sense of organic chemistry (but not polymer science) and they are poly(ethylene-*co*-α-olefin)s. The commonly used chemical names of plastics thus often do not indicate the true chemical structure of the monomeric units of the polymers on which they are based.

The English spelling of poly(monomer) names is not systematic. Simple polymer names are usually written as one word (e.g., polyethylene, polystyrene, etc.), more complex names as two words (e.g., polyvinyl chloride, polyethylene terephthalate, etc.), and highly complex polymer names with the name of the monomer(s) in parentheses (e,g., poly(2,6-dimethyl-1,4-phenylene ether)). Many exceptions exist to this "rule": Polyethyleneimine is, for example, written as one word. In order to avoid misunderstandings, especially with the "poly" nomenclature of organic chemistry, this book will always write the monomer in poly(monomer) with parentheses.

Characteristic group names are used for families of polymers with *characteristic groups* in their repeating units. Polyamides are thus polymers with amide groups $-NHCO-$ in their repeating units, e.g., ~$[NHCO(CH_2)_5]_n$~ = polyamide 6 = nylon 6 = poly(ε-caprolactam). Other examples are polyesters with ester groups $-COO-$ or polyurethanes with urethane groups $-NH-CO-O-$ in the chains. A disadvantage is that this naming scheme is identical with that of organic chemistry where, e.g., "polyisocyanate" denotes a low molar mass compound with more than one isocyanate group per molecule, e.g., $C_6H_3(NCO)_3$. According to this nomenclature, a macromolecular polyisocyanate would thus be a polymer with many intact isocyanate groups per chain, for example, poly(vinyl isocyanate) ~$[CH_2-CH(NCO)]_n$~. The poly(isocyanate)s of polymer chemistry, on the other hand, possess *polymerized* isocyanate groups as, for example, in ~$[N(C_6H_5)-CO]_n$~. Such compounds are unfortunately also often called polyisocyanates.

Constitutive names are recommended by the International Union of Pure and Applied Chemistry (IUPAC); these names are similar to those used in inorganic and organic chemistry. The nomenclature of low and high molar mass inorganic molecules follows the additivity principle; those of low molar mass organic molecules, the substitution principle. The nomenclature of organic macromolecules is a hybrid of both principles: the smallest repeating units are thought of as biradicals according to the substitution principle; then their names are added according to the additivity principle, put in parentheses, and prefixed with "poly". Names of repeating units are written without spaces between words. IUPAC names indicate the chemical structure of macromolecules whereas poly(monomer) names are source-based.

The polymer $\sim[O–CH_2]_n\sim$ from formaldehyde, HCHO, is thus called poly(oxymethylene). The polycondensation of ethylene glycol $HOCH_2CH_2OH$ with terephthalic acid $HOOC(p\text{-}C_6H_4)COOH$ leads to a polymer $\sim[OCH_2CH_2O–OC(p\text{-}C_6H_4)CO]_n\sim$ with the systematic name poly(oxyethyleneoxyterephthaloyl). The trivial names of this polymer are poly(ethylene glycol terephthalate), polyethylene terephthalate, and poly(ethylene terephthalate). It is also known as "saturated polyester" (although it is only one of the many possible saturated polyesters), as PET (an acronym) or PETP (an abbreviation) in the plastics literature, by the acronym PES in the fiber literature, and as PETE or No.1 for recycling purposes.

IUPAC names are rarely used in the plastics literature. They are however important for systematic searches in Chemical Abstracts and other literature services.

1.2. Plastics

1.2.1. Fundamental Terms

Early plastics and their polymers and prepolymers resembled *natural resins*. These natural resins are organic solids that break with a conchoidal fracture in contrast to the planar surfaces created upon the fracture of crystalline materials or the drawn-out zones formed upon the breaking of gums and waxes. The term "natural resin" refers mainly to oleoresins from tree saps but is also used for shellac, insect exudations, and mineral hydrocarbons.

Early polymers and prepolymers were thus sometimes called *synthetic resins*. The term "synthetic resin" (or just "resin") is used today for all prepolymers and polymers which serve as raw materials for plastics, coatings, inks, etc. "Raw material" denotes in this context a material that needs further treatment to become a commercial product. Such a treatment may consist in the physical compounding (formulation, etc.) of the raw material with additives, modifiers, and/or auxiliaries. So-called thermoset(ting) resins furthermore require an additional chemical hardening (cross-linking of primary (pre)polymer molecules). "Resins" may thus not only be prepolymers for thermosets (such as amino resins, phenolic resins, etc.) but also polymers for thermoplastics (such as acetal resins, nylon resins, etc.).

In many countries, the term "(synthetic) resin" is used however exclusively for prepolymers for thermosets. "Resin" is furthermore not be confused with *rosin*, which refers to mixtures of C_{20} fused-ring monocarboxylic acids such as pine oil, tall oil, and kauri resin. Rosin is the main component of naval stores; it is today mainly used for varnishes (world production ca. 900 000 tons per year).

Plastics are usually divided into two groups according to their physical or chemical hardening processes: thermoplastic and thermosetting resins.

Thermoplastics yield solid materials by simply cooling a polymer melt (a physical process); they soften upon heating. They are normally composed of fairly high molar mass molecules since it is only above a certain molar mass that many physical properties become effectively molar mass independent. Examples are melting temperatures and moduli of elasticity. Other properties increase however with increasing molar mass, e.g., melt viscosities.

Thermosetting resins, on the other hand, harden through chemical cross-linking reactions between polymer molecules and become *thermosets* since their shapes and properties are "set" by this process. Upon heating, they do not soften but decompose chemically. The shaping of a thermoplastic is thus a reversible process: the same material can be melted and processed again. Thermosets cannot be remelted and reshaped; their formation is irreversible.

Thermosets are usually generated from fairly low molar mass polymers, called oligomers (science) or prepolymers (industry). High molar masses are here unnecessary since the chemical reactions between prepolymer molecules lead to interconnections ("cross-linking") of these primary molecules and thus to one giant molecule at 100 % conversion of the prepolymers. The properties of these networks are obviously independent (or nearly so) of the molar mass of the prepolymers.

1.2.2. Groups of Plastics

Plastics are usually divided into five groups: (1) commodity thermoplastics, (2) engineering thermoplastics, (3) high-performance thermoplastics, (4) functional thermoplastics (specialty plastics), and (5) thermosets. Some thermosets are sometimes also classified as engineering or high-performance plastics. Fluoroplastics are often considered a special group of plastics because of their outstanding surface properties.

Commodity thermoplastics are manufactured in great amounts, hence the terms "bulk plastics", "volume plastics", or "standard plastics". They include poly(vinyl chloride), poly(ethylene)s (high density, low density, very low density), isotactic poly(propylene), and standard poly(styrene).

Engineering plastics (technical plastics, technoplastics) are in general thermoplastics (ETPs) that possess improved mechanical properties compared to commodity plastics; some thermosets may also classify as engineering plastics They have load-bearing characteristics that permit them to be used in the same manner as metals and ceramics. Such improved properties may be higher moduli of elasticity, smaller cold flows, higher impact strengths, etc. Engineering plastics are also often defined as those thermoplastics that maintain dimensional stability and most mechanical properties above 100°C or below 0°C.

Engineering thermoplastics comprise poly(ethylene terephthalate), poly(butylene terephthalate), polyamides (aliphatic, amorphous, aromatic), polycarbonates, poly-(oxymethylene)s, poly(methyl methacrylate), some modified poly(styrene)s such as styrene/acrylonitrile (SAN) and acrylonitrile/butadiene/styrene (ABS) copolymers, and high-impact poly(styrene)s (SB), as well as various blends such as poly(phenylene oxide)-(poly(styrene) and polycarbonate-ABS.

High-performance plastics, on the other hand, are engineering plastics with even more improved mechanical properties. They comprise liquid crystalline polymers (LCPs), various polyetherketones, different polysulfones, poly(phenylene sulfide), various polyimides, etc.

No sharp dividing lines exist between commodity plastics and engineering plastics on the one hand, and engineering plastics and high-performance plastics on the other; nor are there generally agreed upon property levels beyond which polymers are designated as engineering or high-performance plastics. All these resins have in common that they have many different applications. Poly(ethylene), a commodity plastic, may according to its types or grades be used for containers, as packaging film, as agricultural mulch, etc.

Functional plastics, however, have only one very specific use. Poly(ethylene-*co*-vinyl alcohol) with a high content of vinyl alcohol units is a functional plastic that is only used as a barrier resin in packaging. Other functional plastics are employed in optoelectronics, as piezoelectric materials, as resists, etc. "Functional plastic" is not synonymous with "functional polymer" or "functionalized polymer", because the latter terms refer to polymers with functional chemical groups, i.e., groups that can be reacted. A "reaction polymer", on the other hand, is a polymer generated by polymerization of liquid monomers under simultaneous shaping; such polymers may be either thermoplastics or thermosets.

Fluoroplastics are specialty plastics because of their surface properties. They comprise poly(tetrafluoroethylene) PTFE (Teflon), poly(chlorotrifluoroethylene) PCTFE, poly(vinylidene fluoride) PVDF, and many other fluorinated polymers.

Thermosets are sometimes also classified as engineering or high-performance plastics. In general, thermosets are, however, considered a separate group of plastics. They comprise alkyd resins, phenolic resins, amino resins, epoxides (epoxies), unsaturated polyesters (incl. so-called vinyl esters), polyurethanes, and allylics.

1.2.3. Designations

The plastics trade and technical literature indulges in trivial names, abbreviations, and acronyms. "Acetal polymer" is, for example, the name given to polymers with $-OCH_2-$ as the major repeating unit. "Styrenics" comprise homopolymers of styrene as well as copolymers of styrene with other monomers. Polymers of methacryl esters $\sim[CH_2-C(CH_3)(COOR)]_n\sim$ are known as "methacrylics" and often only as "acrylics" but poly(acrylonitrile) with the structural formula $\sim[CH_2–CH(CN)]_n\sim$ is not considered an "acrylic" in the plastics literature (the textile literature calls this an acrylic fiber). Nylon, originally a protected trade name for fibers from poly(hexamethylene adipamide), is now used in the English language as a generic term for polyamides, especially for aliphatic ones, although its legal use is confined to certain aliphatic polyamide fibers (U.S. Federal Trade Commission).

Acronyms and abbreviations often have more than one meaning. PBT may, for example, denote poly(butylene terephthalate) but it is also used for poly(p-phenylene-benzbisthiazole). Exhaustive lists of commonly used abbreviations and acronyms of plastics, elastomers, fibers, polymers, additives, and auxiliaries have been compiled (see Chapter 1, Literature, and Chapter 15, Appendix).

Companies have long used internal classification systems and the military has issued plastics specifications. An industry-wide classification system for thermoplastic materials recently introduced by ISO is based on an alphanumeric "line call-out", a special code. In this system, a material is characterized by several data blocks which indicate the composition and certain property data and/or ranges. These property data refer to essential criteria of the plastics. Since different thermoplastics have different applications that require different essential criteria, property data in the data blocks are restricted to different "leading properties" (designatory properties) for each type of thermoplastic.

The property data covered by this classification scheme are thus not comprehensive. Rather, various plastics types are characterized by one to four sets of leading criteria which are especially important for the particular plastic material. These leading criteria are selected from the following designatory properties:

Chemical structure data:
- composition such as content of vinyl acetate (VAC) or acrylonitrile units (AN),
- configuration as measured by isotacticity ("isotaxy") (IT),
- branching as revealed by density (D);

Molar mass indicators:
- intrinsic viscosity (IV) ("viscosity number"),
- Fikentscher's K value (FK), another viscosity-based quantity;

Macrostructure data:
- bulk density (BD);

Rheological data:
- melt-flow rate (MFR) (melt-flow index);

Thermal data:
- Vicat temperature (VST),
- torsional stiffness temperature (TST);

Mechanical properties:
- modulus of elasticity (E),
- tensile stress at 100 % strain (TS),
- Shore hardness (SH),
- impact strength (notched) (ISN).

Recipes (contents of plasticizers, etc.) are specified through the properties of the plastics. The following leading criteria are used for the various polymers:

poly(ethylene)	D, MFR
poly(propylene)	I, MFR
poly(styrene) and acrylonitrile/styrene copolymers	VST, MFR
styrene/acrylonitrile copolymers	VST, MFR, AN
styrene/butadiene copolymers	VST, MFR, ISN
acrylonitrile/butadiene/styrene, and acrylonitrile/styrene/acrylic acid copolymers	VST, MFR, ISN, AN
ethylene/vinyl acetate copolymers	VAC, MFR
poly(vinyl chloride)	IV, (FK), BD
poly(vinyl chloride), unplasticized	VST, E, ISN
poly(vinyl chloride), plasticized	TS, SH, TST
polyamides	IV, E
polycarbonates	IV, MFR, E
poly(methyl methacrylate)	IV, VST
poly(ethylene terephthalate)	IV

The **standard designation** of a thermoplastic material consists of a *description block* (giving the type of material), a *standard number block* (consisting of the number of the ISO standard or a national standard such as DIN), followed by a hyphen and an *individual item block* (with five data blocks). An example is:

Molding material DIN 7744-PA12, XF, 22-030, GF 30

Data block 1 of the individual item block contains the abbreviation of the chemical name of the material, e.g., PA 12 = polyamide 12 (see example). It may be followed by analytical data such as the content of vinyl acetate units in ethylene-vinyl acetate copolymers (EVAC). These data are, however, not the exact analytical data but rather code numbers for the range (called "cell") which will permit this material to substitute for a similar one. Separated by a hyphen, supplementary information on this material may be given, e.g., H = homopolymer, P = plasticized, E = polymerized in emulsion, Q = mixtures of different polymers, etc. (see Appendix, Table 15-8).

Data block 2 may comprise up to four letters which give qualitative information. The first letter denotes the intended application, e.g., B = blow molding, G = general purpose, X = no indication (see example), etc. The following letters 2-4 can code up to three essential additives or supplementary information, for example, D = powder (dry blend), F = special burning characteristics (see example), L = light and weather stabilizer, etc. (see Appendix, Table 15-9).

Data block 3 contains quantitative information about the designated properties. The encoding of these data is different for each plastic material and each testing standard. For example, the code "22-030" of data block 3 for the polyamide 12 (example) indicates a polymer with an intrinsic viscosity of 210 cm^3/g (cell 22) and a modulus of elasticity of 280 MPa (cell 030). On the other hand, "20-D050" in data block 3 for a poly(ethylene) tested according to DIN 16 776 means a material with a density within the cell "020" (e.g., 0.918 g/cm^3) and a melt flow rate within the cell "050" (e.g., 4.2 g/min), measured under condition D (temperature of 190°C under a load of 2.16 kg). The definition of the cells can be found in special cell tables.

Data block 4 informs about the type and content of fillers or reinforcing materials. The first letter gives the type of filler: C = carbon, G = glass, etc. The second letter indicates the shape of the filler: F = fiber, S = spheres, etc. (see Appendix, Table 15-10). The third position is a code for the mass content of the filler, e.g., 30 for the range 27.5-32.5 wt% (see Appendix, Table 15-11).

Data block 5 is reserved for specifications based on individual agreements between supplier and customer. It may code additional requirements, restrictions, or supplemental information (none in the example).

A designation system similar to the above ISO/DIN system has been recommended by ASTM. A typical ASTM designation, for example, would read

Molding material ASTM D-4000 PI000G42360

This molding material is a polyimide (PI), whose properties have been specified via the cell table G of the ASTM standard D-4000. This particular cell table identifies five different properties by designating cell limits to the five digits following the letter G. According to cell table G, the "4" indicates a tensile strength of at least 85 MPa. The measured flexural modulus is characterized by "2", the Izod impact strength of 50 J/m by "3", and the heat deflection temperature of 300°C by "6". The fifth digit is undetermined; the "0" indicates an unspecified property.

1.3. History of Plastics

The first plastics were prepared long before their macromolecular nature was discovered. Decorative coatings based on polymers such as egg-white or blood proteins were used in the cave paintings of Altamira, Spain, as early as 15 000 B.C. Later painting methods utilized gelatin, a polymer from the collagen of hides and bones, and the polymers from naturally drying vegetable oils. The use of these materials resulted from the eternal desire of the artists for easy-to-handle materials that give an optimal artistic effect and have an infinite stability.

These requirements were not restricted to surface coatings, which are two-dimensionally applied plastics. Cow horns were used in the medieval ages to prepare transparent windows for lanterns and dyeable intarsia in wood. To this purpose, the horns had to be flattened by steaming, a very difficult process. The flattened horns also tended to curl up after a while. An early, easy-to-process substitute for natural horn was reported by a Bavarian monk, Wolfgang Seidel (1492-1562); it undoubtedly derived from far older recipes. This imitation horn was based on the protein casein, the white material from skim milk. Casein was extracted from skim milk with hot water, treated with warm lye, and shaped while being warm; the desired shape was then fixed by immersion in cold water.

Unknown inventors later discovered that addition of inorganic fillers increased the mechanical stability of this early thermoplastic. The same material was subsequently used for children's building blocks by the German aviation pioneer Otto Lilienthal (1848-1896) and his brother Gustav. In 1885, a patent based on the same physical process was granted to the American Emery Edward Childs. The properties of these plastics were markedly improved by chemical reaction of casein with formaldehyde, as described in 1897 by the German inventors Wilhelm Krische and Adolf Spitteler. The resulting thermoset was called Galalith (Greek: *gala* = milk, *lithos* = stone); it is still used today for haberdasheries.

Another early thermoset used natural rubber as raw material. The American Charles Goodyear discovered in 1839 the cross-linking ("vulcanization") of natural rubber to an elastomer by sulfur under the action of "white lead" (basic lead carbonate) and heat. His brother Nelson used larger amounts of sulfur and in 1851 obtained Ebonite, a hard, black thermoset.

The first fully synthetic thermoset was invented in 1906 by the Belgian-born American chemist Leo H. Baekeland who heated various phenols with formaldehyde under pressure and produced insoluble hard masses. Bakelite™ was recognized in 1909 as excellent electrical insulator and thus became one of the foundations of the modern electrical industry. Baekeland was not the first to prepare phenolic resins though; they were already observed by Adolf von Baeyer in 1872. Baeyer's substances were however only resinous materials; it took Baekeland's "heat and pressure" process to produce commercially useful materials.

The first *semisynthetic thermoplastics* originated from the cellulose of cotton. Cotton fibers have been used by man since prehistoric times. Since cotton is relatively easily grown, many attempts were made to improve its textile properties. The Englishman John Mercer discovered in 1844 that the treatment of cotton with aqueous solutions of caustic soda (sodium hydroxide) leads to fibers with increased strength, higher luster, and improved dyeability. The Frenchman L. Figuier demonstrated in 1846 that cellulose paper was strenghtened similar to cotton when it was treated with sulfuric acid. In 1853, W. E. Gaine received an English patent for the same process, which delivered a parchment-like material. Another English patent by Thomas Taylor described the formation of very resistant materials from layers of paper sheets by the combined action of zinc chloride and pressure. These products were called "vulcanized fiber" since they resembled the vulcanization products of natural rubber.

Mercerized cotton, artificial parchment, and vulcanized paper result from physical transformations of cellulose. If however cellulose materials are treated not by sulfuric acid alone but by a mixture of sulfuric acid and nitric acid, a chemical reaction to cellulose nitrate (gun cotton, "nitrocellulose") occurs. This chance discovery by the German Christian Friedrich Schönbein in 1846 paved the way for the invention of the first semisynthetic thermoplastic. In 1862, Alexander Parkes, an Englishman, found an easy method for the processing of cellulose nitrate into thermoplastic masses by the addition of castor oil, camphor, and dyes. The manufacture of the resulting "Parkesine" was however difficult and the production ceased in 1867. The use of alcoholic solutions of camphor by Daniel W. Spill was equally unsuccesful. In 1869, a patent was granted to the American John Wesley Hyatt for the use of camphor without camphor oil and without alcohol. The resulting Celluloid™ is generally considered to be the pioneer thermoplastic.

Fully synthetic thermoplastics have a far longer history, albeit not as industrial materials. Polymers of formaldehyde were already found by Justus von Liebig in 1839; yet extensive scientific investigations by Hermann Staudinger in the 1920s and 1930s and major industrial development work by the Du Pont Company were necessary before the "acetal homopolymer" became an engineering plastic in 1956.

Vinylidene chloride was synthesized in 1838 by H.V. Regnauld who also observed the formation of solids upon exposure of the gas to sunlight. Vinyl chloride was polymerized in 1912 by F. Klatte in Germany and I. Ostromislensky in England, but it was 1931 when the first poly(vinyl choride) was produced commercially by I.G. Farbenindustrie in Germany and B.F. Goodrich in the United States. The material was difficult to process until the discovery of plasticization in the 1930s.

Styrene was discovered by a chemist called Neuman as reported in W. Nicholson's 1786 "Dictionary of Practical and Theoretical Chemistry". The conversion of styrene to a solid mass was found by E. Simon in 1839, who considered the material a styrene oxide. A.W. von Hofmann and J. Blyth showed in 1845 that the alleged styrene oxide was an isomer of styrene; they called it metastyrene. It was however only in 1920 that the polymeric nature of poly(styrene) was recognized by H. Staudinger; it was first produced commercially by I.G.Farbenindustrie (Germany) and by Dow Chemical (United States) in 1930.

A highly branched poly(ethylene) is found in nature as the "mineral" elaterite. Lightly branched synthetic poly(ethylene) was first obtained in 1933 by high pressure free radical polymerization of ethylene at ICI (United Kingdom); its commercial production began in 1939. Linear poly(ethylene)s were first synthesized in 1953 by low pressure polymerizations using catalysts based on transition metals (Karl Ziegler, Max-Planck-Institute for Coal Research, Germany), chromium oxide (Phillips Petroleum, United States), and molybdenum oxide (Standard Oil of Indiana, United States). For the first time, Ziegler-Natta catalysts also enabled the polymerization of α-olefins such as propylene to solid, stereoregular polymers. The production of isotactic poly(propylene) began in Italy, the Federal Republic of Germany, and the United States in 1957. The use of these catalysts and the availability of inexpensive feedstocks from petroleum refining led to rapid growth of synthetic polymers (Table 1-1).

Table 1-1 World production of polymers (in million tons per year). * Estimated.

Type		1940	1950	1960	1970	1980	1990
Plastics		0.36	1.62	6.7	31.0	59.0	98
Textile fibers,	synthetic	0.005	0.069	0.70	5.0	11.5	15.7
	semisynthetic	1.1	1.6	2.6	3.4	3.3	3.2
	natural	8.7	8.0	12.8	14.0	17.7	18.0*
Elastomers,	synthetic	0.043	0.54	1.9	5.9	8.7	9.9
	natural	1.44	1.90	2.02	3.1	3.9	5.2

Plastics are the leader in this growth, followed by fibers and elastomers. The world production of semisynthetic fibers (rayon and cellulose acetate) has, however, declined, mainly because of cost and environmental concerns in the developed countries and despite an increased production of semisynthetic fibers in developing countries.

Compact and expanded plastics based on synthetic polymers are used in building and construction, as packaging materials, in the electrical and electronic industries, in toys, furniture, appliances, etc. Many other applications of synthetic polymers are known; most noteable are fibers, elastomers, coatings, adhesives, and thickeners. For example, the global demand for water-soluble synthetic polymers is ca. $5 \cdot 10^6$ t/a. Two thirds of water-soluble polymers are used in the manufacture of ca. $200 \cdot 10^6$ t/a cellulose paper; others in oil recovery or as sizes. Oil-soluble polymers serve as VI improvers in gasoline.

1.4. Economic Importance

Plastics are major industrial goods. The rapid growth of their production is caused by three factors: (1) growth of world population, (2) average increase of living standards, and (3) replacement of older materials by plastics. The growth pattern may be economy driven by the price of materials (substitutes), shortages ("ersatz") and government intervention (environmental concerns), and/or technology driven by technological requirements, internal know-how, and accidental discoveries.

The world population grew to $5.32 \cdot 10^9$ in 1990 from $2.53 \cdot 10^9$ in 1950, while the plastics production (approximately equal to consumption) climbed to ca. 18.8 kg per capita (1990) from 0.6 kg per capita (1950) (Table 1-2). These numbers are approximate since statistics are incomplete and may differ considerably from source to source. For example, the 1987 world plastics production is given either as $37.5 \cdot 10^6$ t (excluding South Africa and the Peoples Republic of China) or $55.4 \cdot 10^6$ t, in part, because some statistics relate only to plastics raw materials (resins) and some to plastics themselves (resins plus fillers, modifiers, etc.).

Great differences in plastics consumption exist between the various countries. In 1991, the apparent annual per capita consumption as calculated from production plus import minus export (i.e., including stockpiling) was 150 kg in Belgium, 133 kg in Germany, and 102 kg in the United States or the United Kingdom, but only 5 kg in Ecuador and 1 kg in India. Since the world population continues to grow and the people in less developed countries justifiably aspire to attain higher living standards, the world plastics production is most likely to increase further despite some environmental concerns in highly industrialized countries (see Chapter 14). Although all regions will share in this growth, it will be most pronounced in "other" countries, e.g., Saudi Arabia, Central and South America, Southeast Asia, the Indian subcontinent, and, as soon as planned economies are replaced by free-market type ones, also in the former Comecon countries.

Table 1-2 Plastics production, world population, and apparent plastics consumption per capita.

Region or country	Production in million tons per year				
	1950	1960	1970	1980	1990
European Community	0.3	2.3	11.5	19.7	30
United States	1.1	2.8	9.1	16.1	28
Japan	0.04	0.7	5.3	7.5	13
Eastern Europe	0.1	0.5	4.0	6.9	13
Other	0.07	0.3	1.1	8.8	14
World production (in million tons)	1.6	6.7	31.0	59.0	98
World population (in millions)	2532	3062	3730	4498	5320
Per capita consumption (in kilogram)	0.6	2.2	8.3	13.1	18.8

The replacement of other materials by plastics is difficult to judge. Since developing countries possess a great, unsatisfied need for goods that have long been taken for granted in more affluent societies, an analysis must be based on the production of materials in highly developed countries such as the United States (Table 1-3).

Table 1-3 gives the apparent consumption of selected important materials, calculated from production plus import minus export, i.e., including stockpiling but excluding the import/export of materials in manufactured goods such as machinery (metals), garments (fibers), tires (cars), newsprint (paper), etc. The largest per capita consumption is in construction materials (nonmetallic minerals, lumber, partly steel and aluminum), followed by packaging materials (partly steel, aluminum, paper), and information carriers (mainly paper).

Plastics constitute a small percentage of all consumed materials (Table 1-3). Their production grows however faster than that of other materials as shown by the data for the United States (Fig. 1-1). The data in this figure are given on a volume basis and not on the customary weight basis since materials are used by volume although they are sold by mass.

Table 1-3 Per capita consumption of selected materials in the United States in 1988 (estimated population: 246 million).
Codes for Type: Mt = Metals, P = polymers, O = other.
Codes for Source: N = Natural product, M = modified natural product, S = synthetic product.
Consumption: * Estimated. (Numbers in parentheses): net exports.

Materials	Type	Source	Consumption in kg/a	Imports in % of consumption
Nonmetallic minerals				
Sand, gravel	P, O	N	3336	0
Stones	P, O	N	4492	0
Cement	P	M	339	6
Clays	P	M	165	0
Gypsum	O	N	92	39
Limestone	O	M	62	0
Metals				
Raw steel	Mt	S	417	15
Aluminum, primary	Mt	S	21.9	6
Copper	Mt	S	5.1	31
Lead	Mt	S	1.26	10
Zinc	Mt	S	0.88	39
Polymers				
Wood, wood products				
Lumber	P	N	509	22
Plywood	P	M	96	4
Paper, cardboard	P	M	315	6
Other wood products	P	?	24	?
Fuel wood	P	N	236	?
Fibers (excl. export/import of fabrics and garments)				
Cotton	P	N	5.5	(50)
Rayon, acetate	P	M	3.6	66
Wool, silk	P	N	1.6	92
Synthetic fibers	P	S	16.9	?
Elastomers (excl. export/import of rubber goods)				
Natural rubber	P	M	2.6	100
Synthetic rubbers	P	S	6.7	(12)
Plastics (excl. manufactured goods)				
Synthetic plastics, resins	P	S	120	(9)
Semisynthetic resins	P	M	0.2*	?
Leather	P	M	6.4	?
Polysaccharides (soluble; excl. of foods)				
Starch	P	N, M	49	?
Gums	P	N, M	0.3	?

The United States plastics production has already surpassed the United States steel production and it is continuing to grow, albeit with smaller growth rates. The production was somewhat affected by the strong increases of oil prices during the first and second oil crises but not as dramatic as the production of other materials during the depression years starting with the stock market crash at the Black Friday in 1929. Although the share of plastics in terms of total materials produced, consumed, and disposed is quite low (see Table 1-3), it causes some environmental concern because of the very visible proportion of plastic packaging materials in household refuse (see Chapter 14).

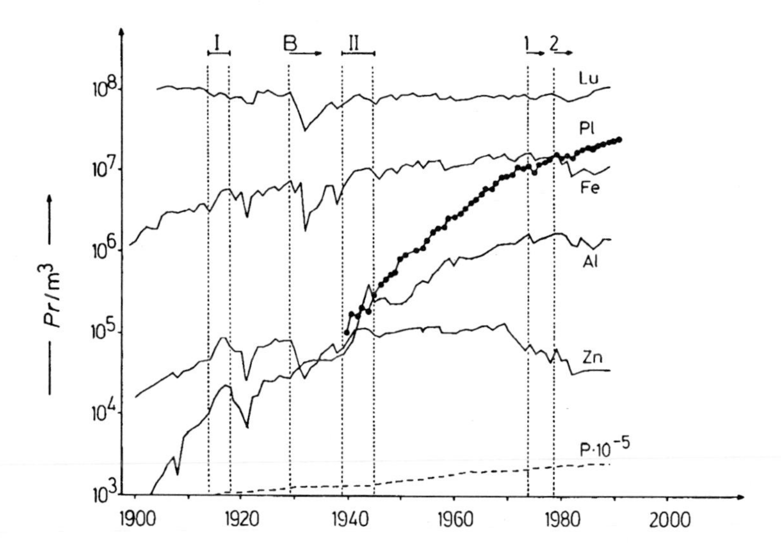

Fig.1-1 Annual production *Pr* in cubic meters of lumber Lu, raw steel (Fe), primary aluminum (Al), and primary zinc slab (Zn) (smelter basis), and consumption of plastics Pl (●) in the United States during 1900-1990 (solid lines) compared to the U.S. population P (broken line).

I = World War I; II = World War II; B = Black Friday in 1929 (beginning of world economic crisis); 1, 2 = first and second oil crisis.

Commerce uses traditional units for lumber (1 board foot = 0.00236 m^3), steel (1 short ton = 0.115 m^3), aluminum (1 short ton = 0.336 m^3), and zinc (1 short ton = 0.127 m^3). Plastics (1 t ≈ 0.91 m^3) are weighted according to annual production.

The appropriate quantity for the use of materials is, of course, the cost per volume and relevant property. Many plastics compare reasonably with conventional materials in this respect. For example, costs per yield strength and volume are 0.027 \$/(GPa L) for ABS plastics, 0.28 \$/(GPa L) for copper, 0.069 \$/(GPa L) for aluminum, and 0.024 \$/(GPa L) for concrete (1987 data).

Since building/construction and packaging are major applications for lumber, steel, and aluminum, it can be expected that plastics are important replacements for these materials. Packaging and building/construction are indeed the most important applications of plastics in the United States (Table 1-4). They also have the highest growth rate of ca. 15 % per year.

Table 1-4 Accounted domestic U.S. plastics applications in 1988 (in 1000 tons) as compared to total sales and percentages of consumption in the United States and in Western Europe (WE).

A = Appliances, B = building/construction, E = electrical/electronic, F = furniture, H = household goods, P = packaging, T = transportation, Y = toys, U = unaccounted.

Polymer	Domestic sales	A	B	E	F	H	P	T	Y	U
		Accounted domestic applications								
Thermoplastics										
Poly(ethylene)s, LD	4479		127	168	4	191	1092	20	74	2803
Poly(ethylene)s, HD	3743	9	251	53	5	143	1692	19	77	1494
Poly(propylene)s, it	3316	67	18	21	30	92	470	170	18	2430
Poly(styrene)s	2329	77	127	186	55	115	108		108	1553
SAN	62	10		2		10	8			32
ABS	562	91	69	53	2		2	95	17	233
Other styrenics	553									553
PPE alloys	82	20	2	1						59
Poly(vinyl chloride)s	3779	64	2301	232	41	29	199	101	15	809
Other vinyls	435		2				53			380
Poly(methyl methacrylate)	316	4	110					25		177
Fluoropolymers	13									13
Poly(oxymethylene)s	58	5	7	2				12		32
Thermoplastic elastomers	225									225
Polyamides	253	8		36			25	69		115
Polycarbonates	195	16	52	25				28		74
PET, PBT	911	3	3	20			450	20		415
Cellulosics	41	1	2	2			3		1	32
Thermosetting polymers										
UP (w/o glass fibers)	623	30	132	17	9			64		371
Epoxides	213	4	19	25			15	45		105
Phenolic resins	1376	15	411	45	77	18	6	10		794
Amino resins	688		121	14	10	6	3			534
PUR (raw materials)	1319	73	309		257				331	349
Alkyd resins	145									145
Other										
Miscellaneous polymers	500	16		21	3	29	86	52	41	252
Total Sales or Applications (in thousand tons per year)										
	26216	513	4063	923	493	633	4212	730	682	13979
Applications as % of Accounted Plastics										
		4.2	33.2	7.6	4.0	5.1	34.4	6.0	5.5	
Applications as % of Domestic Sales (= 100 %)										
United States	100	2.0	15.5	3.5	1.9	2.4	16.0	2.8	2.6	53.3
Applications as % of Consumption (+ 10 % in paints and adhesives), 1990 data										
Western Europe	100	?	25	15	7	3	21	7	?	12

Plastics are consumed in far lesser amounts in electric/electronic applications, transportation, toys, housewares, appliances, and furniture; the United States use in agriculture and for paints and coatings has not been reported. Plastics are critical and irreplaceable components in certain fields such as the electric/electronic industry. Consumption patterns are similar for nations with comparable affluence; the United States and Western Europe applications of plastics do not differ significantly when the fact is considered that ca. 53 % of United States applications are not accounted for.

Literature

1.1. Polymers: Encyclopedias and Dictionaries

W.A.Kargin, ed., Enciklopedia Polimerov, Sovietskaya Enciklopedia Publ., Moscow 1972 (3 vols., in Russian)
H.Mark, N.G.Gaylord, C.Overberger, G.Menges, N.M.Bikales, eds., Encyclopedia of Polymer Science and Engineering, Wiley, New York, 2nd ed. (1985), 20 vols.
G.Allen, J.C.Bevington, eds., Comprehensive Polymer Science, Pergamon, Oxford 1989 (7 vols.)
M.S.M.Alger, Polymer Science Dictionary, Elsevier Appl.Sci., Barking, Essex 1989
R.J.Heath, A.W.Birley, Dictionary of Plastics Technology, Blackie and Son, Glasgow 1992

1.1. Polymers: Selected Textbooks of Polymer Science

F.W.Billmeyer, Jr., Textbook of Polymer Science, Wiley, New York, 3rd ed. 1984
P.Hiemenz, Polymer Chemistry, Dekker, New York, 1984
F.Rodriguez, Principles of Polymer Systems, Hemisphere Publ., New York, 3rd ed. 1989
R.B.Seymour, C.E.Carraher, Jr., Polymer Chemistry, Dekker, New York, 2nd ed. 1989
H.-G.Elias, Macromolecules, vol. I, Structure and Properties, vol. II. Synthesis, Materials, and Technology, Plenum, New York, 2nd edition 1984; Makromoleküle (in German), vol. I, Grundlagen (5th ed. 1990), vol. II, Technologie (1992), Hüthig and Wepf, Basle, Switzerland
J.M.G.Cowie, Polymers: Chemistry and Physics of Modern Materials, Chapman and Hall, New York, 2nd ed. 1991
R.J.Young, P.A.Lovell, Introduction to Polymers, Routledge, Chapman and Hall, New York, 2nd ed. 1991

1.1. Polymers: Nomenclature

International Union of Pure and Applied Chemistry, Compendium of Macromolecular Nomenclature, Blackwell Sci.Publ., Oxford 1991

1.1. Polymers: Data Collections

O.Griffin Lewis, Physical Constants of Linear Homopolymers, Springer, Berlin 1968
J.Brandrup, E.H.Immergut, eds., Polymer Handbook, Wiley, New York, 3rd ed. 1989
D.W.van Krevelen, Properties of Polymers, Elsevier, Amsterdam, 3rd ed. 1990

1.1. Polymers: History

J.K.Craver, R.W.Tess, eds., Applied Polymer Science, ACS, Washington, DC, 1975
R.B.Seymour, ed., History of Polymer Science and Technology, Dekker, New York 1982
H.Morawetz, Polymers: The Origin and Growth of a Science, Wiley-Interscience, New York 1985
R.B.Seymour, G.A.Stahl, eds., Genesis of Polymer Science, ACS, Washington, DC, 1985

1.2. Plastics: General Information

E.R.Yescombe, Plastics and Rubbers: World Sources of Information, Appl.Sci. Publ., Barking, Essex, 2nd ed. 1976
J.Schrade, Kunststoffe (Hochpolymere), Bibliography of German plastics books, c/o J.Schrade, Swiss Aluminum, Zurich, Switzerland, First Section for the years 1911-1969 (1976); Second Section (1980)
G.W.Becker, D.Braun, eds., Kunststoff-Handbuch (in German), Hanser, Munich, 2nd ed. 1983 ff. (many vols.)
H.Saechtling, Kunststoff-Taschenbuch (in German), Hanser, Munich, 24th ed. 1989; International Plastics Handbook, Hanser, Munich, 2nd ed. 1987
D.V.Rosato, Rosato's Plastics Encyclopedia and Dictionary, Hanser, Munich, 2nd ed. 1993
W.J.Roff, J.R.Scott, Handbook of Common Polymers, Butterworths, London 1971
-, Modern Plastics Encyclopedia, McGraw-Hill, New York (vol. 10A of Modern Plastics, annual)
-, Encyclopédie Francaise des Matières Plastiques (in French), Les Publicateurs Techniques Association, Paris (annual)

1.2. Plastics: Textbooks of Plastics

J.H.DuBois, F.W.John, Plastics, Van Nostrand-Reinhold, New York, 5th ed. 1976
D.C.Miles, J.H.Briston, Polymer Technology, CRC, New York 1979
C.Hall, Polymer Materials, Wiley, New York 1981
J.A.Brydson, Plastics Materials, Butterworths, London, 4th ed., 1982
H.Batzer, ed., Polymere Werkstoffe, Thieme, Stuttgart 1984 (3 vols., in German)
H.Ulrich, Introduction to Industrial Polymers, Hanser Publ., Munich, 2nd 1992
A.W.Birley, B.Haworth, J.Batchelor, Physics of Plastics, Hanser, Munich 1992

1.2. Plastics: Designation

K.Wiebusch, Kunststoffe-German Plastics **72** (1982) 22, 167
F.O.Swanson, Plastics Engng. (January 1983) 31
R.H.Wehrenberg, II, Materials Engng. (February 1984) 48

1.2. Plastics: Data Banks

-, The International Plastics Selector, San Diego, CA, 1977
W.W.Flick, Industrial Synthetic Resins Handbook, Noyes Data Corp., Park Ridge, NJ 1985
M.B.Ash, I.A.Ash, Handbook of Plastic Compounds, Elastomers, and Resins, An International Guide by Category, Tradename, Composition, and Supplier, VCH, New York 1992

POLYMAT, German Plastics Institute, Darmstadt, and Fachinformationszentrum Chemie, Berlin (databank with 30-50 properties for each of the 7000 plastics from 80 manufacturers);
Plastics Databank ®CAMPUS (= Computer Aided Material Preselection by Uniform Standards) (total of ca. 5000 plastics with up to 50 of the most important properties each); computer disks (mainly for IBM and IBM compatibles) available from each of the 23 European manufacturers (incl. European subsidiaries of U.S. companies), which joined the system.
PLASPEC, FIZ Chemie, Berlin (up to 80 properties for each of ca. 12 000 polymers)
-, Plastics Databases (four different databases for engineering thermoplastics, engineering thermosets, nylons, and composites), ASM International, Materials Park, OH 44073-0002

1.2. Plastics: Abbreviations and Acronyms

Nomenclature Committee, Division of Polymer Chemistry, Inc., American Chemical Society (H.-G.Elias, et al.), Polymer News **9** 1983) 101-110; **10** (1985) 169-172.

1.3. History of Plastics

J.H.DuBois, Plastics History - U.S.A., Cahners, Boston 1971
J.K.Craver, R.W.Tess, eds., Applied Polymer Science, Am.Chem.Soc., Washington, D.C., 1975
F.M.McMillan, The Chain Straighteners: Fruitful Innovation. The Discovery of Linear and Stereoregular Polymers, MacMillan, London 1981
R.B.Seymour, ed., History of Polymer Science and Technology, Dekker, New York 1982 (reprints of articles in J.Macromol.Sci.-Chem. **A 17** (1982) 1065-1460)
R.Friedel, Pioneer Plastic. The Making and Selling of Celluloid, University of Wisconsin Press, Madison, WI, 1983
W.Glenz, Kunststoffe - Ein Werkstoff macht Karriere, Hanser, Munich 1985
R.B.Seymour, G.S.Kirshenbaum, eds., High Performance Polymers: Their Origin and Development, Elsevier, New York 1986

1.4. Economic Importance: Trade Names of Plastics

The International Plastics Selector, Commercial Names and Sources, Cordura Publ., San Diego, CA, 1978
Fachinformationszentrum Chemie, Parat Index of Polymer Trade Names, VCH, Weinheim, 2nd ed. 1992 (also with computer disk)

1.4. Economic Importance: Production and Consumption Statistics

United Nations, Statistical Yearbook (annually)
United States, Statistical Abstracts (annually); Historical Statistics from Colonial Times to 1970
Modern Plastics, Modern Plastics International (annually, January issue)
Kunststoffe-German Plastics (every four years, October issue)
-, Chemical Economics Handbook, SRI International, Menlo Park, CA (USA) (restricted to participating clients)
Plastics Technology (monthly prices)
Chemical Marketing Reporter (weekly prices)

2. Molecular Structure

2.1. Constitution

2.1.1. Homopolymers

A homopolymer is defined as a polymer derived from one type of monomer. The term homopolymer thus refers to the *source* of the mers in a polymer and not to the actual constitutional units. The homopolymerization of ethylene $CH_2{=}CH_2$ leads, for example, to poly(ethylene) with the structure $\sim[CH_2{-}CH_2]_n\sim$ with ethylene units $-CH_2{-}CH_2-$ as mers and methylene units $-CH_2-$ as constitutional units and repeating units. The structure $\sim[CH_2{-}CH_2]_n\sim$ is idealized since all known ethylene polymerization deliver small proportions of branch units $-CH_2{-}CH(CH_2CH_2)_iR'-$ ($i > 0$) in addition to ideal chain units $-CH_2{-}CH_2-$ (see Chapter 3). The proportion of branch units in these and other polymers depends on the polymerization conditions.

The polymerization of monomers $CH_2{=}CHR$ leads to monomeric units (mers) $-CH_2{-}CHR-$. Mers, constitutional units, and repeating units are identical in this case. However, the mers $-CH_2{-}CHR-$ may be connected to each other not only in head-to-tail positions but also in head-to-head and tail-to-tail arrangements, another deviation from ideality:

$-CH_2{-}CHR{-}CH_2{-}CHR-$	$-CH_2{-}CHR{-}CHR{-}CH_2-$	$-CHR{-}CH_2{-}CH_2{-}CHR-$
head-to-tail	tail-to-tail	head-to-head

Polymer molecules with head-to-tail, tail-to-tail, and head-to-head arrangements of monomeric units are said to show *regioisomerism*. The proportion of head-to-head and tail-to-tail regioisomeric units is usually not very large. It increases with decreasing size of substituents R and with reduced resonance stabilization of the growing species. For example, tail-to-tail and head-to-head connections in free-radically polymerized polymers amount to 1-2 % in poly(vinyl acetate) $\sim[CH_2{-}CH(OOCCH_3)]_n\sim$, 6-10 % in poly(vinyl fluoride) $\sim[CH_2{-}CHF]_n\sim$, and 10-12 % in poly(vinylidene fluoride) $\sim[CH_2{-}CF_2]_n\sim$. Tail-to-tail and head-to-head connections may constitute "weak links" in degradation.

Another source of nonideal structures may be the polymerization of monomer molecules via "wrong" groups. The proportion of the resulting isomeric units depends on the constitution of the monomer and on the mode of polymerization. Acrolein (acrylaldehyde, propenal) I, for example, polymerizes free radically via the carbon-carbon double bond to units II or via the aldehyde group to units III, gives branched units IV by polymerization of both the carbon-carbon double bond and the aldehyde group of the same monomer molecule, forms cross-linked units via IV, and may even generate intramolecular rings V from two units IV:

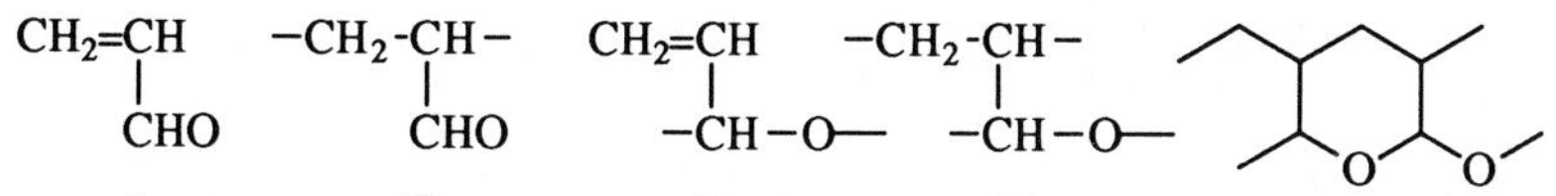

$$\underset{\text{I}}{\begin{matrix}CH_2{=}CH \\ | \\ CHO\end{matrix}} \qquad \underset{\text{II}}{\begin{matrix}-CH_2\text{-}CH- \\ | \\ CHO\end{matrix}} \qquad \underset{\text{III}}{\begin{matrix}CH_2{=}CH \\ | \\ -CH-O-\end{matrix}} \qquad \underset{\text{IV}}{\begin{matrix}-CH_2\text{-}CH- \\ | \\ -CH-O-\end{matrix}} \qquad \text{V}$$

Prospective monomers may also isomerize before or during polymerizations to true monomers. Heptene-3, $C_2H_5{-}CH{=}CH{-}C_3H_7$, isomerizes with certain transition metal catalysts via heptene-2 to the true monomer heptene-1, $CH_2{=}CH(C_5H_{11})$, which polymerizes. Isomerization during propagation are especially prevalent in cationic polymerizations. 4,4-Dimethylpentene-1 (VI) polymerizes via the carbon-carbon double bond (VII) at low temperature (-130°C), but leads at a higher temperature (0°C) to a *phantom polymer* (VIII) with three carbon atoms per monomeric unit:

$$\text{(2-1)} \quad \underset{\text{VII}}{\begin{matrix}\sim CH_2-CH\sim \\ | \\ CH_2-C(CH_3)_3\end{matrix}} \xleftarrow{-130°C} \underset{\text{VI}}{\begin{matrix}CH_2{=}CH \\ | \\ CH_2-C(CH_3)_3\end{matrix}} \xrightarrow{0°C} \underset{\text{VIII}}{\begin{matrix}\sim CH_2-CH_2-CH\sim \\ | \\ C(CH_3)_3\end{matrix}}$$

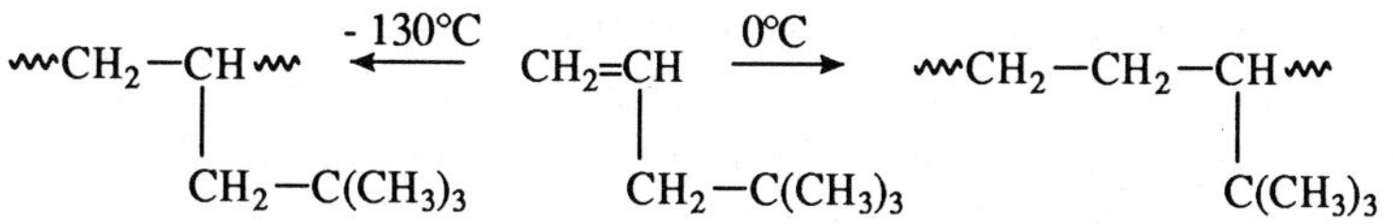

The presence of such isomeric units can often neither proved nor disproved. Polymer structures are therefore in general depicted by the constitution of their most prevalent constitutional repeating units; they are always idealized structures. It is important to distinguish between source-based mers and structure-based constitutional units and constitutional repeating units (Table 2-1).

Table 2-1 Mers, constitutional units, and constitutional repeating units of some polymers.

	Mers	Constitutional units	Constitutional repeating units
Poly(ethylene)	$-CH_2{-}CH_2-$	$-CH_2-$	$-CH_2-$
Poly(propylene)	$-CH_2{-}CH(CH_3)-$	$-CH_2{-}CH(CH_3)-$	$-CH_2{-}CH(CH_3)-$
Poly(hexamethylene adipamide); polyamide 6.6	$-NH(CH_2)_6NH-$ + $-OC(CH_2)_4CO-$	$-NH(CH_2)_6NH-$ + $-OC(CH_2)_4CO-$	$-NH(CH_2)_6NH{-}OC(CH_2)_4CO-$

The groups at the ends of polymer chains are usually not shown, in part, because they are often unknown, and in part because their structure does not influence most of the polymer properties. *Endgroups* may be unreacted functional groups of monomers (polycondensations, polyadditions) and initiator fragments from initiation (and termination!) reactions or groups that are generated by transfer reactions (chain polymerizations). These groups rarely affect mechanical properties; they may, however, negatively influence the thermal and/or photochemical stability of polymers, i.e., the weatherability of plastics.

2.1.2. Copolymers

Copolymers are generated from more than one type of monomers; they are called bipolymers, terpolymers, quaterpolymers, etc., according to the number of monomer types (source-based). Copolymers were named interpolymers in the past; today, they are sometimes called multipolymers.

The copolymerization of ethylene (ethene), $CH_2=CH_2$, and propylene (propene), $CH_2=CH(CH_3)$, thus leads to the bipolymer poly(ethylene-*co*-propylene). Polymers of 1,3-butadiene, $CH_2=CH-CH=CH_2$, that contain 1,4-units $-CH_2-CH=CH-CH_2-$ and 1,2-units $-CH_2-CH(CH=CH_2)-$ are not called copolymers because they are generated from only one monomer type. They are rather *pseudocopolymers* such as the ones resulting from the partial saponification of poly(vinyl acetate) $\sim[CH_2-CH(OOCCH_3]_n\sim$ to polymers with new vinyl alcohol units $-CH_2-CH(OH)-$ and remaining vinyl acetate units $-CH_2-CH(OOCCH_3)-$. For historical reasons, AABB condensation polymers are also not considered copolymers although they are derived from two different monomers, for example, poly(hexamethylene adipamide) (polyamide 6.6; nylon 6.6) from $H_2N(CH_2)_6NH_2$ and $HOOC(CH_2)_4COOH$.

The succession of monomeric units in copolymer chains is known as their sequence. Monomeric units "a" and "b" alternate in *alternating copolymers* $\sim[a-b]_n\sim$ (Table 2-2). Alternating copolymers are limiting cases of *periodic copolymers* such as $\sim[a-b-b]_n\sim$, $\sim[a-a-b-b]_n\sim$, and $\sim[a-b-c]_n\sim$.

Table 2-2 Types of copolymers from monomers A, B and/or C. Monomeric units are characterized in "Structure" by small letters, in the IUPAC "Shorthand name" by capital ones.

Name	Structure	Shorthand name
Copolymer without specified sequence	~(a/b)~	poly(A-*co*-B)
Statistical copolymer (example)	~a-b-b-b-a-a-b-a-a-a-a-b~	poly(A-*stat*-B)
Random copolymer (with Bernoulli statistics of sequence)	ditto	poly(A-*ran*-B)
Alternating copolymer	~a-b-a-b-a-b-a-b-a-b-a-b~	poly(A-*alt*-B)
Periodic copolymer (example)	~a-b-b-a-b-b-a-b-b-a-b-b~	poly(A-*per*-B-*per*-B)
(example)	~a-b-c-a-b-c-a-b-c-a-b-c~	poly(A-*per*-B-*per*-C)
Diblock copolymer	~a..........a-b...........b~	poly(A)-*block*-poly(B)
Triblock copolymer	~a.......a-b.......b-c......c~	poly(A)-*block*-poly(B)-*block*-poly(C)
Graft copolymer	~a-a-a.........a-a-a.........a~ (with side chains b_m and b_n attached)	poly(A)-*graft*-poly(B)

The sequence of monomeric units in *statistical copolymers* is determined by the statistics of copolymerizations (e.g., Markov statistics of zeroth, first, etc., order). *Random copolymers* are special cases of statistical copolymers: the sequence of monomeric units follows a Bernoulli statistics (i.e., zeroth order Markov statistics).

Graded copolymers (= tapered copolymers) exhibit compositional gradients along the chain: one chain end is enriched in "a" units, the other in "b" units. *Block copolymers* are extremal cases of such graded copolymers; they consist of blocks of homosequences that are joined via their ends. Linear multiblock copolymers with short blocks are called segment copolymers or *segmented copolymers*. In *graft* copolymers, "b" side-blocks are connected to "a" main chains via center monomeric units.

Properties of statistical copolymers poly(A-*stat*-B), alternating copolymers poly(A-*alt*-B), block copolymers poly(A)-*block*-poly(B), graft copolymers poly(A)-*graft*-poly(B), etc., of monomers A and B are quite different from each other and also from blends of homopolymers poly(A) and poly(B).

2.1.3. Branched Polymers

Open chain structures of the type R–m_n–R exhibit the simplest chain type. They possess n monomeric units "m" and 2 endgroups R. They are also called *linear chains* (Fig. 2-1, L) because they were originally (and wrongly) assumed to be completely stretched out and the term "linear" now refers to their one-dimensional connectivity (their unbranched structure). Spiropolymers and ladder polymers are also unbranched polymers (see below).

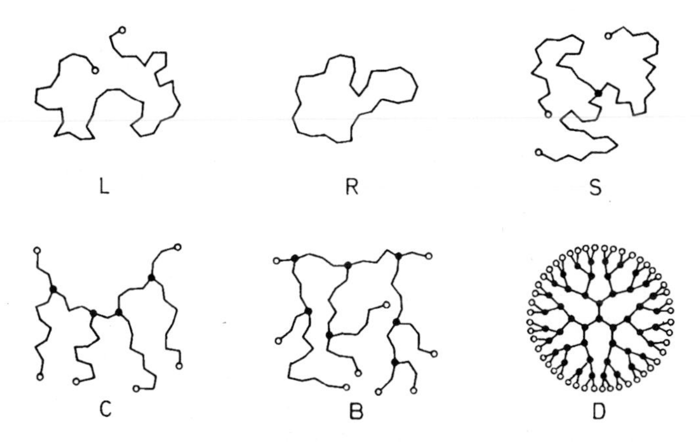

Fig. 2-1 Schematic representation of two-dimensional projections of linear and branched macromolecules (projected bond angles shown only). O Endgroups, ● branching points.
B randomly branched chain with subsequent branching;
C comb-like molecule: a main chain with side chains of different constitution;
D dendrimer with 3-functional branch points and five generations of subchains per core;
L linear (unbranched) chain in the macroconformation of a random coil;
R ring (cyclic) molecule;
S star-like molecule with three subchains.

Cyclic polymers (ring polymers, Fig. 2-1, R) consist of linear polymer molecules that are intramoleculary joined via their own ends; they do not have endgroups. Cyclic macromolecules are not called macrocycles because that term is traditionally used by organic chemistry for "big rings" of ca. 15-20 chain atoms.

Branched polymers contain branch points that connect three or more subchains (sometimes called subunits). Branched polymers comprise the following types:

Star polymers possess one branching point from which 3 or more subchains radiate (Fig. 2-1, S).

Dendrimers (dendritic polymers) are star polymers in which the subchains are themselves star-like branched (Fig. 2-1, D). They are also called cascade polymers (because of the cascade-like sequence of branching points), isotropically branched polymers or iso-branched polymers (since each generation of newly added monomer molecules is more or less "isotropically" arranged around the central core), or, much more poetically, "starburst polymers™".

Comb polymers have "long" side chains, whose chemical structures usually differ from those of the main chains (Fig. 2, C). Comb polymers whose side chains exhibit liquid-crystalline (LC) behavior are called *LC side-chain polymers*. These two types of comb polymers are not considered branched polymers if the "teeth" were originally part of the monomer molecules. The term *graft copolymer* refers to those comb polymers that were synthesized by grafting of monomers or polymer chains onto or from a primary chain; by definition, graft copolymers are branched.

Randomly branched polymers have random distributions of branching points (Fig. 2-1, B); the subchains may be further branched (Christmas tree branching) or not. Short-chain and long-chain branches are distinguished. Short-chain branches are often generated by intramolecular transfer reactions, for example, butyl side chains of poly(ethylene) formed by backbiting during free radical polymerizations of ethylene:

(2-2) $\sim\!C(H)(H)-CH_2-CH_2-CH_2-\dot{C}H_2$ (six-membered ring transition state) $\longrightarrow$ $\sim\!\dot{C}H-CH_2-CH_2-CH_2-CH_3$ $\xrightarrow{+\,CH_2=CH_2}$ $\sim\!CH(CH_2-\dot{C}H_2)-CH_2-CH_2-CH_2-CH_3$

Long-chain branches are usually generated by *inter*molecular transfer reactions. In the polymerization of vinyl acetate, side-chain radicals formed by transfer reactions

(2-3) $$\sim\!CH_2-\dot{C}H(OOC-CH_3) + CH_3-COO-CH\!\sim \longrightarrow \sim\!CH_2-CH_2(OOC-CH_3) + \dot{C}H_2-COO-CH\!\sim$$

may start the polymerization of further vinyl acetate molecules. Such side-chain radicals are also formed by radical transfer from certain initiator radicals.

Side chains introduced by the inherent structure of a (co)monomer are not called branches in polymer science. Copolymers of ethylene with small proportions of α-olefins $CH_2=CH(CH_2)_iH$ ($i \geq 4$) are even called *linear* low density poly(ethylene)s.

Branching decreases the ordering of chains, which in turn lowers crystallinities and densities of polymers. The "characteristic density" of a (10·10·4) mm^3 test specimen at 23°C is used as a measure of the branching of poly(ethylene)s in the CAMPUS® system (ISO 1183; DIN 53 479 A).

2.1.4. Ordered Chain Assemblies

Two or more chains may also be joined in an ordered way by chemical bonds at more than two points per chain. The chains are thought to be fully extended; for classification purposes; a distinction is made between one-, two- or three-dimensional assemblies or, according to IUPAC, between catena ("1" in Fig. 2-2), phyllo ("2" in Fig. 2-2) and tecto ("3" in Fig. 2-2) structures.

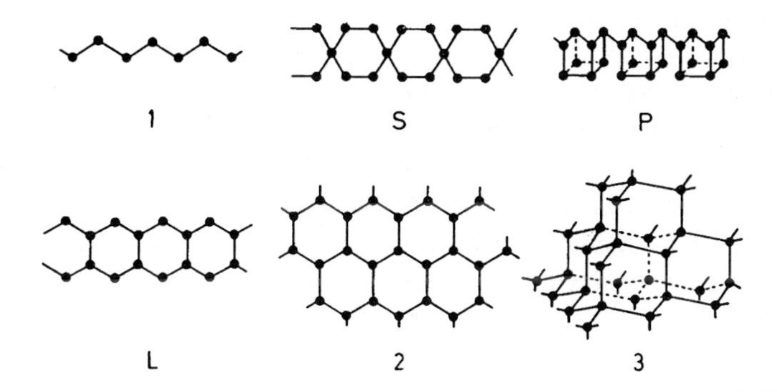

Fig. 2-2 Schematic representation of some ordered chain assemblies. ● Chain atoms (substituents not shown). 1 = Catena polymer in all-trans conformation, 2 = phyllo polymer, 3 = tecto polymer, S = spiro polymer, P = pearl string-like polymer, L = ladder polymer.

In the IUPAC nomenclature, linear polymers are catena polymers (Lat.: *catena* = chain) since the fully extended chains extend in only one direction. Spiro polymers can be considered either as catena polymers from rings as units that are connected to each other at one point or as two-chain polymers whose chains share common joint points at regular intervals (S in Fig. 2-2).

Ladder polymers consist of two chains joined by bonds in regular intervals (for example, L in Fig. 2-2); they are also called *double-strand polymers*. Such ladder polymers are usually generated by first forming a linear chain with pendent polymerizable side groups that are subsequently polymerized. An example is the polymerization of 1,3-butadiene to 1,2-poly(butadiene) with subsequent cyclization to cyclo-poly(butadiene):

(2-4) $CH_2{=}CH{-}CH{=}CH_2 \longrightarrow \sim\sim CH_2{-}CH(CH{=}CH_2)\sim\sim \longrightarrow$

Ladder polymers may also be synthesized by two-step polycondensation or polyaddition reactions, in general, however, only fairly short ladder-like units are formed this way (see below). One-step ladder syntheses are very rare. The polymerization of the silicon analogue of cubane, $(C_6H_5)_8Si_8O_{12}$, does not lead to a ladder polymer of type L in Fig. 2-2 but to a pearl-string like polymer P. Ladder polymers and double-ladder polymers are common amongst silicates.

Phyllo polymers are also called parquet or *layer polymers* ("2" in Fig. 2-2). Graphite and the silicates montmorillonite, bentonite, and mica are well known examples. Cell walls of bacteria consist of bag-shaped polymers, i.e., layer polymers. Diamond and quartz are examples of tecto polymers ("3" in Fig. 2-2).

2.1.5. Unordered Networks

Multifunctional oligomer and polymer molecules can be interconnected via chain groups or endgroups to cross-linked polymers (networks). The cross-linking points must be at least trifunctional, i.e., the cross-linkable groups in the polymer must be at least bifunctional. The cross-linking generates bridges between chains. The bridges can be short or long; the distribution of cross-linking points, at random (homogeneous networks) or in clusters (heterogeneous networks) (Fig. 2-3). Both chemical and physical bonds can be employed for cross-linking; chemically and physically cross-linked polymers are thus distinguished.

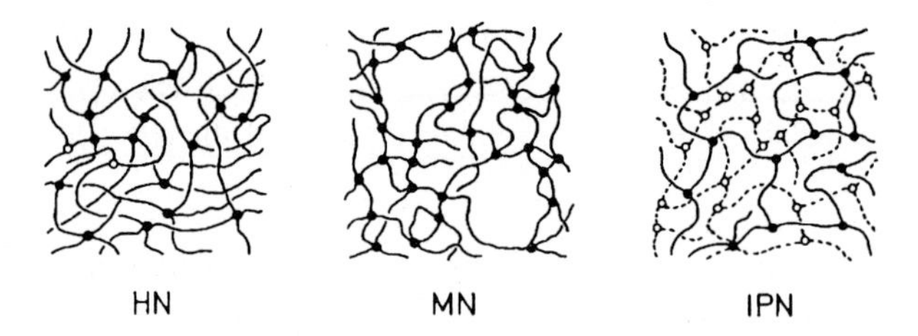

Fig. 2-3 Schematic representation of chemically cross-linked polymers. HN = Homogeneous network with branching points O and tetrafunctional cross-linking points ●; MN = macroreticular network with tetrafunctional cross-linking points; IPN = interpenetrating network of polymer chains A (solid lines) with trifunctional cross-linking points ● and polymer chains B (broken) with trifunctional cross-linking points O.

Chemical Networks. The structure of chemical networks depends on the relative amount, functionality, and distribution of cross-linking sites and branches generated by the cross-linking reaction. Structures of chemical networks are in first approximation independent of the chemical structure of the cross-linking sites.

Networks are fairly homogeneous with respect to the distribution of cross-links if they are prepared by polymerization in bulk or from homogeneous solutions and if the corresponding linear chains are soluble in their own monomers or the applied solvent (HN in Fig. 2-3). Inhomogenous (heterogeneous) networks are formed if phase separation occurs during polymerization, e.g., if the polymers or oligomers generated during the early stages of the polymerization process are insoluble in the applied solvent or the remaining monomer. The resulting *macroreticular networks* show cavities (MN in Fig. 2-3).

In principle, two chemically different networks poly(a) and poly(b) can co-exist independent of each other in *interpenetrating networks* (IPN's). In reality, these networks are not molecularly interdispersed; rather, each network forms interconnected domains with higher cross-link densities. Semi-interpenetrating networks consist of cross-linked poly(a) in uncross-linked poly(b).

The proportion of the cross-linking sites with respect to the total number of mers determines the cross-link density. Light cross-linking (low cross-linking density) does not change the mobility of chain segments between cross-links and the resulting networks behave as elastomers if the temperature is above the glass temperature of the chain segments (see Chapter 5). Chemical networks with high cross-linking densities and/or chain segments at temperatures below the glass temperature are thermosets.

Chemical cross-links are in most cases irreversible. These cross-linked polymers with their "infinitely high" molar masses do not dissolve in solvents. With increasing temperature, subchains would in general rather decompose than revert to the original monomers. Chemically reversible cross-linked networks are presently investigated since they would allow the recycling of thermosets.

Physical Networks. Physical cross-links are formed by topological restraints between two polymer chains or by assemblies of many polymer chains held together by nonchemical bonds (bonds with low bond strength).

Topological restraints are responsible for the physical cross-links of catenanes, rotaxanes, and entanglements. *Catenanes* consist of two intertwined rings which are not bound to each other (Fig. 2-4). Such structures are well known for deoxyribonucleic acids. They can be prepared synthetically but have yet to find industrial applications.

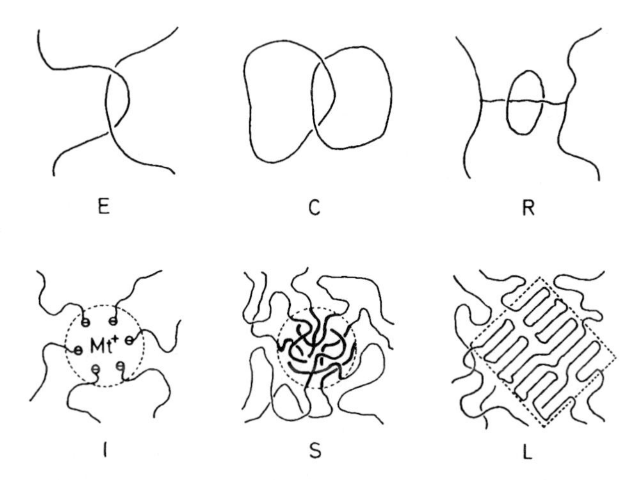

Fig. 2-4 Schematic representation of physical cross-links generated by two chains (upper row) or segments of many chains (lower row). E = Entanglement; C = catenane; R = rotaxane; I = ion multiplet in ionomers; S = spherical domain in diblock copolymers; L = lamellae of folded chains in semicrystalline polymers.

Rotaxanes are rings on chains that are sealed on both ends by branched or cross-linked structures in such a way that the ring cannot slip out. Such structures are now being synthesized; they exist probably in some cross-linked poly(dimethylsiloxane)s.

Catenanes and rotaxanes are irreversible physical cross-links. *Entanglements* form reversible physical crosslinks between two high molar-mass chains. They behave under strain as cross-links since the two chains cannot diffuse away rapidly enough; the chains disentangle, however, if enough time is allowed.

Reversible physical cross-links can be also formed by assemblies of segments from various polymer chains. Examples are ion multiplets around metal ions with high coordination numbers, ion clusters, microdomains generated by immiscible blocks of block copolymers, or crystallites in partly crystallized polymers. Such microdomains disassemble (dissociate, melt) at higher temperatures; physically cross-linked polymers can also often be dissolved in suitable solvents.

2.2. Molar Mass and Molar Mass Distribution

Molar masses of polymers can be varied over a wide range; they may extend from several hundreds to several millions, albeit not for every polymer type. The industrially used molar mass ranges depend on the limitations of the available polymerization procedures, on the restrictions imposed by processing methods, and on the desired properties of the plastics.

Presently available commercial polymerization methods all employ statistical approaches to initiation and propagation of growing chains, as well as to termination and transfer, where applicable. The polymer molecules of the resulting polymers have a distribution of molar masses; the polymers are not molecularly uniform and the experimentally determined molar masses are averages. Synthetic copolymers are, in addition, nonhomogeneous with respect to the distribution of the different types of monomer units; they are polydisperse with respect to both constitution and molar mass. These features allow great variations of physical properties at constant chemical composition because individual physical properties rely differently on the various averages and distributions of monomeric units and molar masses of polymers.

In contrast, biopolymers are synthesized via matrix polymerizations. Many (if not all) biopolymers are monodisperse in their biological environments: each macromolecule of a polymer has the same constitution and the same molar mass. Examples are nucleic acids and enzymes. On isolation from living cells and further processing, degradation may occur. Industrial celluloses thus have molar mass distributions.

The recently discovered Starburst® dendrimers™ are molecularly uniform synthetic polymers. Their molecular homogeneity can approach that of isotope distributions in atoms and molecules; it is caused by a self-limitation of divergent growth due to surface crowding after a certain number of growth generations if the polymerization is started from a core.

2.2.1. Molar Mass Averages

Polymer properties often depend on the type and width of molar mass distributions; it is thus desirable to know these distributions. In many cases, molar mass averages are determined instead, in part, because it is experimentally easier, in part, because a number is easier to grasp than a function.

Most molar masses M are determined by physical methods which deliver the mass per amount of substance (physical unit: g/mol). These molar masses are traditionally called "molecular weights" although they are not "weights" (a weight is a force; usually measured in newtons). Chemical methods such as endgroup determinations usually determine the molecular mass of a substance relative to the molecular mass of a standard. These quantities with the physical unit "1" are "dimensionless"; they are *relative molecular masses* M_r and can be called *molecular weights*.

Molar mass averages. The averages employed in polymer science are usually arithmetic averages. Polymer molecules of size i (degree of polymerization X_i or molar mass M_i) are counted according to their statistical weights g_i, which may be their mole fraction $x_i \equiv n_i/\Sigma_i n_i$, mass fraction $w_i = m_i/\Sigma_i m_i = x_i M_i / \overline{M}_n$, or z-fraction defined as $Z_i \equiv z_i/\Sigma_i z_i = w_i M_i/\overline{M}_w = x_i M_i^2/(\overline{M}_n \overline{M}_w)$. In these expressions, n is the amount of substance (in e.g. moles of molecules), m the mass, M the molar mass, and $\overline{M}_n$ and $\overline{M}_w$ the number- and mass-average molar masses, respectively.

The number-, mass-, and z-average molar masses are defined as

$$\overline{M}_n \equiv \sum_i x_i(\overline{M}_n)_i = \frac{1}{\sum_i (w_i / (\overline{M}_n)_i)} \tag{2-5}$$

$$\overline{M}_w \equiv \sum_i w_i(\overline{M}_w)_i = \frac{\sum_i x_i(\overline{M}_n)_i(\overline{M}_w)_i}{\sum_i x_i(\overline{M}_n)_i} \tag{2-6}$$

$$\overline{M}_z \equiv \sum_i Z_i(\overline{M}_w)_i = \frac{\sum_i w_i(\overline{M}_w)_i(\overline{M}_z)_i}{\sum_i w_i(\overline{M}_w)_i} = \frac{\sum_i x_i(\overline{M}_n)_i(\overline{M}_w)_i(\overline{M}_z)_i}{\sum_i x_i(\overline{M}_n)_i(\overline{M}_w)_i} \tag{2-7}$$

Equations (2-5)-(2-7) are valid for polymers with negligible contributions of molar masses of endgroups to the molar mass of the molecules. For molecular homogeneous components, molar mass averages of the components i become identical, i.e., $(\overline{M}_z)_i = (\overline{M}_w)_i = (\overline{M}_n)_i = M_i$. Polymers with identical i exhibit $\overline{M}_n = \overline{M}_w = \overline{M}_z$. Polymers with nonidentical i are often characterized by ratios of molar masses, for example, the polymolecularity index ("polydispersity index") $\overline{M}_w / \overline{M}_n \geq 1$.

Each of the three molar mass averages $\overline{M}_z$, $\overline{M}_w$, and $\overline{M}_n$ is essentially a single moment of the distribution of molar masses present. The so-called *viscosity average molar mass*, on the other hand, is the ath root of the ath moment of the mass distribution of molar masses (see below). There are also molar mass averages that are composed of more than one moment, for example, molar masses from sedimentation and diffusion coefficients for certain types of molecule shapes and solvent interactions.

2.2.2. Determination of Molar Masses

Many methods exist for the determination of molar masses. Some are absolute methods that require neither a knowledge of the chemical structure of the polymer nor (in principle) a calibration of the instruments; examples are membrane osmometry and static light scattering. Relative methods such as viscometry demand calibrations with similar specimens. Equivalent methods need a knowledge of chemical structures, for example, in endgroup determinations.

Membrane osmometry is the most important absolute method for the determination of number-average molar masses. It measures the osmotic pressure Π of polymer solutions at various polymer concentrations c against the neat solvent. Solution and solvent are separated by a semipermeable membrane that is permeable to the solvent but not to the polymer molecules.

Older osmometers determined osmotic pressures in true thermodynamic equilibrium, which was reached only after days and weeks because great volumes of solvent had to be moved through the membrane. Modern instruments compensate for the increase in pressure difference between solution and solvent caused by the flow of solvent into the solution chamber by a change of solvent pressure via a servomechanism; equilibrium pressures are established after 10-30 min. Reduced osmotic pressures, Π/c are extrapolated to zero concentration to give the number-average molar mass according to the van't Hoff equation, $\overline{M}_n = RT/[\lim_{c\to 0} (\Pi/c)]$ where R is the gas constant and T the thermodynamic temperature. Number-average molar masses can be determined by osmometry in the range 10^4 g/mol (lower limit of semipermeability of membranes) to ca. 10^6 g/mol (upper limit of sensitivity). Membrane osmometry is an absolute method, that is, number-average molar masses are obtained from the experimental parameters Π, c, and T without any additional assumption.

Ebullioscopy and **cryoscopy** also deliver number-average molar masses. They are rarely used in polymer science because they are insensitive to higher molar masses (upper limit of M ca. 20 000 g/mol) and often are experimentally difficult.

Static light scattering of dust-free dilute solutions determines mass average molar masses (molecular Tyndall effect). The scattering intensity i_s is normalized by the number concentration N/V of the solute and the intensity I_o of the incident light to give the so-called Rayleigh ratio $R_\vartheta = i_s(N/V)/I_o$. Modern instruments measure the Rayleigh ratio $R_0 = Kc(M_w)_{app}$ at very low angles $\vartheta \approx 0$ relative to the incident light beam (*laser low angle light scattering*, LLALS), where K is an optical constant and $(M_w)_{app}$ an apparent mass-average molar mass which is calculated from R_0, c, and K. Values of $Kc/R_0 \equiv 1/(M_w)_{app}$ are determined at various polymer concentrations and extrapolated to zero concentration. The limiting value $\lim_{c\to 0} (Kc/R_0)$ delivers the inverse mass-average molar mass, $1/\overline{M}_w$. Molar masses from several hundred g/mol to several million g/mol can be determined by light scattering, the upper limit being given by experimental problems such as multiple scattering. From measurements at a wide range of scattering angles ϑ, important additional information about the radii of gyration of the molecules can be gained from the angular dependence of the Rayleigh ratio.

Dynamic light scattering (quasielastic light scattering) measures relaxation times of translational motions of the center of mass (Brownian motion) and the segmental motions within a flexible molecule. The Brownian motion determines the diffusion coefficient D, which is obtained as a z-average; D is related to the molar mass M by $D = KM^{-b}$. The exponent b depends on the shape of the solute particles or molecules. Dynamic light scattering is thus a relative method, which requires knowledge of the calibration function $D = f(M)$.

Viscometry is fast and simple and therefore the most often applied method for the determination of molar masses. From solvent viscosity η_1 and viscosities η of dilute solutions of various concentrations c (measured at low shear rates), reduced viscosities (i.e., viscosity numbers) are calculated via $\eta_{red} = (\eta - \eta_1)/(\eta_1 c)$. The reduced viscosities are extrapolated to zero concentration to give the *intrinsic viscosity* (limiting viscosity number), $[\eta]$; DIN recommends the symbol J_0. This quantity has the physical unit of a specific volume since the concentration c is measured as mass per volume. The recommended physical units are mL/g (cm^3/g) but most literature data in English-speaking countries are reported in 100 mL/g = dL/g; in the old German literature, $[\eta]$ was given as Z_η in L/g. Instead of intrinsic viscosities $[\eta]$ at zero polymer concentrations, viscosity numbers $(\eta_{red})_c$ or *inherent viscosities* $\{[\ln (\eta/\eta_1)]/c\}_{c = \mathrm{const}}$ at standard concentrations c are often used to characterize molar masses.

Older thermoplastics such as poly(vinyl chloride) are still characterized industrially by Fikentscher's *K-values*, which are calculated from viscosities and concentrations with the help of empirical conversion tables. *K-values* were assumed to be independent of polymer concentrations; however, they are insensitive to variations in high molar masses and should not be confused with the K of Equation (2-9).

Intrinsic viscosities, if given in mL/g, measure the volume in mL occupied by 1 g of polymer at $c \to 0$, i.e., the specific volumes (inverse densities) of isolated polymer molecules. Since the molecule density varies with molecular size for most molecule shapes, intrinsic viscosities can be used to measure molar masses for a homologous series of polymer molecules; they do not measure molar masses *per se*.

The viscosity increment η_i ("specific viscosity") of unsolvated, compact ("hard") spheres in dilute solutions of solvent 1 is given by the Einstein equation

$$(2\text{-}8) \qquad \eta_i = (\eta - \eta_1)/\eta_1 = (5/2)\,\phi = (5/2)\,(N_A V_H c/M)$$

where $\phi = V_2/V$ = volume fraction of polymer 2 in the volume V of the solution, $V_2 = N_2 V_H$ = volume of N_2 spheres with hydrodynamic volumes V_H, $N_2/V = cN_A/M$ = number concentration of spheres, N_A = Avogadro number, $c = m_2/V$ = mass concentration of spheres, M = molar mass of spheres. Compact ("hard") spheres from the same material have the same density $\rho_H = mV_H$. Since $m \sim M$ and $\eta_i/c \approx [\eta]$, intrinsic viscosities of hard spheres are independent of their molar masses, i.e., $[\eta] = K \cdot M^0 = K$.

The hydrodynamic densities and thus also $[\eta]$ of rods, coils etc., change with the molar mass according to the empirical Kuhn-Mark-Houwink-Sakurada equation

$$(2\text{-}9) \qquad [\eta] = KM^a$$

"Viscosity coefficients" are thus used to characterize molar masses according to ISO 1628 + 1191 and DIN 53 728 T4 in the CAMPUS® system.

Viscometry delivers a peculiar molar mass average ("viscosity average") as can be derived from Equation (2-9) with $[\eta] = \Sigma_i w_i[\eta]_i$:

$$(2\text{-}10) \quad \overline{M}_v = ([\eta]/K)^{1/a} = (K^{-1} \Sigma_i w_i[\eta]_i)^{1/a} = (\Sigma_i w_i M_i^a)^{1/a}$$

Numerical values of viscosity-average molar masses, $\overline{M}_v$, lie between number and mass averages for $0 < a < 1$ and become identical with mass averages if $a = 1$.

Theory and experiment indicate a value of $a = 1/2$ for random coils in the unperturbed state (Section 2.4.4); this is a limiting case for very flexible chains in theta solvents. A theta solvent (Θ solvent) is a solvent at a temperature at which a polymer adopts the theta state, i.e., where coil molecules are in their unperturbed states with Gaussian segment distributions (no excluded volume). A theta solvent is a thermodynamically bad solvent, i.e., a solvent which delivers low dilute solution viscosities (this is a "good solvent" in the paint industry).

Theory predicts $a = 0.764$ for nondraining coils of high molar mass in thermodynamically good solvents (Fig. 2-5); at very low molar masses, $a = 1/2$ is expected. Experimentally, intermediate values of $1/2 < a < 0.764$ are often found for flexible chains (such as most thermoplastics in solution) since their molar masses are not high enough for the assumptions of the theory to be fulfilled (see Fig. 2-5).

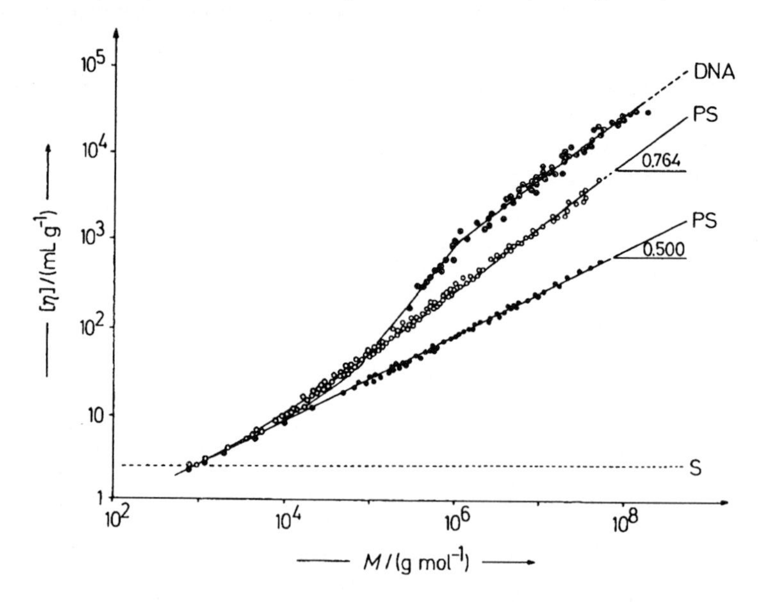

Fig. 2-5 Molar mass dependence of intrinsic viscosities of molecules with various shapes.
PS = Flexible chain molecules (coils) of poly(styrene) in the theta solvent cyclohexane at 34.5°C (●) and in the thermodynamically good solvent benzene at 25°C (O). The slopes at high molar masses adopt the theoretical values of 0.764 (good solvent) and 0.500 (theta solvent).
DNA = Worm-like chains of double helices of deoxyribonucleic acids in dilute salt solutions at room temperature (O).
S = Compact spheres with density of 1 mL/g (- - - -) (e.g., highly branched poly(lysine)).

The exponent a equals 2 for rigid rods of infinite length. Worm-like molecules may adopt values of $0 < a < 2$ at intermediate molar masses since they resemble spheres and ellipsoids at low molar masses, rigid rods at medium ones, and random coils at very high molar masses (cf. the behavior of short and very long garden hoses). Such a behavior is found for stiff molecules such as the double helices of deoxyribonucleic acids in aqueous solutions or certain poly(α-amino acid)s in helicogenic solvents (see Fig. 2-5 and Section 4.3.3.).

2.2.3. Molar Mass Distributions

The types and characteristic parameters of molar mass distributions of polymers are determined thermodynamically or kinetically by the synthesis conditions, and, to a smaller extent, by degradations during the processing of plastics. The molar mass distributions are mathematically described by distribution functions which may be discontinuous (discrete) or continuous and, in addition, differential or integral (cumulative) (Fig. 2-6).

In these functions, polymer molecules of size i (molar mass M_i, degree of polymerization X_i) are considered according to their statistical weights, which may be mole fractions x_i, mass fractions w_i, etc. Theoretical distribution functions describe the mole fractions x_i of degrees of polymerization X_i because they relate to the thermodynamically or kinetically controlled formation of molecules; mole fractions may be converted into mass fractions by $w_i = x_i(\overline{X}_n)_i/\overline{X}_n$. Experimental distribution functions are usually expressed by mass fractions w_i of molar masses M_i, however; molar masses can be converted into degrees of polymerization by $M_i = M_u X_i + M_e$ where M_u = molar mass of monomeric units and M_e = total molar mass of endgroups.

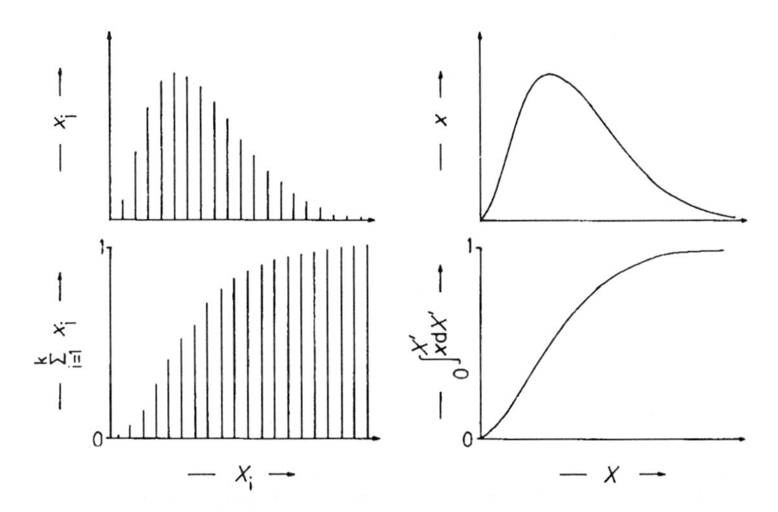

Fig. 2-6 Representations of a distribution of mole fractions x of degrees of polymerization X. Left: discontinuous (discrete); right: continuous; top: differential; bottom: integral (cumulative). Reproduced with permission by Hüthig and Wepf Publ., Basle [1].

All molar mass distributions are discrete since successive polymer molecules of a homologous series differ from each other by a degree of polymerization of 1 and the molar mass M_u of one monomeric unit. Discrete distribution functions can be normally replaced by continuous ones since there are so many different degrees of polymerization and the difference in molar masses between molecules with M_i and M_{i+1} is small compared to the average molar mass.

The various types of distribution functions are usually named after their discoverers. The *Schulz-Zimm distribution* function derives from processes in which reactive chains add indiscrimately monomer or polymer molecules until the chains are deactivated. Such deactivation processes may be chain termination reactions (chain polymerizations, polyeliminations) or the "freezing-in" of equilibrium reactions by lowering the temperature (polycondensation, polyaddition). The mass fraction of polymer molecules of degree of polymerization X_i of polymers with Schulz-Zimm distribution functions of degrees of polymerization, X_i, is given by

$$(2\text{-}11)\quad w_i = \frac{(k/\overline{X}_n)^k X_i^k \exp(-kX_i/\overline{X}_n)}{\Gamma(k+1)}$$

where k is the degree of coupling of chains (e.g., $k = 2$ for recombination of 2 growing chains to 1 dead chain, and $k = 1$ if 2 growing chains react by disproportionation to 2 dead chains), and $\Gamma(k+1)$ the gamma function of $(k+1)$.

The Schulz-Zimm distribution function is called *Schulz-Flory distribution* (sometimes only *Flory distribution*) for the special case of $k = 1$ (polycondensation; radical chain polymerization with termination by disproportionation). Equation (2-11) reduces for $k = 1$ and high degrees of polymerization to

$$(2\text{-}12)\quad w_i = (k/\overline{X}_n)^2 X_i[1-(k/\overline{X}_n)]^{X_i}$$

The various average degrees of polymerization are connected via

$$(2\text{-}13)\quad \overline{X}_n/k = \overline{X}_w/(k+1) = \overline{X}_z/(k+2)$$

Poisson distributions are formed if a constant number of chains starts to grow simultaneously and if monomer molecules are added to the chains at random and independent of previous additions; an example are so-called living polymerizations. The mass fraction of molecules of degree of polymerization X_i is given by

$$(2\text{-}14)\quad w_i = \frac{X_i(\overline{X}_n-1)^{X_i-1}\exp(1-\overline{X}_n)}{(X_i-1)!\,\overline{X}_n}$$

and the interrelationship between the mass- and number-averages by

$$(2\text{-}15)\quad \overline{X}_w/\overline{X}_n = 1+(1/\overline{X}_n)-(1/\overline{X}_n)^2$$

The ratio of mass to number average approaches 1 at infinitely high molar masses. The Poisson distribution is thus a very narrow distribution, in contrast to the Schulz-Zimm distribution where the ratio $\overline{X}_w/\overline{X}_n$ equals 2 for k = 1 and 3/2 for k = 2.

The Schulz-Zimm distribution is a special case of the *Kubin distribution*, an empirical generalized exponential distribution (GEX distribution) with three empirical, adjustable parameters γ, ε, and β:

$$(2\text{-}16) \qquad w_i = \frac{\gamma\beta^{(\varepsilon+1)/\gamma} X_i^{\varepsilon} \exp(-\beta X_i^{\gamma})}{\Gamma[(\varepsilon+1)/\gamma]}$$

This expression converts to the Schulz-Zimm distribution if $\gamma = 1$, $\varepsilon = k$ and $\beta = k/\overline{X}_n$. The Kubin distribution also includes the *Tung distribution* ($\varepsilon = \gamma$-1) and the *logarithmic normal distribution* ($\gamma = 0$; $\varepsilon = \infty$), both of which describe frequently molar mass distributions of polymers from Ziegler-Natta polymerizations. The Kubin distribution is a very adoptable distribution since it contains a "stretched exponential", i.e., a variable X with an exponent γ in the exponential term.

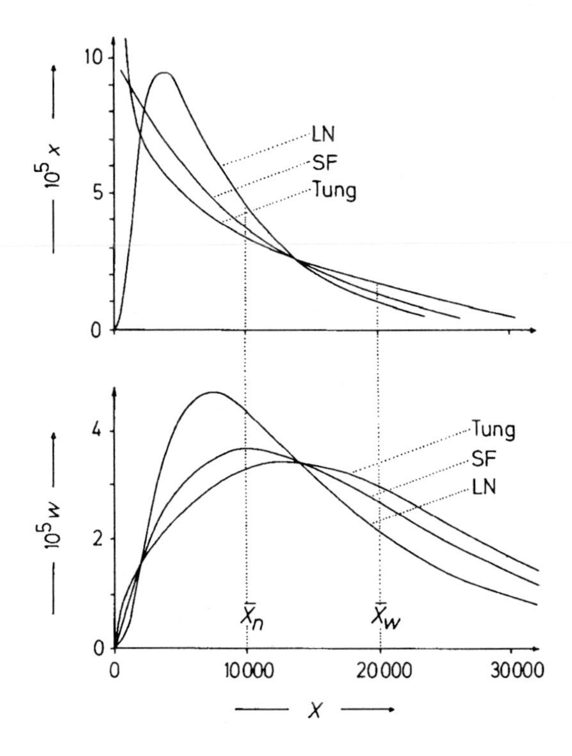

Fig. 2-7 Continuous differential distributions of degrees of polymerization, X. The logarithmic normal distribution (LN), the Schulz-Flory distribution (SF), and the Tung distribution (Tung), are shown as distributions of mole fractions (above) and mass fractions (below) for a polymer with $\overline{X}_n$ = 10 000 and $\overline{X}_w$ = 20 000. The Poisson distribution for a polymer with $\overline{X}_n$ = 10 000 is so narrow ($\overline{X}_w$ = 10 000.9999) that it is practically identical with the vertical line for $\overline{X}_n$ = 10 000. Reproduced with permission by Hüthig and Wepf Publ., Basle [2].

Some of these distribution functions are compared in Fig. 2-7. Note that only the logarithmic normal distribution shows a maximum in the distribution of mole fractions and that none of the maxima in the distributions of the mass fractions corresponds to simple molar mass averages.

2.2.4. Determination of Molar Mass Distributions

Molar mass distributions can be determined by preparative *fractionations* of polymers from solutions since the various species of a polymer-homologous series exhibit small differences in solubility. Fractionation occurs upon the change of temperature or the addition of a nonsolvent. The molar masses of the resulting fractions are determined in separate experiments. Preparative fractionations use inexpensive instrumentation but are very time-consuming.

The method of choice for the fast determination of molar mass distributions is *size exclusion chromatography* (SEC). Dilute polymer solutions are placed on the top of a column filled with a porous carrier. Low molar mass molecules can enter the pores but high molar mass molecules cannot. Medium-sized molecules enter with difficulty and remain for shorter times than low molar mass molecules. Higher molar masses are thus eluted first, and elution curves are observed that give the concentration of eluted molecules as a function of the eluted volume V_e. The maximum of the elution curve is called the retention volume.

The retention volume depends on the SEC system (carrier, solvent, temperature) and the investigated polymer. The carrier consists of materials with pores of 5-500 nm diameter. The materials may be rigid such as porous glass beads or may be swollen, cross-linked polymers such as cross-linked poly(styrene)s or dextrans. In the latter case, the method is also called gel permeation chromatography GPC (synthetic polymers) or gel filtration (biopolymers). Carrier and eluting solvent should have matched solubility parameters in order to avoid distribution effects. The concentration of the leaving solution is determined by refractive index or spectroscopy; the molar mass of the solute may be measured simultaneously by LALLS or viscosity.

Retention volumes are constant below and above certain molar masses of the polymers (Fig. 2-8). The volume V_0 at lower molar masses gives the total volume available for the flow of solvent, the volume V_i at upper molar masses indicates the interstitial volume between carrier particles. Polymer molecules can thus be separated only in the range $V_i < V_e < V_0$, where V_e is the elution volume.

Elution volumes depend to a first approximation on the logarithms of molar masses (Fig. 2-8). A function $V_e = K \ln M$ is thus often used to determine unknown molar masses with the help of a constant K derived from calibrations with narrow-distribution poly(styrene)s. This procedure does not give absolute molar masses because poly(styrene)s and test polymers with the same molar masses possess different chromatographically effective volumes. A frequently used "universal" calibration method thus employs a function $V_e = f(\lg [\eta]M)$ since $[\eta]$ measures the specific hydrodynamic volumes of solute molecules and $[\eta]M$ has the physical unit of a molar volume.

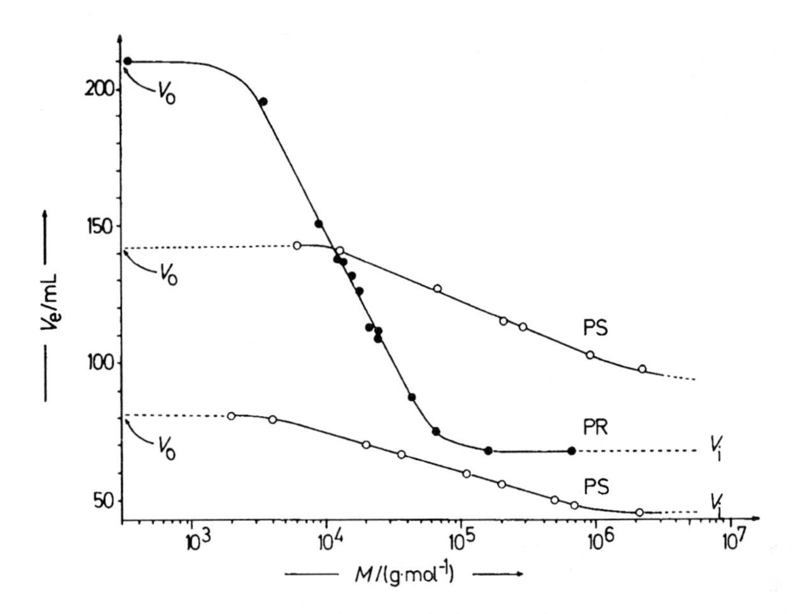

Fig. 2-8 Molar mass dependence of elution volumes, V_e, for (O) coil-like linear poly(styrene)s PS on SEC columns with two different sets of cross-linked poly(styrene)s as carriers and (●) various spheroidal proteins PR on cross-linked dextran. V_o = Total volume, V_i = interstitial volume.

2.3. Configuration

Polymers with symmetric repeating units such as $-CH_2-CH_2-$, $-CH_2-C(CH_3)_2-$, $-O-CH_2-$, $-NH-CO-(CH_2)_5-$, etc., do not possess isomeric configurational units. Isomeric configurational units do exist, however, for polymer molecules with nonsymmetric repeating units such as poly(methylmethylene) $\sim[CH(CH_3)]_n\sim$, poly(propylene) $\sim[CH(CH_3)-CH_2]_n\sim$, poly(propylene oxide) $\sim[O-CH(CH_3)-CH_2]_n\sim$, etc. Poly(propylene), for example, has two configurational repeating units I and II with two different monomeric units each:

$$-C(H)(CH_3)-CH_2- \quad \text{Ia} \qquad -C(CH_3)(H)-CH_2- \quad \text{Ib} \qquad -CH_2-C(CH_3)(H)- \quad \text{IIb} \qquad -CH_2-C(H)(CH_3)- \quad \text{IIa}$$

The configurational units Ia and Ib are enantiomeric; they belong to the same constitutional unit $-CH(CH_3)-CH_2-$. The configurational units Ia and IIa, on the other hand, are based on two different constitutional units. Several configurational units may be joined head-to-tail to give steric repeating units; the three simplest steric repeating units for poly(propylene) are:

$$\underset{\text{IT}}{-\overset{\overset{\displaystyle H}{|}}{\underset{\underset{\displaystyle CH_3}{|}}{C}}-CH_2-} \qquad \underset{\text{ST}}{-\overset{\overset{\displaystyle H}{|}}{\underset{\underset{\displaystyle CH_3}{|}}{C}}-CH_2-\overset{\overset{\displaystyle CH_3}{|}}{\underset{\underset{\displaystyle H}{|}}{C}}-CH_2-} \qquad \underset{\text{HT}}{-\overset{\overset{\displaystyle H}{|}}{\underset{\underset{\displaystyle CH_3}{|}}{C}}-CH_2-\overset{\overset{\displaystyle H}{|}}{\underset{\underset{\displaystyle CH_3}{|}}{C}}-CH_2-\overset{\overset{\displaystyle CH_3}{|}}{\underset{\underset{\displaystyle H}{|}}{C}}-CH_2-\overset{\overset{\displaystyle CH_3}{|}}{\underset{\underset{\displaystyle H}{|}}{C}}-CH_2-}$$

The repetition of these units leads to polymer chains that are called isotactic (IT), syndiotactic (ST), and heterotactic (HT) (from Greek: *isos* = equal, *syn* = together, *dios* = two, *taxis* = arrangement). A heterotactic unit ht in a polymer chain consists of *three* monomeric units (the first or last three of HT), but the repetition of such triads does not lead to a completely heterotactic polymer HT since it would consist of alternating heterotactic and syndiotactic triads. It is immaterial whether Ia or IIa is used as the simplest configurational repeating unit since infinitely long poly(propylene) chains of Ia differ from those of IIa only by the orientation of these units.

The term "tacticity" thus refers to the *relative* arrangements of configurational units in a chain. Relative configurations are classified by starting at one end of the polymer chain and considering the configuration around a central atom relative to the preceding one. This classification is different from that of the absolute configuration of organic chemistry where the configuration of each central atom is determined relative to the ligand with the lowest seniority.

The ligands R of *isotactic polymer* molecules are always "on the same side" if these molecules are shown in Fischer projections (projection of three-dimensional structures unto two-dimensional planes) or in other stereo formulas with hypothetical cis-conformations of the chains (see Fig. 2-9). In stereo formulae based on trans-conformations, ligands are only on the "same side" for isotactic molecules consisting of two chain atoms per monomeric unit, e.g., in $\sim[CHR{-}CH_2]_n\sim$, but not in units with odd numbers of chain atoms such as in $\sim[CH(CH_3)]_n\sim$ or $\sim[O{-}CH(CH_3){-}CH_2]_n\sim$.

Real polymer chains of the types $\sim[CHR]_n\sim$, $\sim[CHR{-}CH_2]_n\sim$, $\sim[O{-}CHR{-}CH_2]_n\sim$ etc., contain configurational mistakes; they are neither 100 % isotactic nor 100 % syndiotactic. The tacticity of such chains is expressed by the fraction x_J of their isotactic or syndiotactic J-ads (diads, triads, etc.). A diad consists of two monomeric units, a triad of three monomeric units, etc. Each monomeric unit of a polymer chain belongs to two tactic diads, three tactic triads, four tactic tetrads, and so on.

The sum of the mole fractions of all J-ads of given J must equal 1 (for diads: $x_i+x_s \equiv 1$, for triads: $x_{ii}+x_{is}+x_{si}+x_{ss} \equiv 1$, etc.), where i = isotactic diad, s = syndiotactic diad, ii = isotactic triad of two isotactic diads, etc. The mole fraction x_i of isotactic diads is given by the mole fraction x_{ii} of isotactic triads ii plus 1/2 of the mole fraction of the sum of the two heterotactic triads is and si, that is, $x_i = x_{ii} + (1/2)(x_{is} + x_{si})$, etc. The

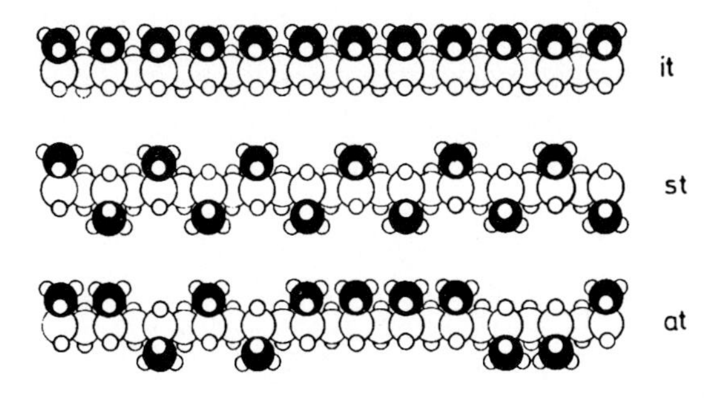

Fig. 2-9 Isotactic (it), syndiotactic (st) and atactic (at) poly(propylene)s ~$[CH_2–CH(CH_3)]_n$~ in (hypothetical!) trans conformations. O Carbon atoms of polymer chains in methylene (CH_2) and methine (CH) groups; ● carbon atoms C in methyl substituents (CH_3); o hydrogen. Reproduced with permission by Springer-Verlag, Heidelberg [3].

number-average sequence length of isotactic sequences (= average number of isotactic diads between two syndiotactic diads) is thus given by $(\overline{X}_I)_n = 2\, x_i/(x_{is} + x_{si})$. A similar expression applies to the number-average sequence length of st-sequences.

The presence and/or proportion of tactic J-ads can be investigated by various experimental methods. 2D-NMR spectroscopy allows the absolute determination of the types of J-ads and their amounts up to pentads whereas conventional high resolution NMR spectroscopy requires the prior knowledge of the J-ad type by X-ray crystallography or other methods. IR spectroscopy detects only diads whereas crystallinity, solubility, chemical reactions, and glass or melting temperatures may or may not indicate the presence of shorter or longer tactic sequences.

Highly isotactic polymers such as it-poly(propylene) or it-poly(butene-1) can be obtained by Ziegler-Natta polymerizations of α-olefins. Special Ziegler-Natta catalysts lead to st-poly(propylene) and st-poly(styrene); these syndiotactic polymers have no commercial use so far. Highly syndiotactic polymers are also generated by most very low-temperature free radical polymerizations of vinyl and acryl monomers. Conventional vinyl and acryl polymers are, however, synthesized at ambient temperatures or above, and consist, at best, of slightly syndiotactic polymers; in general, they are considered atactic polymers. "Atactic" is conventionally used in the sense of "not highly tactic" and not in the strict sense of "random distribution of equal numbers of the possible configurational base units" (Bernoulli statistics), which would give for diads $x_i \equiv x_s \equiv 1/2$, for triads $x_{ii} \equiv x_{is} \equiv x_{si} \equiv x_{ss} \equiv 1/4$, etc. Atactic polymers are thus also called *random polymers* although they may not have a truly random distribution of configurational base units. It is incorrect to name them amorphous polymers because amorphicity and atacticity are not synonyms (see below).

Polymers with units of different geometric isomerism are also tactic. Depending on the configuration of the chain segments relative to the double bonds in the chain, *cis*-tactic (ct) and *trans*-tactic (tt) structures are distinguished. The cis-tactic structure of 1,4-polyprenes ~$[CH_2–CR=CH–CH_2]_n$~ from a polymerization via both double bonds of 1,3-dienes $CH_2=CR–CH=CH_2$ corresponds to an E isomer; the trans-tactic structure, to a Z isomer. If the polymerization proceeds via one double bond only,

1,2-units $-CH_2-CR(CH=CH_2)-$ or 3,4-units $-CH_2-CH(CR=CH_2)-$ are formed that may be arranged in isotactic or syndiotactic diads, triads, etc.

cis-1,4 (E)	trans-1,4 (Z)	1,2 (it or st)
$R(CH_2\sim)C=C(H)(CH_2\sim)$	$(\sim H_2C)(R)C=C(H)(CH_2\sim)$	$\sim CH_2-CR(CH=CH_2)\sim$

2.4. Conformation

2.4.1. Microconformations

Rotations of atoms or groups of atoms around single bonds create spatial arrangements called conformations in organic chemistry and local conformations or microconformations in macromolecular chemistry. The sequence of these microconformations determines the shape of the macromolecule, i.e., the macroconformation (in statistical mechanics: the "configuration").

In principle, an infinite number of conformations is possible around each single bond. In practice, certain positions are energetically preferred; only these are called (micro)conformations. In two joined tetrahedrons, such as ethane H_3C-CH_3, two extremes of energetically different positions are possible: the staggered position corresponds to a minimum of energy and the eclipsed position to a maximum of energy if repulsive forces are prevalent (Fig. 2-10).

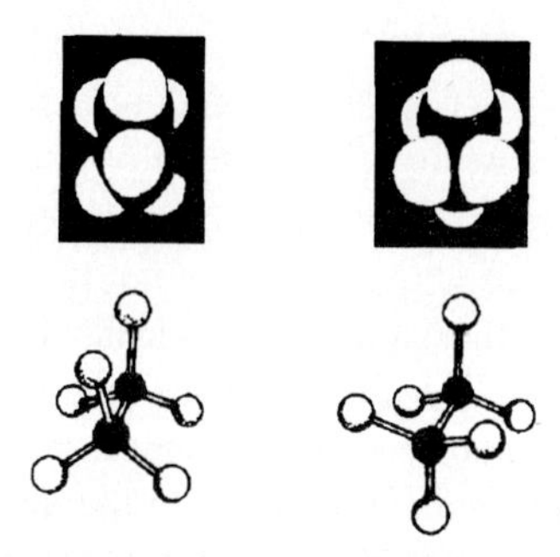

Fig. 2-10 Microconformations of an ethane molecule. ● Carbon atoms, o hydrogen atoms. E = Eclipsed position (cis or synperiplanar); S = staggered position (trans or antiperiplanar). Reproduced with permission by Springer-Verlag, Heidelberg [4].

On rotation by 360° around the C-C axis, three energetically equivalent eclipsed positions (at +120°, 0°, -120°) and three energetically equivalent staggered positions (at +60°, ±180°, -60°) may be occupied since the three H atoms bound to the same C atom ("bonded atoms") are equivalent. The rotational angles are also called conformational angles or torsion angles. Small molecules, such as ethane, can be considered as definite species in conformations with energy minima (+120°, 0°, -120°); they are called conformers, rotamers, or rotational isomers. Conformers can be observed at low temperatures and/or with fast experimental methods.

The number of types of conformers increases if the three bonded atoms are not equivalent, for example, in butane $CH_3–CH_2–CH_2–CH_3$ where one CH_3 group and two H atoms are bonded to each of the methylene carbon atoms participating in the central C–C bond. In polymer chains, one of the bonded atoms is never equivalent since it is part of the chain; in poly(ethylene) $\sim[CH_2–CH_2]_n\sim$, the bonded atoms are two H atoms and a C atom of the chain. There are two energetically different eclipsed positions (*cis* and *anti*) and two energetically different staggered ones (*trans* and *gauche*). Each gauche and each anti position can occur in two spatially different positions that are energetically equivalent (plus and minus) if two of the three bonded atoms are equivalent. These (micro)conformers have different names in macromolecular and organic chemistry (Fig. 2-11).

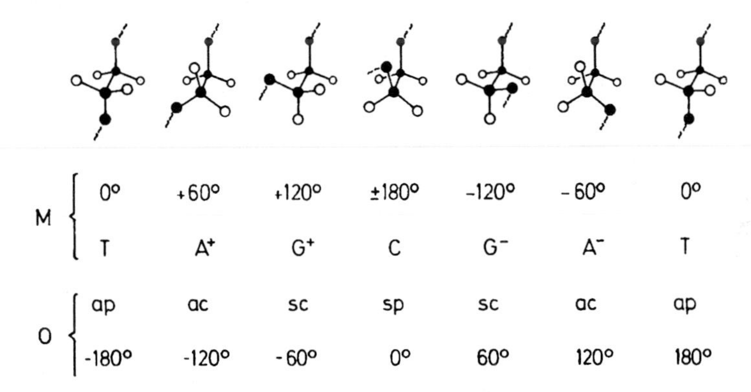

Fig. 2-11 Chain conformations and their names and symbols in macromolecular (M) and organic (O) chemistry. ● Chain atoms, O substituents. T = Trans, A = anti, G = gauche, C = cis, ap = antiperiplanar, ac = anticlinal, sc = synclinal, sp = synperiplanar. The torsion angles between three chain atoms refer to the two conventions.

Eclipsed positions are sterically hindered in polymer chains (see Fig. 2-11) and therefore only trans and gauche positions must be considered for polymers with dominating repulsive forces unless electrostatic forces are more prevalent (see below). The conformational energy is the energy difference between the energies of trans and gauche conformers. An activation energy (potential energy) is necessary to overcome the rotational barrier between trans and gauche conformations. This rotational barrier increases with decreasing length of the central bond and increasing number and size of bonded atoms. Its value is, for example, 12.1 kJ/(mol bond) for the C–C bonds in poly(ethylene) chains ~$[CH_2–CH_2]_n$~ but only 2.1 kJ/mol for the CH_2–CO bonds in polyester chains with units –CH_2–CO–O–CH_2–. Since these potential energies can be easily overcome by thermal energy, (hindered) rotations are possible around the chain bonds and the molecules may adopt many chain conformations (Section 2.4.3). Isolated aliphatic polyester chains ~$[O–OC–(CH_2)_m]_n$~ (with $m > 3$) are thus more flexible than carbon chains; such polymers are utilized as polymeric plasticizers.

Two different types of flexibility are distinguished. A chain molecule is said to be *statically flexible* if it possesses many accessible conformational minima. *Dynamic flexibilities* are characterized by low barriers between conformational minima.

2.4.2. Conformations in Ideal Polymer Crystals

A regular sequence of microconformations leads to regular macroconformations of polymer chains. A chain in all-*trans* conformation (zig-zag chain) is said to be linearly or fully extended. The end-to-end distance of such a chain is commonly called the contour length $r_{cont} = Nb \sin(\tau/2) = N_e b_e$ of the chain (Fig. 2-12), where N_e is the number of "effective bonds" with the bond length b_e and τ is the bond angle (valence angle). The contour length originally referred to the contour of the chain along the individual chain bonds, that is, to the length $L = Nb$ given by the number N of bonds with the bond length b.

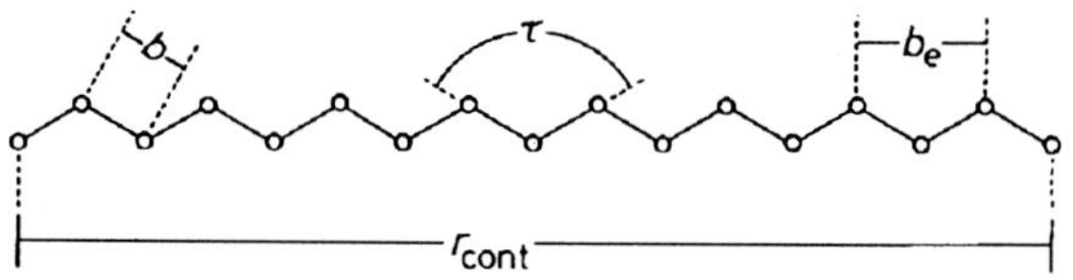

Fig. 2-12 Conventional contour length of a chain ~$[CH_2\text{-}CHR]_n$~ with $N_a = 17$ chain atoms, $N_b = 16$ chain bonds with bond length b, $N_e = 8$ effective bonds with effective bond lengths b_e (crystallographic lengths), and τ = bond angle (valence angle).

Macroconformations of polymer chains are effected by steric and electrostatic potentials. *Repulsive forces* dominate in poly(ethylene) ~$[CH_2–CH_2]_n$~, which crystallizes ideally in an all-trans conformation (zig-zag chain) because the shortest distance between nonbonded hydrogen atoms (0.254 nm) is greater than the sum of the van der Waals radii of H-atoms (0.24 nm). The size of substituents R in isotactic poly(α-olefin)s ~$[CH_2–CHR]_n$~, however, forces the microconformation around each second chain bond to adopt a gauche position (see Fig. 2-11). All gauche conformations must be alike for steric reasons; conformational diads G^+G^- and G^-G^+ are forbidden. The chain thus adopts either a ...$TG^+TG^+TG^+$... or a $TG^-TG^-TG^-$... macroconformation; i.e., it becomes helical (Fig. 2-13). Both macroconformations have the same energy: left-handed and right-handed helices exist in equal amounts.

The number of monomeric units per complete helix turn is determined mainly by the size of the immediate substituent groups. it-Poly(propylene) ~$[CH_2–CH(CH_3)]_n$~ has three propylene units per one turn (3_1 helix), it-poly(4-methylpentene-1) P4MP

~$[CH_2–CH(CH_2CH(CH_3)_2)]_n$~ has seven units per two turns (7_2 helix = 3.5 helix), and poly(3-methylbutene-1) P3MB ~$[CH_2–CH(CH(CH_3)_2)]_n$~ has four units per one turn (4_1 helix) (Fig. 2-13). The conformational angles are not necessarily the ideal ones of 0° for *trans* and 120° for *gauche* as found for it-poly(propylene); they are rather -13°/110° for P4MP and -24°/96° for P3MB. Conformational positions with deviations up to ± 30° from the ideal conformational angles are still named after the ideal microconformations.

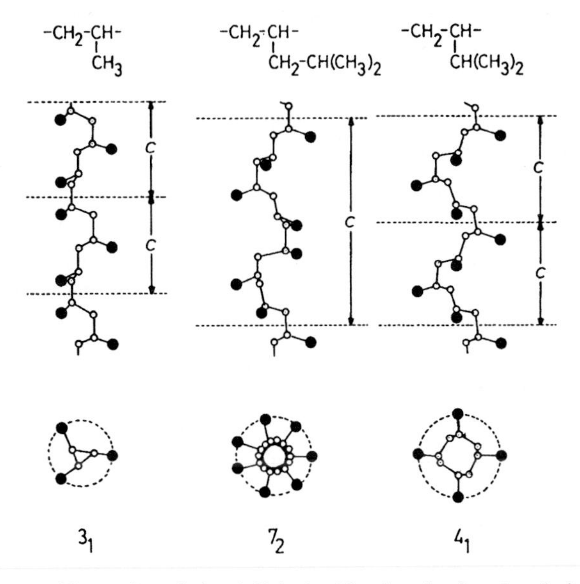

Fig. 2-13 Helix types of isotactic poly(α-olefin)s in side view (top) and end view (bottom). o Chain atoms; ● substituents; hydrogen atoms are not shown. c = Chain identity period.

The R substituents of syndiotactic vinyl polymers ~$[CH_2–CHR]_n$~ are farther apart than those of their isotactic counterparts (see Fig. 2-9). In general, *trans* conformations have thus the lowest energy in those st-polymers where only repulsive (steric) forces operate. *Attractive forces*, such as intramolecular hydrogen bonds between neighboring OH groups in poly(vinyl alcohol) ~$[CH_2–CH(OH)]_n$~, (PVAL), lead to different conformational sequences: isotactic PVAL chains crystallize in all-*trans* conformation whereas syndiotactic PVAL forms helices.

Electrostatic forces dominate in other polymers, which can lead to energy minima for gauche instead of trans and even to cis microconformations instead of trans or gauche. The most stable conformation of poly(dimethylsiloxane) ~$[O–Si(CH_3)_2]_n$~ is cis-trans; the trans-trans conformation corresponds to the energy maximum. Chain atoms with free electron pairs lead to *gauche effects*. Chains of crystallized poly(oxymethylene) ~$[O–CH_2]_n$~ exist in all-gauche conformations (9_5 helix). Poly(oxyethylene) ~$[O–CH_2–CH_2]_n$~ and poly(glycine) ~$[NH–CO–CH_2]_n$~, on the other hand, possess the conformational sequence TTG. The resulting 7_2 helices of poly(glycine) are stabilized by intramolecular hydrogen bonds between the first, fourth, seventh, etc., peptide bond of a chain.

Polypeptides ~[NH–CO–CHR]$_n$~ have chiral monomeric units. An L-polymer may thus form right-handed and left-handed helices. Since the helices are diastereomers with different energy contents, one handedness is preferred over the other in chiral polymers. Poly(L-α-amino acid)s form usually right-handed helices, whereas most polysaccharides from D-sugars and poly([S]-α-olefin)s exist as left-handed ones.

2.4.3. Conformations in Polymer Melts and Solutions

Melts. Helices can survive melting and dissolution processes only if the helical structure is stabilized by intramolecular attractive forces or if the substituents are so tightly packed along the chain that no rotations are possible around chain bonds. Examples of the former are hydrogen bonds in helices of poly(α-amino acid)s or hydrogen bonds plus base stacking in double helices of deoxyribonucleic acids. These forces may be so strong that the molecules decompose upon heating rather than melt. A delicate balance between intramolecular bonding and solvation of substituents is necessary to preserve helical structures in solution; examples are poly(α-amino acid)s in helicogenic solvents such as poly(γ-L-benzyl glutamate) in dimethylformamide.

The packing of helices of apolar polymer chains in crystals stabilizes the microconformations that originate from repulsive forces between nonbonded substituents. Such helices do not survive the melting process intact. Each macromolecule can form many macroconformers that equilibrate rapidly. Only very short helical sequences in very low concentrations may thus exist in melts of such polymers.

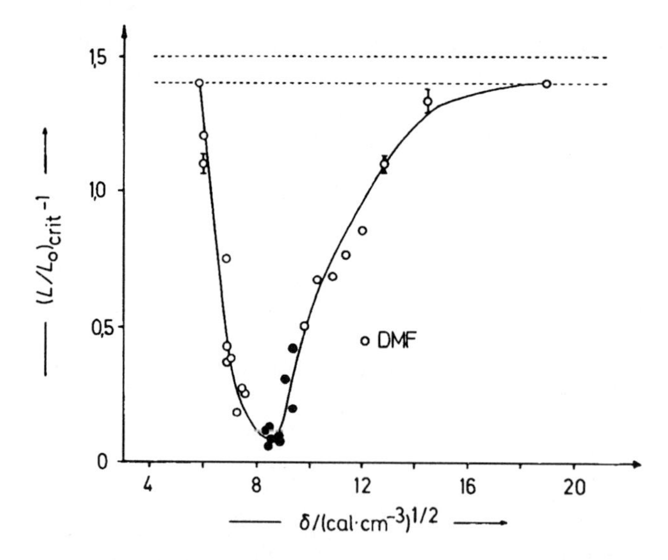

Fig. 2-14 Dependence of critical elongation $\varepsilon_{crit} = (L/L_o)_{crit} - 1$ of poly(oxy-2,6-dimethyl-1,4-phenylene) PPE for fracture (O) and crazing (●) on the solubility parameter δ of the solvent in traditional units (1 $(cal\ cm^{-3})^{1/2}$ = 2.045 $(J/cm^3)^{1/2}$. DMF = N,N-Dimethylformamide. Dotted lines indicate critical elongation in air.
Reprinted with permission by American Chemical Society, Washington (D.C.) [5].

Table 2-3 Solubility of unstressed high molar mass polymers with solubility parameters δ_P in solvents 1, 2, or 3 (see table head) with solubility parameters δ_S. Solubility parameters are given in traditional units: 1 $(cal/cm^3)^{1/2}$ = 2.04 $(J/cm^3)^{1/2}$.

Symbols: - completely or relatively unaffected; ( ) swollen at room temperature and soluble at elevated temperatures; [ ] soluble at elevated temperatures; * stress corrosion.

Abbreviations: HC = hydrocarbon; Tol. = toluene; Xyl. = xylenes; BuAc = butyl acetate; EtAc = ethyl acetate; MeAc = methyl acetate; But. = butanone (methyl ethyl ketone); Acet. = acetone. For abbreviations and acronyms of polymer names see Appendix.

Polymer			Solvents and their solubility parameters					
Name	δ_P		*Gasoline*	*Chloro-HC*	*Aromatics*	*Esters*	*Ketones*	*Alcohols*
		1 →	C_5H_{12} 7.0	CCl_4 8.6	C_6H_6 9.2	BuAc 8.5	But. 9.3	Butyl 11.4
		2 →	C_7H_{16} 7.4	C_2Cl_6 9.3	Tol. 8.9	EtAc 9.1	Acet. 9.9	Ethyl 12.7
		3 →	C_8H_{18} 7.6	$CHCl_3$ 9.3	Xyl. 8.8	MeAc 9.6		Methyl 14.5
Amorphous polymers								
PIB	8.0		1 2 3	1 2 3	1 2 3	1	-	-
PS	9.3		*	1 3	1 2	2 (3)	1	*
PMMA	9.3		-	(1) 3	1 2 (3)	1 2 3	1 2	(1)
PVAC	11.2		-	(1) 3	1 2 3	1 2 3	1 (2)	1 (2) (3)
Slightly crystalline polymers								
PC	8.3		-	(1) 3	(1) 2* 3*	1* 2* 3*	1* (2)	-
PVC	11.0		-		(1) (2) (3)		1 (2)	-
Semicrystalline polymers								
PE	8.1		[1 2 3]	[1 3]	[3]	[1]		-
PP	8.7		1 2 3	[3]	(3)			(1)
PET	10.2		-	-	-	-	-	-
PA 6.6	13.6		-	-	-	-	-	-

Solubility. The solubility range of an amorphous polymer P in a solvent S can be estimated from the difference $\delta_P - \delta_S$ between their so-called solubility parameters δ. These parameters are the square roots of cohesion energy densities. Apolar polymers dissolve in apolar solvents if the absolute difference $|\delta_P - \delta_S| <$ ca. 1.6 $(J\ cm^{-3})^{1/2} \approx$ 0.8 $(cal\ cm^{-3})^{1/2}$ and polar polymers in polar solvents if $|\delta_P - \delta_S| <$ 6.8 $(J\ cm^{-3})^{1/2} \approx$ 3.3 $(cal\ cm^{-3})^{1/2}$. The situation is more complex for semicrystalline polymers because the solvent has also to overcome the heat of crystallization (see also Table 2-3).

Solubility parameters can be used to estimate the effect of solvents on fracture and crazing (Chapter 7) of polymers (Fig. 2-14). Polymers in solvents with solubility parameters $\delta_S \approx \delta_P$ craze and crack at considerably smaller strains than polymers in solvents outside that range.

Macroconformations. The macroconformation of polymer molecules in solution depends on the interaction between polymers and solvents. The interaction itself can be described by thermodynamic model parameters such as the Flory-Huggins interaction parameter χ from the concentration dependence of apparent molar masses.

Two extreme cases must be considered for the dissolution of polymers. Only weak interactions (or none at all) exist between monomeric units of apolar polymers and apolar solvent molecules. Conformational changes are thus entropy driven; the sequence of microconformations becomes irregular and the polymer molecule adopts the macroconformation of a coil.

In apolar enantiomeric polymers in apolar solvents, long conformational sequences are conserved although fast conformational transformations from left-handed to right-handed helices and vice versa may occur. At any given time, only a few "wrong" microconformations exist. The surviving helical sequences may be stabilized by association processes. The macroconformation of such a polymer molecule is still a coil but the regio-conformation is that of a helix.

Polar solvents, on the other hand, interact strongly with polar polymers; these solvents cause considerable changes in the proportion of microconformations. Since the ligands around each chain bond can adopt various microconformations and since these microconformations can change rapidly, a given macromolecule may exist in time in many macroconformations. Figures generated by the rolling of a die may serve as crude models for macroconformations of an individual chain (Fig. 2-15).

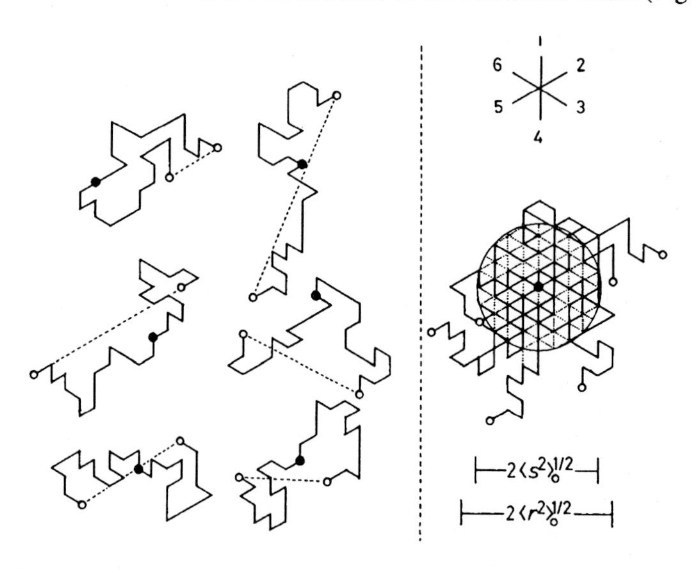

Fig. 2-15 Left: Two-dimensional projections of the three-dimensional spatial arrangements of six chains with thirty chain bonds each, each generated by rolling a die thirty times and placing the bonds in directions according to the numbering convention (upper right). Central chain atoms are indicated by ●, endgroups by O, and end-to-end distances by broken lines.

Right: All six chains are projected on top of each other at the position of their central atoms ●; the circle indicates the range of the radius of gyration s (all chain segments outside the circle fit into the empty spaces (···) inside the circle).

Reprinted with permission by Springer-Verlag, Heidelberg [6].

A poly(methylene) $\sim[CH_2]_n\sim$ of degree of polymerization 20 001 possesses n = 20 000 chain bonds, each of which can adopt three microconformations: *trans*, *gauche* (plus) and *gauche* (minus). According to statistics, such a chain can exist in $3^n = 3^{20\,000} \approx 10^{9542}$ different macroconformations if all three types of microconformations are equally probable! Conversely, a collection of chains may have many macroconformations at any given time. None of these macroconformations adopts a simple geometric shape such as a rod or a sphere, not even instantaneously (Fig. 2-15).

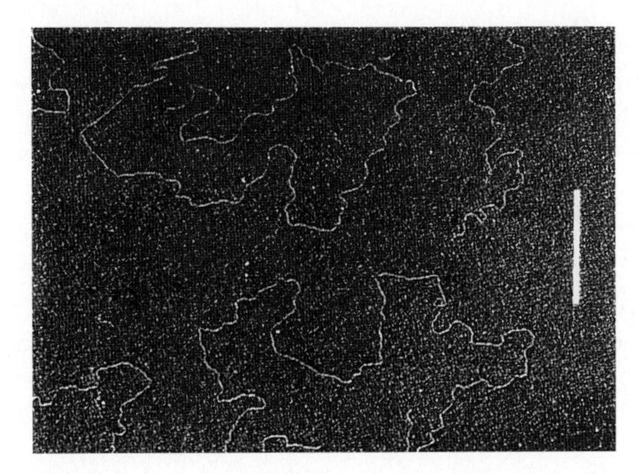

Fig. 2-16 Electron micrograph of the double helix molecules of deoxyribonucleic acid chains showing two-dimensional random coils (projection of three-dimensional coils on a plane). Reprinted with permission by Academic Press, London [7].

The chains are rather forming rapidly changing coil structures, which can be made visible by electron microscopy for chains of sufficiently large chain diameter (such as the double helices of deoxyribonucleic acids) as snapshots of two-dimensional projections of the three-dimensional coil shapes (Fig. 2-16).

The coil size in dilute solution depends on the thermodynamic interaction of the polymer with the solvent. Coils adopt their unperturbed dimensions in so-called theta solvents since the net polymer-polymer and polymer-solvent interaction is zero. The coil size does not change with the polymer concentration.

Coils swell in thermodynamically good solvents because of the strong polymer-solvent interaction. The size of the coil depends on the concentration regime. Dilute, semidilute, and concentrated solutions can be distinguished. In dilute solutions, individual coils are widely separated from each other; a variation of the concentration c does not change the coil dimension. At a certain (higher) concentration c^*, the whole solution volume is occupied by coils. Coils retain their identity at still higher concentrations $c^* < c < c^{**}$ but coil dimensions decrease with increasing c of these semidilute solutions. At another critical concentration c^{**}, coils loose their identity and start penetrating each other; the coil dimensions in these concentrated solutions approach those of unperturbed coils in melts.

2.4.4. Unperturbed Coils

The instantaneous shapes of coils cannot be determined by presently known methods. Their time-averaged radius of gyration can be measured, however, by light scattering, small angle X-ray scattering, and small angle neutron scattering. The end-to-end distances of coiled linear chains can be furthermore modeled for various types of chains and calculated for individual chains via the rotational isomeric state method.

The simplest model is that of a freely jointed chain (no fixed bond angles). This model assumes infinitely thin segments without interactions, i.e., a *phantom chain.* The mean square end-to-end distance $\langle r^2 \rangle_\infty$ of such a coil is given by the number N of its unspecified segments with segment lengths b and molar masses M_b:

$$\text{(2-17)} \quad \langle r^2 \rangle_\infty = Nb^2 = M/(b^2 M_b)$$

Real chains possess fixed bond angles τ between chain atoms and finite torsion angles (dihedral angles) θ ($\theta = 0°$ for *trans* conformations). The mean square end-to-end distance $\langle r^2 \rangle_o$ for such a chain in the unperturbed state (zero net polymer-polymer and polymer-solvent interactions) is given by

$$\text{(2-18)} \quad \langle r^2 \rangle_o = Nb^2 \left(\frac{1 - \cos \tau}{1 + \cos \tau} \right) \left(\frac{1 + \cos \theta}{1 - \cos \theta} \right)$$

$$\text{(2-19)} \quad \langle r^2 \rangle_o = Nb^2 \left(\frac{1 - \cos \tau}{1 + \cos \tau} \right) \sigma^2$$

The factor σ is called *hindrance parameter* or steric factor. Very often hindrance parameter and bond angle term are combined to give the characteristic ratio C_N:

$$\text{(2-20)} \quad C_N = \frac{\langle r^2 \rangle_o}{Nb^2} = \left(\frac{1 - \cos \tau}{1 + \cos \tau} \right) \sigma^2$$

The characteristic ratio C_N (and thus $\langle r^2 \rangle_o / M$) becomes approximately independent of the chain link number N (and thus of the molar mass M) for chains with more than 50-100 chain atoms ($C_N \to C_\infty$) because M_b, b, τ, and σ are all constant for a given polymer. The characteristic ratios C_∞ vary widely (Table 2-4); they can be calculated by the RIS method (rotational isomeric state) and determined experimentally in melts by neutron scattering and in dilute theta solutions by light scattering.

Table 2-4 Calculated (RIS) and experimentally observed characteristic ratios C_∞ at temperatures T.

Polymer	T/°C	C_∞ (calc.)	C_∞ (exp.)
Poly(ethylene)	140	6.7	6.7
Poly(propylene), isotactic	145	5.7	5.7
Poly(propylene), syndiotactic	150	12	12
Poly(butadiene), 1,4-cis	25	4.9	4.9
Poly(butadiene), 1,4-trans	25	5.8	5.8
Poly(oxymethylene)	25	9.0	10.5
Poly(ε-caprolactone)	25	6.1	5.9
Poly(ethylene terephthalate)	25	4.2	4.2
Poly(glycine)	25	2.2	-
Poly(L-alanine)	25	9.3	9.0

A chain can also be characterized by its *Kuhn length*, L_K, which can be calculated from its unperturbed mean square end-to-end distance $\langle r^2 \rangle_o$, its degree of polymerization X, and the effective length, b_e, of monomeric units (i.e., length of units in chain direction projected on a plane) (see Fig. 2-11; $r_{cont} = Xb_e = N_K L_K$):

(2-21) $\langle r^2 \rangle_o = N_K L_K^2$

End-to-end distances can rarely be measured directly. They are however related to the experimentally accessible radii of gyration, s, via

(2-22) $\langle r^2 \rangle_o = 6 \langle s^2 \rangle_o$

This relationship applies to all chains with random flight statistics, e.g., unperturbed chains and freely jointed chains. It is not valid for perturbed chains (Section 2.4.5).

The segments of an unperturbed coil possess a Gauss distribution; the coil density decreases with increasing chain length. Coil densities are very small (cf. Fig. 2-13). The volume fraction of monomeric units is, for example, only 0.012 at the center of gyration of a poly(ethylene) chain with a molar mass of $1.2 \cdot 10^6$ g/mol; 99.8 % of the space of these poly(ethylene) coils is occupied by solvent molecules (in very dilute solutions) or by units of other polymer chains (in melts).

A chain in a melt cannot distinguish between its own segments and segments of other chains. Because the interactions between the segments of various chains are also the same, such chains adopt their unperturbed dimensions, which has been confirmed experimentally by small angle neutron scattering. Most of the coil volume is filled by segments of other chains to avoid empty space. Long chains may thus become entangled, which manifests itself in properties such as slow polymer self-diffusion, high melt viscosity, good melt stability, and certain mechanical properties.

2.4.5. Perturbed Coils

Coils adopt their unperturbed dimensions in melts or in certain solvents at certain temperatures (so-called theta solvents). In such theta solvents, polymer-solvent and polymer-polymer interactions cancel each other and the chain behaves as if it is infinitely thin. The mean square radius of gyration increases with the molar mass according to Equations (2-19) and (2-22) with $M = NM_u$

(2-23) $$\langle s^2 \rangle_o = \left(\frac{b^2}{6 M_u} \right) \left(\frac{1 - \cos \tau}{1 + \cos \tau} \right) \sigma^2 M = K_o M$$

In thermodynamically good solvents, polymer-solvent interactions dominate and the coil is swollen. The large volume requirements of such coils must thus lead to interpenetrations by other coils even at low concentrations, approximately starting at concentrations of $c^* > 1/[\eta]$. Since segments are not infinitely thin, a part of the total space is excluded for segments of other chains (and also for other segments of the own chain). These excluded volumes cause the coils to expand by an expansion factor $\alpha = (\langle s^2 \rangle / \langle s^2 \rangle_o)^{1/2}$, which depends on the molar mass and thus can be incorporated in the constants K and v:

$$(2\text{-}24) \quad \langle s^2 \rangle^{1/2} = K_o^{1/2} M^{1/2} \alpha = K M^{v}$$

The exponent $v = 1/d$ is the inverse of the fractal dimension d of the coils. According to mean field theory, the fractal dimension of random coils is $d \approx 5/3$ (renormalization theory: 1.7007) in good solvents and $d = 2$ in theta solutions. Both numbers have been confirmed by experiment.

2.4.6. Worm-Like Chains

Some polymer chains are not totally flexible. Their finite thickness and the partially hindered rotation around chain bonds prevent them from adopting all possible positions in space. Such chains can be described by the Kratky-Porod model of a worm-like chain.

The characteristic parameter of this model is the *persistence length a*. This parameter is defined as the average of the projection of the end-to-end distance of an infinitely long and infinitely thin chain in the direction of the first segment. It can be calculated from the radius of gyration and the conventional contour length via

$$(2\text{-}25) \quad \langle s^2 \rangle_o = a^2 \left[\left(\frac{y}{3}\right) - 1 + \left(\frac{2}{y}\right) - \left(\frac{2}{y^2}\right)(1 - \exp(-y)) \right]$$

where $y = r_{cont}/a$. For flexible chains, it is related to the unperturbed chain end-to-end distance via

$$(2\text{-}26) \quad \langle r^2 \rangle_o / 6 = \langle s^2 \rangle_o = a r_{cont}/3 = a N_K L_K/3$$

Equation (2-25) reduces for infinitely stiff chains to

$$(2\text{-}27) \quad \langle s^2 \rangle_o = a^2 y^2/12 = (r_{cont})^2/12 = (1/3)(r_{cont}/2)^2$$

An infinitely stiff chain behaves like an infinitely thin rod with a length r_{cont} because the radius of gyration of a rod is directly proportional to its length. Worm-like chains thus describe the whole transition from rod-like molecules (small y; Eqn. (2-27)) to random coils (large y; Eqn. (2-26)) (see Fig. 2-5). The model is strictly valid for infinitely thin chains only but the error produced by this assumption is negligible if the persistence length is much greater than the chain diameter.

Literature

2.1. Constitution: Analysis of Polymers and Plastics

J.Urbanski, W.Czerwinski, K.Janicka, F.Majewska, H.Zowall, Handbook of Analysis of Synthetic Polymers and Plastics, Wiley, New York 1977
H.-W.Sisler, K.Holland-Moritz, Infrared and Raman Spectroscopy of Polymers, Dekker, New York 1980
L.S.Bark, N.S.Allen, eds., Analysis of Polymer Systems, Appl.Sci.Publ., Barking, Essex 1982
Q.T.Pham, R.Petiaud, H.Waton, Proton and Carbon NMR Spectroscopy of Polymers, Wiley, New York 1983 (2 vols.)
A.Krause, A.Lange, M.Ezrin, Plastics Analysis Guide, Hanser, Munich 1984
W.Klöpfer, Introduction to Polymer Spectroscopy, Springer, Berlin 1984
E.Komoroski, ed., High Resolution NMR Spectroscopy of Synthetic Polymers in Bulk, VCH, New York 1986
J.Mitchell, jr., ed., Applied Polymer Analysis and Characterization, Hanser, Munich 1987
T.R.Crompton, Analysis of Plastics, Pergamon, Oxford 1989
A.E.Tonelli, NMR Spectroscopy and Polymer Microstructure, VCH, New York 1989
G.Allen, J.C.Bevington, eds., Comprehensive Polymer Science, Vol. 1, C.Booth, C.Price, eds., Polymer Characterization, Pergamon Press, Oxford 1989
D.O.Hummel, Atlas of Polymer and Plastics Analysis, VCH, Weinheim, 3rd ed. 1991 (3 volumes in several parts)
H.G.Barth, J.W.Mays, eds., Modern Methods of Polymer Characterization, Wiley, New York 1991

2.1. Constitution: Chemical Structure

J.L.Koenig, Chemical Microstructure of Polymer Chains, Wiley, New York 1980
L.H.Sperling, Interpenetrating Polymer Networks and Related Materials, Plenum, New York 1981
J.M.G.Cowie, ed., Specialty Polymers, Vol. 1, Alternating Copolymers, Plenum, New York 1984
J.A.Semlyen, ed., Cyclic Polymers, Elsevier, New York 1986
N.A.Plate, V.P.Shibaev, eds., Comb-Shaped Polymers and Liquid Crystals, Plenum, New York 1987
A.E.Tonelli, NMR Spectroscopy and Polymer Microstructure, VCH, New York 1989
J.E.Mark, H.R.Allcock, R.West, Inorganic Polymers: An Introduction, Prentice-Hall, Englewood Cliffs (NJ) 1992

2.2. Molar Mass and Molar Mass Distribution

L.H.Peebles, Molecular Weight Distribution in Polymers, Interscience, New York 1971
M.B.Huglin, ed., Light Scattering from Polymer Solutions, Academic Press, New York 1972
P.E.Slade, Jr., Polymer Molecular Weights, Dekker, New York 1975 (2 vols.)
H.-G.Elias, Polymolecularity and Polydispersity in Molecular Weight Determinations, Pure Appl.Chem. **43**/1-2 (1975) 115
N.C.Billingham, Molar Mass Measurements in Polymer Science, Halsted Press, New York 1977
L.H.Tung, ed., Fractionation of Synthetic Polymers, Dekker, New York 1977
J.Cazes, ed., Liquid Chromatography of Polymers and Related Materials, Dekker, New York 1977

W.W.Yan, I.J.Kirkland, D.D.Bly, Modern Size-Exclusion Liquid Chromatography, Wiley, New York 1979
M.Bohdanecky, J.Kovar, Viscosity of Polymer Solutions, Elsevier, Amsterdam 1982
J.Janca, ed., Steric Exclusion Chromatography of Polymers, Dekker, New York 1984
C.G.Smith, N.E.Skelly, C.D.Chow, R.A.Solomon, Chromatography: Polymers, CRC Press, Boca Raton (FL) 1982
G.Glöckner, Polymer Characterization by Liquid Chromatography, Elsevier, Amsterdam 1986
J.Janca, Field-Flow Fractionation, Dekker, New York 1987
K.S.Schmitz, An Introduction to Dynamic Light Scattering by Macromolecules, Academic Press, San Diego (CA) 1990
B.Chu, Laser Light Scattering, Academic Press, San Diego (CA) 1990
W.Brown, ed., Dynamic Light Scattering, Oxford University Press, New York 1993

2.3. and 2.4. Configuration and Conformation

A.D.Ketley, ed., The Stereochemistry of Macromolecules, Dekker, New York 1967 (3 vols.)
F.A.Bovey, Polymer Conformation and Configuration, Academic Press, New York 1969
A.Hopfinger, Conformational Properties of Macromolecules, Academic Press, New York 1973
J.L.Koenig, Chemical Microstructure of Polymer Chains, Wiley, New York 1980

2.4.3. Solubility

A.F.M.Barton, CRC Handbook of Solubility Parameters and Other Cohesion Parameters, CRC Press, Boca Raton (FL), 2nd ed. 1992

2.4.4.-2.4.6. Polymer Coils in Solution

H.Yamakawa, Modern Theory of Polymer Solutions, Harper and Row, New York 1971
H.Morawetz, Macromolecules in Solution, Interscience, New York, 2nd ed. 1975
M.Kurata, Thermodynamics of Polymer Solutions, Harwood Academic Publ., Chur 1982

References

[1] H.-G. Elias, Makromoleküle, vol. I: Grundlagen, Hüthig and Wepf, Basle, 5th ed. 1990, Fig. 3-1.
[2] H.-G. Elias, Makromoleküle, vol. I: Grundlagen, Hüthig and Wepf, Basle, 5th ed. 1990, Fig. 3-3.
[3] H.-G. Elias, Grosse Moleküle, Springer-Verlag, Berlin 1985; Mega Molecules, Springer-Verlag 1987; both Fig. 21.
[4] H.-G. Elias, Grosse Moleküle, Springer-Verlag, Berlin 1985; Mega Molecules, Springer-Verlag 1987; both Fig. 11.
[5] G.A.Bernier, R.P.Kambour, Macromolecules **1** (1968) 393, Fig. 4
[6] H.-G. Elias, Grosse Moleküle, Springer-Verlag, Berlin 1985; Mega Molecules, Springer-Verlag 1987, both Fig. 14.
[7] D.Lang, H.Bujard, B.Wolff, D.Russell, J.Mol.Biol. **23** (1967) 163, Plate II.

3. Polymer Manufacture

3.1. Raw Materials for Polymers

Most plastics are based on synthetic organic polymers or prepolymers; a few are synthetic semiorganics (inorganic chains with organic ligands). These polymers are derived from monomers that are in turn mainly manufactured from intermediates. The main raw material for intermediates is petroleum, followed by natural gas. Coal and certain natural oils deliver relatively small amounts of monomers.

Few plastics are derived from naturally products such as wood, starch, natural oils, etc. Renewable raw materials are of industrial interest if their structure leads directly to polymers or to easy to produce and convert intermediates and monomers. As a source of chemical base products, they have several disadvantages with respect to petroleum. Renewable raw materials are, in general, chemical compounds or mixtures of chemical compounds with often considerable proportions of molecularly bound oxygen; their C/H ratio is less favorable for the manufacture of base chemicals. Their conversion into intermediates or monomers requires also more energy than that of petroleum.

3.1.1. Wood

Wood is a naturally occuring composite of oriented cellulose fibers in a continuous matrix of cross-linked lignins; it is plasticized by water and "foamed" by air (vacuoles). The water content varies from 40-60 % in green wood to 10-20 % in air-dried wood. Solids include cellulose (42 %); hemicelluloses (28-38 %); lignins (19-28 %); and proteins, resins, and waxes (2-3 %), depending on the type and age of the plant. Most of the harvested wood is used directly as fuel or for construction purposes but ca. 1/6 is converted into wood pulp or cellulose pulp for the manufacture of paper, cardboard, and rayon. A very small amount of wood is filled with monomers that are subsequently polymerized to give polymer wood.

Delignified wood is the main source for industrial celluloses; a very small amount is nowadays produced industrially by bacteria. Wood celluloses are used for the manufacture of rayon fibers. They are also chemically transformed into various cellulose derivatives, some of which are utilized as plastics (Chapter 12.7.).

Lignins are a highly cross-linked polymers based on various phenol alcohols such as coniferyl alcohol. The chemical composition of lignins depends on their origin. Lignins themselves are not commercial products. Wood pulp production generates huge amounts of lignin sulfonates. Most of these sulfonates are burnt; small amounts are utilized for street pavements, as binders for foundry sands, or as drilling agents. Very small amounts of lignins are used for the synthesis of ion exchange resins and other polymers or as raw materials for organic intermediates.

3.1.2. Coals

Coals are polymerized and partly cross-linked hydrocarbons with average compositions between $C_{75}H_{140}O_{56}N_2S$ (peat) and $C_{240}H_{90}O_4NS$ (anthracite coal).

Coals deliver coke and coal tar. Coke can be converted into acetylene, which was the main source for a number of aliphatic monomers during the first half of this century; examples are chloroprene and various vinyl monomers. Coal tar is the raw material for aromatics such as benzene, xylene, phenol, and phthalic anhydride.

3.1.3. Natural Gas

Natural gas has a high proportion of aliphatic hydrocarbons. European gas is rich in methane whereas American and Saudi Arabian gases are relatively rich in higher hydrocarbons. Natural gas is processed to synthesis gas (mixture of carbon monoxide and hydrogen), ethylene, and acetylene. These gases are used to produce a variety of monomers for polymers; examples are vinyl monomers, styrene, ethylene oxide, ethylene glycol, trimethylol propane, acrylonitrile, and hexamethylene diamine.

3.1.4. Petroleum

Petroleum (crude oil) is the main feedstock for monomers. It consists of 95-98 % hydrocarbons and 2-5 % of oxygen, nitrogen, and sulfur compounds. The hydrocarbons are largely aliphatic, partly naphthenic, and to a small extent aromatic, depending on the source. Petroleum distillation leads to saturated hydrocarbons (gas, naphtha, gasoline, kerosene, diesel fuel, heating oil, etc.). These hydrocarbons are subsequently cracked to olefin mixtures that are fractionated by distillation. The resulting compounds are used as such or are converted further into the desired monomers.

The polymer industry is a major petrochemical customer: ca. 60 % of the cracking and subsequent products of naphtha are used to produce polymers (ca. 50 % for plastics and elastomers, ca. 10 % for fibers). However, the final yields of polymers and converted plastics, respectively, are fairly small: 1000 kg of petroleum deliver only 30 kg of poly(ethylene) and subsequently 16 kg of poly(ethylene) film, in addition to 360 kg of useful byproducts (liquid gas, gasoline, propylene, etc.).

3.1.5. Other Natural Raw Materials

Vegetable oils are a relatively large source of raw materials. They consist of mixtures of fatty acid triglycerides and are usually subdivided into drying oils with high linolenic and linoleic acid content, semidrying oils with high linoleic and oleic acid content, and nondrying oils with high oleic acid content. Some oils are used directly for paints; others are chemically converted into monomers.

Castor oil is converted into methyl ricinoleate, which is thermally cracked into methyl undecenate and heptaldehyde. Methyl undecenate is the raw material for 11-aminoundecanoic acid, which is subsequently polycondensed to form polyamide 11 (PA 11). Alkali scission of castor oil or ricinoleic acid produces sebacic acid, a monomer used to prepare PA 6.10. Castor oil is also used directly for alkyd resins.

Soybean oil is converted by a series of chemical reactions into the methyl ester of the C_{10} amino acid $H_2N(CH_2)_9COOCH_3$, which is the monomer for PA 10. The plant *Crambe abyssinica* contains about 55 % of erucic acid, which leads to the C_{13} amino acid and subsequently to PA 13, $\sim[NH(CH_2)_{12}CO]_n\sim$ (not commercially available).

Corn husks and other agricultural wastes are rich in *pentosans* (= polyoses with pentose sugar units; also called *hemicelluloses*). These polymers can be converted into tetrahydrofuran and further into poly(tetrahydrofuran), $HO[CH_2CH_2CH_2CH_2O]_nH$, which serves as a soft block in certain polyurethanes and polyesters.

The supply of raw materials from vegetable sources is relatively small. It is also fairly insecure since quality and quantity may depend on weather conditions or be influenced by political turmoil. Most monomers for the polymer industry are thus obtained from fossil feedstocks.

3.2. Polymer Syntheses

3.2.1. Classifications

Polymers may be synthesized from monomers by polymerization or from other polymers by chemical transformation. Classifications of polymerizations have grown and changed with the times; most are not very systematic and some are confusing.

All reactions of small molecules (monomers) to polymer molecules are called *polymerizations*. Polymerizations are classified according to the

1) origin of polymers (biopolymerizations, synthetic polymerizations);
2) chemical structure of monomers (vinyl, diene, ring-opening, etc.);
3) chemical structure of polymers (linear, branching, cross-linking, ring-forming, etc.);
4) relative composition of monomers and monomer units (chain polymerization, condensation polymerization, etc.);
5) formation of low molar mass byproducts (polycondensation, polyaddition, etc.);
6) type of polymerization start (thermal, initiated, catalytic, photochemical, enzymatic, electrolytic, etc.);
7) type of propagating species (free radical, anionic, cationic, etc.);
8) type of mechanism (equilibrium, addition, insertion, living, stepwise, etc.);
9) reaction media (bulk, solution, emulsion, suspension, crystal, mesophase, etc.);
10) state of matter (gaseous, homogeneous, heterogeneous, etc.).

The most common classification is according to the relative chemical composition of monomer vs. monomeric unit and the type of molecules which may react with the growing polymer chain. The resulting four types of polymerization can be depicted schematically as

(3-1)	$P_n + P_m \rightarrow P_{n+m}$ (polyaddition)	$P_n + M \rightarrow P_{n+1}$ (chain polymerization)
	$P_n + P_m \rightarrow P_{n+m} + L$ (polycondensation)	$P_n + M \rightarrow P_{n+1} + L$ (condensative chain polymerization)

where M is the monomer, P_n, P_m, P_{n+m}, and P_{n+1} are polymers, and L is the leaving molecule (e.g., water in polyesterifications of diacids and diols).

The names of three of these four types have been used with different meanings by different researchers over the past 70 years (see below). The names in Equation (3-1) are those presently recommended by the International Union of Pure and Applied Chemistry (IUPAC).

3.2.2. Functionality

Polymerizable monomers must be at least bifunctional under the reaction conditions so that they can connect to the growing chains on one hand and generate new coupling sites on the other. Styrene and other monomers with ethylenic double bonds are such bifunctional monomers under all polymerization conditions; they lead to linear polymers whereas the tetrafunctional divinyl benzenes generate networks:

(3-2)		monomer	polymer
	bifunctional	$CH_2{=}CH(C_6H_5)$	$\sim\sim CH_2\text{-}CH(C_6H_5) \sim\sim$
	tetrafunctional	$CH_2{=}CH\text{-}C_6H_4\text{-}CH{=}CH_2$	$\sim\sim CH_2\text{-}CH \sim\sim$ / C_6H_4 / $\sim\sim CH\text{-}CH_2 \sim\sim$

The functionality f of polymer molecules is given by the functionality f_o of the monomer molecules and the degree of polymerization X of molecules:

$$f = 2(f_o - 1) + (X - 2)(f_o - 2) = 2 + X(f_o - 2) \quad (3\text{-}3)$$

if no intramolecular rings are formed between functional groups. A monomer with functionality $f_o = 3$ thus gives a dimer ($X = 2$) with a functionality $f = 4$ and a 100-mer ($X = 100$) with a functionality $f = 102$. Since molecules need only to be bifunctional for the formation of linear molecules, the "extra" functionalities lead to interconnections between various molecules (branching) and thus finally to cross-linked, "infinitely large" molecules, which are insoluble in all solvents. Examples are the divinylbenzenes $(CH_2{=}CH)_2C_6H_4$, which behave as tetrafunctional compounds in free radical polymerizations.

The second vinyl group of a divinylbenzene molecule possesses, however, a different reactivity after the polymerization of the first one. The two reactivities may differ so strongly in certain ionic polymerizations of divinylbenzenes that linear polymers, and not cross-linked ones, result (e.g., in the cationic polymerization of 1,4-divinyl benzene in dichloromethane by acetyl perchlorate). The functionality of a chemical group thus depends not only on the chemical structure of that group but also on the reaction conditions.

3.2.3. Cyclopolymerization

Polymerizations of multifunctional molecules may proceed not only intermolecularly to produce branched and cross-linked molecules but also intra-intermolecularly with formation of rings in chains. These polymerizations are called cyclopolymerizations. They occur especially with monomers having carbon-carbon double bonds in the 1,5- or 1,6-position. An example is acrylic anhydride (I), $(CH_2{=}CH{-}CO)_2O$, which polymerizes free-radically to soluble polymers, even at high monomer conversions. The polymer molecules contain 90-100 % six-membered (II) and five-membered (III) ring structures and only 0-10 % of "linear" monomeric units IV:

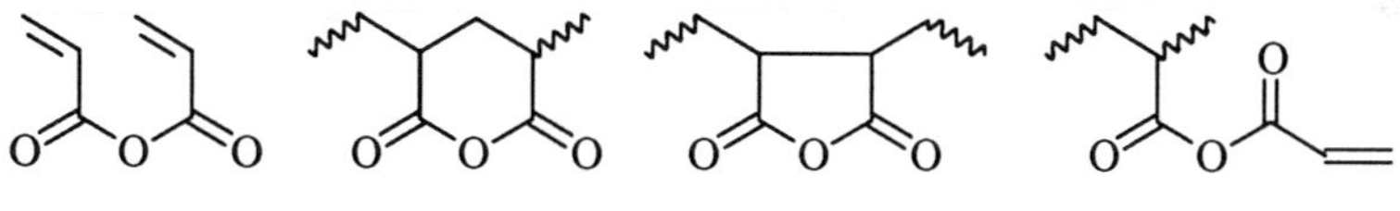

I II III IV

3.2.4. Polymer Formation

Polymerizations are generally accompanied by a decrease of volume of the reacting system because long intermolecular distances by van der Waals forces between monomer molecules (ca. 0.3-0.5 nm) are replaced by covalent bonds between monomeric units (ca. 0.14-0.19 nm). The contraction is higher for smaller monomer molecules since relatively more van der Waals contacts are removed per mass. Ethylene $CH_2{=}CH_2$ contracts 66 % upon polymerization, vinyl chloride $CH_2{=}CHCl$ 34 %, and styrene $CH_2{=}CHC_6H_5$ only 14 %. Larger rings show less contraction upon ring-opening polymerization: the three-membered ethylene oxide ring leads to a volume decrease of 23 % but the five-membered tetrahydrofuran to one of only 10 %. The contraction during polycondensation is smaller, if the leaving molecules are small: reaction of hexamethylene diamine with adipic acid releases water H_2O (22 % contraction) but that of hexamethylene diamine with dioctyl adipate gives the larger octanol molecules $C_8H_{17}OH$ (66 % contraction).

Polymerization can also lead to expansion. This rare case occurs if the polymerization reduces the total number of chemical bonds, σ-bonds are converted into smaller π-bonds, and/or crystalline monomers are transformed into amorphous polymers.

3.3. Chain Polymerization

3.3.1. Overview

Chain polymerization is the name recommended by IUPAC for a polymerization by repeated addition of monomer molecules M to a growing polymer chain without elimination of low molar mass molecules (Eq. (3-1)). An example is the polymerization of vinyl monomers $CH_2{=}CHR$ by initiators Y* which may be free radicals $Y^{\bullet}$, anions Y^-, cations Y^+, or coordination compounds:

$$(3\text{-}4) \qquad Y^* \xrightarrow{+\,CH_2=CHR} Y\text{-}CH_2\text{-}\overset{*}{C}HR \xrightarrow{+\,CH_2=CHR} Y\text{-}CH_2\text{-}CHR\text{-}CH_2\text{-}\overset{*}{C}HR \quad \text{etc.}$$

The initiation reaction Y* + $CH_2{=}CHR$ leads to Y–CH_2–*CHR (monomer radical, monomer anion, monomer cation, etc.). The active centers ~*CHR add more monomer molecules (propagation). They may also be transferred to other molecules where they may or may not start a new chain and/or they may also be deactivated to dead chains (termination).

The name "chain polymerization" refers to the presence of kinetic chains; these reactions are also known as *chain-growth polymerizations*. The traditional name of chain polymerization is *addition polymerization* because monomers are "added" and not "condensed". Addition polymerization is not to be confused with polyaddition.

Condensative chain polymerization is the name recommended by IUPAC for a chain polymerization by repeated addition of monomer molecules *and* elimination of low molar mass molecules L; another name is thus *polyelimination*.

Polyelimination seems to be the process which nature uses to generate polysaccharides. Cellulose of cotton, for example, is not formed by polymerization of glucose but by that of guanosine diphosphate D-glucose attached to a lipid matrix. The reaction proceeds with elimination of guanosine diphosphate via insertion of the glucose moiety into the growing chain.

Polyeliminations are not very common in synthetic macromolecular chemistry. An example is the polymerization of α-amino acid N-carboxyanhydrides (Leuchs anhydrides) with elimination of carbon dioxide:

$$\text{(3-5)} \qquad R\text{-}NH_2 + \begin{matrix} R & & O \\ & C\!-\!C & \\ H & | & O \\ & HN\!-\!C & \\ & & O \end{matrix} \longrightarrow R\text{-}NH\text{-}CO\text{-}CHR\text{-}NH_2 + CO_2 \quad \text{etc.}$$

These polyeliminations are *living polymerizations* consisting only of a fast initiating reaction and the propagation reaction without termination of the growing chains and without side reactions. Many anionic polymerizations can be performed as living polymerizations. An example is the polymerization of styrene $CH_2{=}CH(C_6H_5)$ by the butyl anion $C_4H_9^-$ from lithiumbutyl (butyllithium) LiC_4H_9 as initiator. Such living polymerizations are utilized in the syntheses of block copolymers. True cationic living polymerizations are rare and living radical polymerizations are unknown.

In living polymerizations with fast initiations, the degree of polymerization is directly proportional to the conversion u of monomer molecules or the extent of reaction p of functional groups, respectively (Fig. 3-1, curve CP-L). The molar mass distribution of polymers is quite narrow and approximately of the Poisson type.

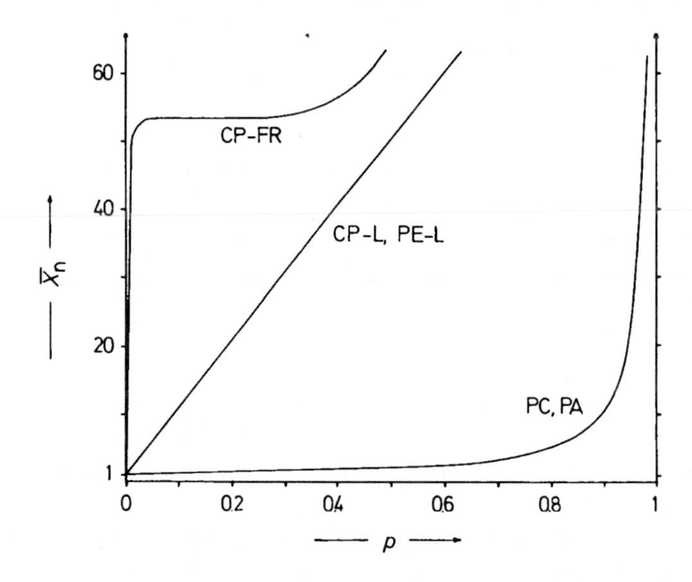

Fig. 3-1 Change of number-average degree of polymerization, $\overline{X}_n$, with the extent of reaction of functional monomer groups, p.

- CP-FR free radical chain polymerizations with gel effect (schematic; high initiator conc.);
- CP-L living chain polymerizations (with initial monomer/initiator ratio 100:1 mol/mol);
- PA equilibrium polyadditions PA (numerically exact);
- PC equilibrium polycondensations PC (numerically exact);
- PE-L living polyeliminations (initial monomer/initiator ratio 100:1 mol/mol).

The variation of the degree of polymerization with monomer conversion is quite different for chain polymerizations where active centers are successively formed and destroyed; examples are *radical polymerizations*. Depending on reaction conditions, the degree of polymerization may remain constant (primarily at low conversions) or increase with monomer conversion (so-called gel effect) (Fig. 3-1, curve CP-FR). Molar mass distributions are fairly broad; they are often Schulz-Zimm distributions.

3.3.2. Thermodynamics

Polymerizations occur only if their molar standard Gibbs energies of polymerization, ΔG_p^o, are negative. Since the Gibbs energy

$$\Delta G_p^o = \Delta H_p^o - T\Delta S_p^o = -RT \ln K \qquad (3\text{-}6)$$

is determined by both the standard polymerization enthalpy, ΔH_p^o, and the standard polymerization entropy, ΔS_p^o, four different cases are possible:

Case 1. Both enthalpy and entropy are negative; the entropy term $-T\Delta S_p^o$ becomes more positive with increasing temperature until finally the Gibbs energy turns zero. At this *ceiling temperature* T_c, enthalpy and entropy term balance each other; no polymerization to high molar mass molecules is possible above this temperature. Oligomers may be formed at the ceiling temperature, however, since polymerization equilibria consist of a series of consecutive equilibria between monomer and growing chains of different degrees of polymerization.

Case 1 is the most common case. It is found for monomers with polymerizable double and triple bonds, such as C=C, C=O, C=S, or C≡N. The polymerization entropies of such monomers are practically determined only by the loss of translational entropy if a monomer molecule is added to a chain; they are thus almost independent of the chemical structure of the monomers.

The polymerization enthalpy of monomers with double bonds is given by the difference in bond energies of π-bonds in monomers and σ-bonds in monomeric units. Steric hindrance, however, strongly decreases the polymerization enthalpy: styrene $CH_2{=}CHC_6H_5$ (T_c = 300°C) homopolymerizes in bulk at 70°C but α-methylstyrene $CH_2{=}C(CH_3)C_6H_5$ (T_c = 60°C) does not.

Differences of bond energies in monomers and polymers are practically zero for cyclic monomers because σ-bonds are opened and formed again. Polymerization enthalpy is determined by delocalization and strain energies; very few 6-membered rings can be polymerized.

Case 2. The polymerization enthalpy is negative (or zero) and the polymerization entropy is positive. Polymerization is possible at all temperatures in this rare case, the polymerization of octamethylcyclotetrasiloxane being an example.

Case 3. The polymerization enthalpy is positive (or zero) and the polymerization entropy is negative. No polymerization is possible at any temperature. This case seems to apply to the polymerizations of acetone or hexafluoroacetone; the latter may be copolymerized, however.

Case 4. Both polymerization enthalpy and polymerization entropy are positive. A *floor temperature* exists below which no polymerization is possible. This rare case is fulfilled for the polymerization of tightly packed cyclic monomers such as oxacycloheptane (oxepane) or cyclooctasulfur since the number of rotatory degrees of freedom increases strongly upon polymerization to open chains.

Polymerization enthalpies and entropies are influenced by the state of monomers and polymers (gaseous, liquid, crystalline, etc.), by pressure, and by the physical interactions between reactants themselves or between reactants and solvents. Especially important for industrial polymerizations is the fact that most (but not all, see above) chain polymerizations are exothermic. The heat of polymerization can be considerable: a complete adiabatic polymerization of gaseous ethylene to crystalline poly(ethylene) would increase the temperature by 1800 K! Polymerization reactors must be designed in such a way that the heat of polymerization can be removed rapidly. Otherwise, inhomogeneous charges may result or the reactor may even explode.

3.3.3. Free Radical Polymerization

Overview. Free radical polymerizations are initiated and propagated by free radicals; most follow the kinetics of true chain reactions. Such polymerizations are fairly easy to control and relatively insensitive to impurities in the reaction mixture. They are thus the methods of choice for the industrial production of polymers if the monomers can be subjected to free radical polymerizations. This is true for ethylene and several ethylene derivatives: many commodity plastics such as poly(vinyl chloride), poly(styrene), and low density poly(ethylene) are obtained by free radical polymerization (Table 3-1). Other industrial polymers by free radical homopolymerizations include poly(vinyl acetate) (coatings, adhesives), poly(acryl ester) (adhesives), poly(acrylamide) and poly(acrylic acid) (thickeners), poly(acrylonitrile) (fibers), poly(chloroprene) (elastomers), and poly(N-vinyl pyrrolidone) (various applications).

Polymerizations can proceed in the gas phase, in bulk, in suspension, in emulsion, in solution, or under precipitation. Most monomers are bifunctional; they lead to linear or slightly branched polymers and thus to thermoplastics. The only thermosets obtained from free radical homopolymerizations are those from diallyl and triallyl compounds that cross-link under polymerization conditions.

Initiation. The polymerization starts with the formation of initiator radicals $R^{\bullet}$, mostly from the thermal decomposition of initiators such as dibenzoyl peroxide (BPO) $C_6H_5COO{-}OOCC_6H_5$, vinylsilane triacetate $CH_2{=}CH{-}Si(OOCCH_3)_3$, N,N-azobisisobutyronitrile $(CH_3)_2C(CN){-}N{=}N{-}C(CN)(CH_3)_2$ (AIBN), or dipotassium disulfate $K_2S_2O_8$. Low temperature emulsion polymerizations are often initiated by *redox initiators* which generate radicals by reaction of a reducing agent with an oxidizing agent (e.g., $Fe^{2+} + H_2O_2$). *Photochemical initiations* are utilized for the production of lithographic plates and for the hardening of lacquers but not in the manufacture of plastics. *Electrolytic polymerizations* find applications in the coating of metal sheets by polymers. Initiator radicals are rarely generated from the monomers themselves; the only known cases are the thermal polymerization of styrene and the polymerizations of *p*-cyclophane (the cyclic dimer of *p*-xylene $CH_2{=}(p\text{-}C_6H_4){=}CH_2$).

Tab. 3-1 Industrial polymers by free radical chain polymerization in the gas phase G, in bulk B, as suspension S, in aqueous emulsion E, in solution L to dissolved polymers, and by precipitation P of the dissolved or neat liquid monomer. + Major processes, (+) minor processes.

Monomers		Polymerizations					
		G	B	S	E	L	P
Living polymerizations							
p-Cyclophane	$[CH_2(p\text{-}C_6H_4)CH_2]_2$	+					
Irreversible polymerizations to linear or slightly branched polymers							
Ethylene	$CH_2{=}CH_2$		+	+	(+)	(+)	(+)
Styrene	$CH_2{=}CHC_6H_5$		+	+	(+)	(+)	
p-Methyl styrene	$CH_2{=}CH(p\text{-}CH_3C_6H_4)$		+	+			
Vinyl acetate	$CH_2{=}CH(OOCCH_3)$		(+)	(+)	+	(+)	
Vinyl chloride	$CH_2{=}CHCl$	+	(+)	(+)	+	(+)	+
Vinyl fluoride	$CH_2{=}CHF$		+				
Vinylidene fluoride	$CH_2{=}CF_2$			+	+		
Trifluorochloroethylene	$CF_2{=}CFCl$			+			
Tetrafluoroethylene	$CF_2{=}CF_2$			+			
Methyl methacrylate	$CH_2{=}C(CH_3)(COOCH_3)$		+	+	+	+	
Cross-linking polymerization							
Diallyl phthalate	$(CH_2{=}CH{-}CH_2{-}OOC)_2(o\text{-}C_6H_4)$		+				

Propagation. Only initiator radicals are formed at the very early stages of free radical polymerizations. These radicals are successively converted into monomer radicals by addition of initiator radicals to monomer molecules and then into polymer radicals by further addition of monomers. At the same time, some radicals are removed by termination reactions. Finally, a *steady state* is established in which as many radicals (initiator radicals, monomer radicals, and polymer radicals) are generated as are removed by termination reactions. This state of constant total radical concentration (ca. 10^{-8} mol/L) is generally observed within seconds at monomer conversions of 10^{-2}-10^{-4} %.

Monomer molecules are usually added head-to-tail to growing macroradicals but tail-to-tail additions do occur (see Section 2.1.). Free radical polymerizations give in general "atactic" or predominantly syndiotactic polymers (exception: very bulky ligands). Decreasing polymerization temperatures lead to higher syndiotacticities.

Termination. Macroradicals are terminated by other radicals: by recombination with another macroradical

(3-7) $2\ R'(CH_2CHR)_n^{\bullet} \rightarrow R'(CH_2CHR)_n(CHRCH_2)_nR'$

by disproportionation by another macroradical

(3-8) $2\ R'(CH_2CHR)_n^{\bullet} \rightarrow R'(CH_2CHR)_{n-1}CH_2CH_2R + CHR{=}CH_2\text{-}(CHRCH_2)_{n-1}R'$

and by addition of initiator radicals $I^{\bullet}$ (at higher initiator concentrations)

(3-9) $R'(CH_2CHR)_n^{\bullet} + I^{\bullet} \rightarrow R'(CH_2CHR)_nI$

Since initiator radicals and thus also polymer radicals are formed successively and are deactivated at random by different termination reactions, distributions of molar masses are formed. The width of these distributions is given by the relative proportion of the different types of termination reactions.

Chain Transfer. Radicals can be transferred to other molecules (e.g., $R^{\bullet} + ClR' \rightarrow RCl + (R')^{\bullet}$). The transfer of radicals to monomers, solvents, and initiators terminates growing polymer chains and generates radicals that may start new polymer chains; these reactions are not termination reactions in the kinetic sense.

Chain transfer often leads to new radicals that have approximately the same reactivity as the disappearing ones. The rate of polymerization is not decreased but the degree of polymerization is lowered. Certain transfer agents with high ratios of rate constants of transfer to rate constants of propagation (transfer constants) are thus often employed to regulate molar masses by this *degradative chain transfer.*

Retardation occurs if the new radical is more sluggish in adding monomers than the transferring polymer radical; *inhibition*, if no new chain is started at all. Such inhibitions are caused often by small amounts of impurities in the polymerization system, for example, oxygen.

Chain transfers to polymers lead to branched polymers; most radically polymerized polymers are thus slightly branched. The chain transfer to nonionic allyl monomers $CH_2{=}CH{-}CH_2R$ generates resonance-stabilized allyl radicals $CH_2{\cdots}CH{\cdots}CHR$. Addition of monomer molecules to these allyl radicals produces macroradicals which are more resonance stabilized than their predecessors. Since these radicals do not like to add additional monomer molecules, degrees of polymerization of noncharged monoallyl polymers are low (ca. 10-20). The formation of cross-linked polymers from diallyl monomers is of course not affected.

Polymerization Kinetics. Initiator radicals are generated by initiator decomposition with a rate of $2\,fk_d[I]$ where f is the radical yield (i.e., the fraction of radicals that start the polymerization). Initiator radicals disappear in the start reaction with a rate of $k_{st}[I^{\bullet}][M]$. In the steady state, the rate R_{st} of the start reaction equals the rate R_t of the termination reaction(s), thus $R_{st} = R_t$. For terminations through deactivation by other polymer radicals, the rate $d[P^{\bullet}]dt$ of formation of polymer radicals is given by

$$(3\text{-}10)\quad d[P^{\bullet}]/dt = R_{st} - R_t = 2\,fk_d[I] - k_t[P^{\bullet}]^2 = 0$$

Monomer is practically only consumed by the propagation reaction with a rate of $R_p = -d[M]/dt = k_p[P^{\bullet}][M]$. For diminishing initiator consumption ($[I] \approx [I]_o$), the rate of propagation is thus with Equation (3-10):

$$(3\text{-}11)\quad R_p = -d[M]/dt = k_p\left(\frac{2fk_d}{k_t}\right)^{1/2}[I]_o^{1/2}[M]$$

The polymerization rate is directly proportional to the monomer concentration; bulk polymerization is thus always faster than solution polymerization. The predicted decrease of propagation rate with decreasing monomer concentration [M] (increasing monomer conversion) is often observed for monomer conversions of ca. 0-20 %.

A "self-acceleration" of the polymerization and a concomitant increase in the degree of polymerization is observed at higher monomer conversion, especially in bulk and in concentrated solutions. This *gel effect* or *Trommsdorff-Norrish effect* is caused by a changing diffusion control of the termination by mutual reaction of two polymer radicals (i.e., a decrease of termination rate), which in turn comes from an increasing entanglement of polymer chains. Because neither the rate of the initiator decomposition nor the rate of monomer addition is affected by increasing entanglements, both overall polymerization rates and molar masses increase. The gel effect broadens molar mass distributions.

Cross-Linking Polymerizations. Free radical chain polymerizations are employed commercially in the manufacture of diallyl thermosets and in styrene/divinylbenzene copolymerizations for polymer beads. Primary macromolecules produced at very low monomer conversions are almost linear and carry pendant polymerizable groups. The likelihood of an attack at these groups increases with increasing mass average molar mass. Branched molecules are formed that convert to cross-linked polymers at a certain monomer conversion. At this conversion, the reaction mixture turns into a gel-like mass consisting of a network of cross-linked polymer molecules that is swollen by still unreacted monomer. At present, the monomer conversion at this "gel point" cannot be predicted theoretically for such free radical polymerizations.

Process Engineering. Free radical polymerizations may be performed in the gas phase, in bulk, solution, emulsion, suspension, or under precipitation (Table 3-1). Each method has advantages and disadvantages.

Bulk Polymerizations. Industrial bulk polymerizations are carried out in *liquid monomers*, occasionally with the addition of 5-15 vol-% solvent as polymerization aid. The process leads to very pure products. However, much heat is generated per unit volume, especially in the presence of a gel effect. Local overheating may cause multimodal molar mass distributions, branching, polymer degradation, discoloring, and/or explosions. Bulk polymerizations are thus often terminated deliberately at 40-60 % monomer conversion. Alternatively, polymerizations may be performed in two steps: first in large batch reactors, then in thin layers. The residual monomer is usually removed by steam and recovered.

Suspension polymerization is a "water-cooled" bulk polymerization. Water-insoluble monomers are suspended as small droplets of 0.001-1 cm diameter with the help of suspending agents. The polymerization is initiated by oil-soluble initiators. The droplets are converted by polymerization into pearls (beads) and the process is thus also called *pearl polymerization* or *bead polymerization.* Suspension polymerization allows easy control of the reaction and delivers the polymer as easy-to-handle beads. A disadvantage is the cost of water removal and cleaning, along with the sometimes deleterious effects of incorporated suspension agents.

Emulsion Polymerization. In emulsion polymerizations, water-insoluble monomers are solubilized in water by micelles of surfactant molecules (micelle diameters: 4-10 nm); the monomers are also suspended in a few droplets (ca. 1000 nm diameter).

Water-soluble initiators decompose into radicals that travel through the water to the micelles where polymerization begins. Diffusion of monomer from the droplets to the micelles replenishes the monomer that has been used up by polymerization in the micelles. The micelles finally grow into latex particles of 500-5000 nm diameter. Since the reaction volume is very small in micelles and in growing latex particles, termination reactions are rare and the polymerization proceeds in a quasi-living manner, leading to much higher molar masses than in bulk polymerization. At high monomer conversions, the polymerization rate becomes very small, however, due to the small proportion of remaining monomer and a decrease of propagation rate constants caused by a diffusion-controlled propagation.

In terms of reaction control, emulsion polymerization has advantages similar to suspension polymerization. In addition, it offers higher polymerization rates by redox initiators and higher molar masses (which can be regulated by transfer agents). The resulting latex can be used directly for paints, adhesives, and coatings.

Gas-phase polymerization. Gas-phase polymerizations can be initiated photochemically above the dew point of monomers. The growing polymer chains aggregate; the polymerization thus proceeds in a mist of polymer particles and not in the gas phase. Each growing particle contains only one radical and the polymerization is thus quasi-living. The polymerization rate is determined by the rate of monomer adsorption by the precipitating particles and later by monomer diffusion to the occluded polymer radicals, provided the polymerization temperature is below the polymer melting temperature. The resulting polymers are very clean.

Solution polymerizations decrease polymerization rates by monomer dilution and allow easy removal of the heat of polymerization. Solvents are however costly to remove, and the process is used only if the resulting polymer solutions can be used directly and/or if monomers and polymers decompose in melts.

3.3.4. Anionic Polymerization

Anionic polymerizations are initiated by bases or Lewis bases such as alkali metals, alkoxides, amines, phosphines, Grignard compounds, and sodium naphthalene, very often in solution. The initiators dissociate "instantaneously" into the initiating species. These dissociations and the subsequent start reactions need little activation energy; anionic polymerizations thus often proceed with high speed, even at -100°C.

The initiating species and the growing macroions are rarely completely dissociated into free anions and counterions. They are rather in equilibrium with various types of ion pairs (nondissociated species ⇄ contact ion pair ⇄ solvent-separated ion pair ⇄ free ions; plus associates of these species). The type and proportion of the initiating species strongly affect the polymerization rates and the tacticity of polymers.

The initiating species are present in their effective concentrations at the beginning of the polymerization; they are not formed one after another as in free radical polymerizations. Termination and transfer reactions are fairly rare; anionic polymerizations are often "living". Very low initiator concentrations are required for living polymerizations to high molar masses (see Section 3.2.1); low initiator/monomer ratios are difficult to control and may lead to low polymerization rates. Since termination reactions are absent, polymers from living polymerizations have very narrow molar mass distributions of the Poisson distribution type if diffusion effects are avoided during the mixing of monomer and initiator solutions.

Very few anionic homopolymerizations are therefore employed for the production of thermoplastics, most notably that of formaldehyde, ε-caprolactam (also for reaction injection molding), and laurolactam (Table 3-2). Anionic polymerizations are however the method of choice for the syntheses of block copolymers (e.g., thermoplastic elastomers of the styrene-butadiene-styrene type).

Tab. 3-2 Industrial anionic homopolymerizations:
D diene polymerization;
C=C polymerization of carbon-carbon double bond;
C=O polymerization of carbon-oxygen double bond;
R ring-opening polymerization of cyclic monomer consisting of 1, 2, 3, or 4 repeating units.

Monomer	Type	Repeating unit	Application
Butadiene	D	$-CH_2-CH=CH-CH_2-$	Elastomer (1,4-cis)
Isoprene	D	$-CH_2-C(CH_3)=CH-CH_2-$	Elastomer (1,4-cis)
Methylcyanoacrylate	C=C	$-CH_2-C(CN)(COOCH_3)-$	Adhesive
Formaldehyde	C=O	$-O-CH_2-$	Engineering plastic
Ethylene oxide	R (1)	$-O-CH_2-CH_2-$	Thickener
Glycolide	R (2)	$-O-CO-CH_2-$	Surgical threads
ε-Caprolactone	R (1)	$-O-CO-(CH_2)_5-$	Polymer plasticizer
ε-Caprolactam	R (1)	$-NH-CO-(CH_2)_5-$	Fiber,thermoplastic
Lauryllactam	R (1)	$-NH-CO-(CH_2)_{11}-$	Fiber, film
Hexamethylcyclotrisiloxane	R (3, 4)	$-O-Si(CH_3)_2-$	Elastomer

3.3.5. Cationic Polymerization

Three groups of monomers can be polymerized cationically: Olefin derivatives $CH_2=CHR$ with electron-rich substituents R; monomers $CH_2=Z$ with double bonds containing heteroatoms or heterogroups Z; and rings with heteroatoms. Initiators are Brønsted acids (perchloric acid, trichloroacetic acid, trifluoromethanesulfonic acid, etc.); Lewis acids ($AlCl_3$, $TiCl_4$, etc.) with "coinitiators" such as water; and carbenium salts (acetyl perchlorate, tropylium hexachloroantimonate, etc.).

The propagating macrocations are thermodynamically and kinetically instable. They attempt to stabilize themselves by addition of nucleophilic species, which leads to very fast polymerizations on one hand and to a host of transfer and termination reactions on the other. Quasiliving cationic polymerizations are obtained if the propagating carbocations are stabilized through nucleophilic interaction; other than true living polymerizations, quasiliving ones have transfer reactions. Very few monomers are polymerized cationically on an industrial scale (Table 3-3).

Table 3-3 Industrial cationic polymerizations. c = cyclo.

Monomer		Application
Isobutylene	$CH_2=C(CH_3)_2$	Elastomers, adhesives, VI-improvers
Vinylethers	$CH_2=CHOR$	Adhesives, textile aids, plasticizers
Formaldehyde	$CH_2=O$	Engineering plastic
Ethyleneimine	c-($NHCH_2CH_2$)	Paper additive, flocculant
Tetrahydrofuran	c-($O(CH_2)_4$)	Soft segment for polyurethanes or polyetherester elastomers

3.3.6. Ziegler-Natta Polymerization

Ziegler-Natta polymerizations are chain polymerizations that are initiated and propagated by Ziegler catalysts, a vast group of coordination compounds from transition-metal compounds and metal alkyls. These polymerizations are also called *coordination polymerizations* or *anionic-coordination polymerizations*, although coordination of a monomer to an initiator need not lead to a Ziegler-Natta type of polymerization and no anions are involved at all.

Typical Ziegler catalysts for industrial polymerizations are shown in Table 3-4. The so-called third-generation catalysts for ethylene polymerizations are composed of $TiCl_3 + AlR_3 + MgCl_2$; their high catalyst yield of more than 50 000 g poly(ethylene) per 1 g $TiCl_3$ makes removal of catalyst residues from the polymer unnecessary.

Table 3-4 Important industrial polymerizations with transition metal catalysts.

Monomer		Initiator	Application
Ethylene	$CH_2=CH_2$	$TiCl_3/AlR_3$; Cr/silica	Thermoplastic
Propylene	$CH_2=CHCH_3$	$TiCl_3/R_2AlCl$	Thermoplastic
Ethylene + propylene + nonconjugated diene		$VOCl_3/R_2AlCl$	Elastomer
Butadiene	$CH_2=CH-CH=CH_2$	TiI_4/AlR_3;	Elastomer
		$R_3Al_2Cl_3/Co(OOCR')_2$	Elastomer
Isoprene	$CH_2=CCH_3-CH=CH_2$	$TiCl_3/AlR_3$	Elastomer
Butene-1	$CH_2=CH(C_2H_5)$	$TiCl_3/Et_2AlCl$	Thermoplastic
4-Methylpentene-1	$CH_2=CH(CH_2CH(CH_3)_2)$	?	Thermoplastic

Various complexes between transition metal compounds and metal alkyls are known to be formed, but which ones are polymerization active is rarely known. In any case, the transition metal atoms of these complexes are bound to the growing chains; Ziegler catalysts are thus not catalysts but initiators. Propagation proceeds via α-insertion of monomer molecules $CH_2=CHR$ between chains and their bound transition metals Mt to $\sim[CHR-CH_2)_nMt$ or via β-insertion to $\sim[CH_2-CHR]_nMt$. The isospecific polymerization of propylene by $TiCl_4/AlR_3$ proceeds via an α-insertion; the syndiospecific polymerization of the same monomer is very probably a β-insertion.

The ability of Ziegler catalysts to regulate the stereocontrol of such polymerizations to a very high degree makes Ziegler-Natta polymerizations extremely useful industrially. Propylene, for example, polymerizes free radically or cationically to highly branched, atactic, low molar mass, noncrystallizable polymers that have low glass temperatures; they can be used only for hot-melt adhesives. Ziegler-Natta polymerizations (Table 3-4) lead, however, to stereoregular polymers with high melting temperatures; isotactic poly(propylene)s find extensive use as thermoplastics.

Similar catalysts are used in *metathesis polymerizations*. Metatheses are exchange and disproportionation reactions of double bonds, mainly carbon-carbon double bonds in olefins and cycloolefins, for example, the metathesis of pentene-2 to butene-2, pentene-2, and hexene-3 (in the ratio 1:2:1) by the catalyst system $C_2H_5AlCl_2$–WCl_6–C_2H_5OH. Industrially, three monomers are polymerized by metathesis polymerization. Cyclooctene gives poly(octenamer) $\sim[CH{=}CH{-}(CH_2)_6]_n\sim$; both the *cis* and the *trans* isomers are elastomers. Norbornene polymerizes to a thermoplastic polymer that is plasticized by mineral oil to give an elastomer. Dicyclopentadiene has been proposed as a monomer for reaction injection molding (RIM) processes.

Group transfer polymerizations are probably also insertion polymerizations. Only a small-scale polymerization of alkyl methacrylates to low molar mass polymers by silylketene acetals and nucleophilic catalysts ($[(CH_3)_3SiF_2]^-$, $[HF_2]^-$, CN^-, etc.) is utilized industrially at present.

3.3.7. Copolymerization

Copolymerizations are joint polymerizations of two or more monomers; they were also called interpolymerizations in the older literature. In the simplest case, both active chain ends ~a* and ~b* react irreversibly with the monomers A and B:

$$\begin{array}{llllll} (3\text{-}12) & \sim a^* + A & \rightarrow & \sim a{-}a^* & ; & R_{aA} = k_{aA}[a^*][A] \\ & \sim a^* + B & \rightarrow & \sim a{-}b^* & ; & R_{aB} = k_{aB}[a^*][B] \\ & \sim b^* + A & \rightarrow & \sim b{-}a^* & ; & R_{bA} = k_{bA}[b^*][A] \\ & \sim b^* + B & \rightarrow & \sim b{-}b^* & ; & R_{bB} = k_{bB}[b^*][B] \end{array}$$

Four rates R_{ij} and four rate constants k_{ii} are to be considered in this terminal model which corresponds to Markov first order statistics. At high molar masses, monomers are consumed only by the four propagation reactions. The relative monomer conversion is given by

$$(3\text{-}13)\quad \frac{-d[A]/dt}{-d[B]/dt} = \frac{R_{aA} + R_{bA}}{R_{bB} + R_{aB}} = \left(\frac{k_{bA} + k_{aA}\{[a^*]/[b^*]\}}{k_{bB} + k_{aB}\{[a^*]/[b^*]\}}\right)\left(\frac{[A]}{[B]}\right) = \frac{d[A]}{d[B]}$$

The ratios of rate constants of homopropagations and cross-propagations are called *copolymerization parameters r*:

(3-14) $r_A \equiv k_{aA}/k_{aB}$; $r_B \equiv k_{bB}/k_{bA}$

Five different cases can be distinguished for each copolymerization parameter:

$r = 0$ No homopropagation; the active center adds only the other monomer.
$r < 1$ The other monomer is preferentially added.
$r = 1$ Both monomers are added in the same amounts.
$r > 1$ The own monomer is preferentially added but not exclusively.
$r = \infty$ No copolymerization, only homopolymerization.

Depending on the relative numerical value of the two copolymerization parameters, five different types of copolymerizations can be distinguished (Table 3-5). In general, one of the monomers is consumed preferentially during copolymerization ($r_A \neq r_B$). This drift of copolymer composition can be avoided if the more reactive monomer is fed into the reactor according to its consumption. A conversion-independent polymer composition also results if the copolymerization is performed under azeotropic conditions. These conditions exist if the relative rate of monomer consumption equals the ratio of monomer concentrations at any time, i.e., if $d[A]/d[B] \equiv [A]/[B]$: ratios of monomer and the polymer concentrations do not drift with the progress of polymerization.

Table 3-5 Special cases in the copolymerization of monomers A and B.

Arrangement of mers	Copolymerization parameter if copolymerization is azeotropic		not azeotropic		
	r_A	r_B	r_A	r_B	$r_A r_B$
Alternating	0	0	0	> 0	0
Statistical	< 1	< 1	< $1/r_B$	> 1	< 1
Ideal (random)	1	1	$1/r_B$	$1/r_A$	1
Block forming	> 1	> 1	> $1/r_B$	< 1	> 1
Blend forming	∞	∞	∞	< ∞	∞

The azeotrope equation $d[A]/d[B] = [A]/[B]$ is a special case of the *Lewis-Mayo equation* which describes the relative instantaneous composition of the copolymer (e.g., x_a/x_b) as a function of the relative change of monomer concentrations:

$$\text{(3-15)} \quad \frac{d[A]}{d[B]} = \frac{1 + r_A\{[A]/[B]\}}{1 + r_B\{[B]/[A]\}} = \frac{x_a}{x_b}$$

The Mayo-Lewis equation describes reasonably well the composition of copolymers at small conversion intervals (e.g., at low conversions). It often fails for polymerization rates because it considers only the effects of the last monomeric units of the growing polymer chains (see Eq. (3-9)). Rate studies have to take into account penultimate monomeric units, that is, ~aa* + A, ~ab* + B, ~ba* + A, and ~bb* + B.

Tab. 3-6 Industrial copolymerizations. B = Bulk, L = solution, P = precipitation, S = suspension, E = emulsion; (1) in t-butanol, (2) in acetone, 1,4-dioxane or hexane, (3) in water.

Polymerization type and monomers	Polymerization	Application
Free radical, linear (or slightly branched)		
Ethylene + 10 % vinyl acetate	B	Shrink films
Ethylene + 10-35 % vinyl acetate	B	Thermoplastics
Ethylene + 35-40 % vinyl acetate	P (1)	Films
Ethylene + > 60 % vinyl acetate	E	Elastomers
Ethylene + < 10 % methacrylic acid	B	Extrusion coatings
Ethylene + trifluorochloroethylene		Thermoplastics
Butadiene + styrene	E	Multipurpose elastomers
Butadiene + 37 % acrylonitrile	E	Oil-resistant elastomers
Vinyl chloride + 3-20 % vinyl acetate	L (2)	Paints
Vinyl chloride + 15 % vinyl acetate	L (2)	HiFi records
Vinyl chloride + 3-10 % propylene	B	Thermoplastics
Acryl esters + 5-15 % acrylonitrile		Oil-resistant elastomers
Acrylonitrile + 4 % of various monomers	P (3)	Fibers with improved dyebility
Acrylonitrile + styrene		Thermoplastics
Acrylonitrile + butadiene + styrene		Thermoplastics
Tetrafluoroethylene + propylene	E	Thermoplastics
Methacrylic acid + methacrylonitrile		Hard foams (after cyclization)
Free radical, cross-linking		
Glycol methacrylate + 2-4 % glycol dimethacrylate		Contact lenses
Unsaturated polyesters + styrene or methyl methacrylate	B	Glass fiber-reinforced thermoset
Anionic		
Styrene + butadiene	L	Elastomers
Cationic		
Isobutene + 2 % isoprene	L	Butyl rubber
Trioxane + ethylene oxide	L	Thermoplastics
Ethylene oxide + propylene oxide	L	Thickener, detergents
Propylene oxide + nonconjugated dienes	L	Elastomers

Most industrial copolymerizations are by free radical initiation (Table 3-6) because of three reasons. Free radical copolymerizations allow to vary the copolymer structure widely (see below); they can be performed with olefin, vinyl, and acrylic type monomers; and these monomers are inexpensive.

The type of copolymerization depends on the type of initiator. For example, the free-radical initiated copolymerization of styrene and methyl methacrylate is azeotropic (Fig. 3-2), whereas cationic and anionic polymerizations are nonazeotropic. Initiation by $Et_3Al_2Cl_3$ and traces of oxygen leads to almost alternating copolymers. Cationic polymerization generates long styrene sequences; anionic polymerization, long methyl methacrylate sequences. The sequence length determines many polymer properties, most notably glass and melting temperatures.

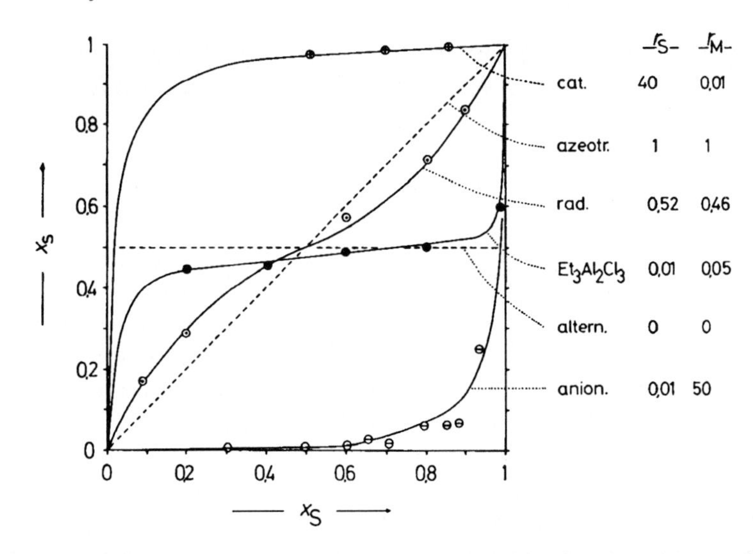

Fig. 3-2 Instantaneous mole fraction, x_s, of styrene units as function of the instantaneous mole fraction, x_S, of styrene in the copolymerization of styrene (copolymerization parameters r_S) with methyl methacrylate (copolymerization parameters r_M) initiated by (⊕) cations, (O) free radicals, (●) $Et_3Al_2Cl_3$, or (⊖) anions.
Solid lines: Calculated with copolymerization parameters shown to the right.
Horizontal broken line: Theory for alternating copolymerizations.
Slanted broken line: Theory for azeotropic polymerizations.
Reprinted with permission by Hüthig and Wepf Publ., Basle [1].

3.4. Polycondensation and Polyaddition

3.4.1. Introduction

Polycondensation is a polymerization by reaction between molecules of *all* degrees of polymerization (not just between polymer molecules and monomer molecules as in chain polymerization) and with the formation of low molar mass leaving molecules. Many polymers with hetero atoms in the chains are generated by polycondensations. An example is the polycondensation of 11-aminoundecanoic acid to polyamide 11 with water as the leaving compound:

(3-16) $$n\ H_2N(CH_2)_{10}COOH \rightarrow H[NH(CH_2)_{10}CO]_nOH + (n-1)\ H_2O$$

Dimers ($n = 2$) are formed first which may then react with monomers to trimers ($n = 3$) or with other dimers to tetramers ($n = 4$), etc. The reaction is thus statistical with respect to the size of the molecules to which the functional groups $-NH_2$ and -COOH are attached. The degree of polymerization of the polymer is a number average, $\overline{X}_n$, over all degrees of polymerization of the reactants, including the monomer ($n \geq 1$).

All amino and carboxyl groups have the same chance to react if their reactivity is independent of the molecule size. The reaction of functional groups attached to two molecules with high molar masses adds, however, considerably more to the increase of the average degree of polymerization than the reaction of two lower molar mass molecules: the degree of polymerization snowballs with increasing extent of reaction of functional groups (Fig. 3-1, curve PC).

Most polycondensations can be easily stopped at any desired conversion of functional groups: cooling the reaction system allows to isolate the steps (in the meaning of organic chemistry). Polycondensations are thus also known as *step-growth polymerizations*. The term *polycondensation* was used in the past and is still used in the technical literature with a different meaning, i.e., as a synonym for condensation polymerization. The latter term includes both the current concepts for polycondensation and for polyelimination (condensative chain polymerization).

Polyaddition is a polymerization by reaction between molecules of *all* degrees of polymerization (as in polycondensation) but *without* the formation of small leaving molecules (as in chain polymerization). The number-average degree of polymerization of the resulting polymers varies with the extent of reaction of functional groups in the same way as it does in polycondensation. An example is the polyurethane formation by polyaddition of a diisocyanate to a diol

$$n\ \mathrm{OCN{-}R{-}NCO} + n\ \mathrm{HO{-}R'{-}OH} \rightarrow \mathrm{OCN[R{-}NH{-}CO{-}O{-}R'{-}O]}_n\mathrm{H} \quad (3\text{-}17)$$

These polymerizations have also been called "polyadduct formations". Polyaddition is not to be confused with addition polymerization, the traditional term used in literature for chain polymerization.

Thermodynamics. Many polycondensations and polyadditions are exothermic equilibrium reactions between A and B functional groups. Equilibrium constants of polycondensations are defined by $K \equiv ([AB][L])/([A][B])$. Shifts to higher polymer concentrations [AB] are thus facilitated by removal of leaving molecules L. Equilibrium constants of polycondensations are low for transesterifications (K = 0.1-1), higher for esterifications (K = 5-10), and highest for amidations (K = 100-1000). They decrease with increasing temperature. The main contribution to the Gibbs energy of polycondensations comes from enthalpy, only a small part from entropy.

3.4.2. Bifunctional Reactions

Polycondensations and polyadditions are generally divided into bifunctional reactions (two functional groups per reactant) and multifunctional ones (three and more functional groups per monomer molecule). Bifunctional reactions lead to linear macromolecules, multifunctional ones to branched and finally cross-linked polymers (exception: A_2B monomers). Bifunctional polycondensations and polyadditions are further subdivided into AB reactions (e.g., that of $H_2N{-}Y{-}COOH$) and AABB reactions (e.g., $H_2N{-}Y{-}NH_2$ + $HOOC{-}Y'{-}COOH$).

All reactants, from monomer to high polymer (but exclusive of leaving molecules such as water in polyesterifications), may react to form macromolecules in bifunctional polycondensations and polyadditions. Only monomers and later oligomers are present at the beginning of the reaction. If the reactivity of the functional groups is independent of the molecule size, then the greater number of smaller molecules leads first to low molar mass polymers and only at high conversions to chemical compounds with high molar masses (see Fig. 3-1). In equilibrium, the number-average degree of polymerization $\overline{X}_n = 1/(1 - p)$ of polymers from AB and stoichiometric AA/BB reactions is dictated by the extent p of reaction of functional groups.

At 50 % conversion ($p = 0.5$), the number-average degree of polymerization $\overline{X}_n$ of reactants i ($1 \leq X_i \leq \infty$) is only 2 (this includes monomer molecules!) and the number-average degree of polymerization of polymers ($2 \leq X_i \leq \infty$) is only 3. Industrial AB and AABB polycondensations rarely deliver number-average degrees of polymerization greater than 100-200 (i.e., $p = 0.99$-0.995) in equilibrium.

The endgroups of equilibrium polymers may condense further during processing by, e.g., extrusion or injection molding. The accompanying strong increase in viscosity is undesirable because processing conditions need to be adjusted continuously for such postcondensations. Endgroups are therefore blocked against further polycondensation by "sealing" them with monofunctional reagents.

Polycondensations for plastics are preferably conducted in the melt. At higher conversions of functional groups, melting temperatures of the resulting polymers may surpass the reaction temperature, however. Higher molar masses may then be obtained by polycondensation in the solid state. For example, poly(ethylene terephthalate) grades for containers require high molar masses that are obtained by "solid stating"

Equilibria are established not only by retroreactions (reversal of polycondensations and polyadditions) but also by trans reactions, i.e., catalyzed exchange reactions between chain segments:

$$(3\text{-}18) \qquad A_n{-}A_p + A_q{-}A_r \rightleftarrows A_n{-}A_r + A_q{-}A_p$$

In these reactions, number-average molar masses remain constant but mass-average molar masses increase if the initial ratio of mass- to number-average molar masses was smaller than dictated by equilibrium conditions.

Higher degrees of polymerization at lower extents of reactions can be obtained by irreversible "activated polycondensations" and especially by heterogeneous reactions of fast reacting monomers. An example is the *interfacial polycondensation*, a Schotten-Baumann reaction of diamines $H_2N{-}Y{-}NH_2$ in water with diacid dichlorides $ClOC{-}Z{-}COCl$ in, for example, chloroform. A polyamide film is formed at the interface of the two solutions. Functional groups of this polymer are buried in the interior of the film; they cannot react with each other. Exchange reactions and retroreactions are also absent. Only surface groups can react; the resulting irreversible reaction leads to high molar masses. Industrially, this reaction is utilized for one of the two industrial syntheses of polycarbonates, that is, from phosgene $COCl_2$ and the sodium salt of bisphenol A $NaO(p{-}C_6H_4){-}C(CH_3)_2{-}(p{-}C_6H_4)ONa$ (Table 3-7).

Most industrial polycondensations are of the AABB type (Table 3-7); they can be described by

(3-19) n A–V–A + n B–W–B → A–(V–W)$_n$–B + (2 n - 1) AB

where A and B are leaving groups (not necessarily endgroups), AB represents leaving molecules L, V and W are monomeric units, and V–W is a repeating unit. Industrial AB polycondensations

(3-20) n A–V–B → A–V$_n$–B + (n - 1) AB

are fairly rare because the presence of two different reactive groups in one monomer molecule may lead to premature polycondensations on storage (short shelf-life of monomers). The types of endgroups (and thus of leaving molecules) used industrially are fairly limited: HO- (NaO-, KO-), H_2N- and occasionally NaS-, H-, or O=C- on one hand, and -COOR (R = H, CH_3, C_6H_5, Cl), $-SO_2Cl$, -NCO, and -F on the other.

Table 3-7 Industrially important linear AB polycondensations and AABB polycondensations. Ar = Aromatic residue, *p*Ph = para-substituted phenylene group, R = organic residue.

Polymer	Chemical structure A	V	W	B	Remarks
AA/BB polycondensations					
Poly(ethylene terephthalate)	H	$O(CH_2)_2O$	CO-*p*Ph-CO	OH or OCH_3	
Poly(butylene terephthalate)	H	$O(CH_2)_4O$	CO-*p*Ph-CO	OH or OCH_3	
Unsaturated polyesters	H	$O(CH_2)_2O$	CO-CH=CH-CO	$O_{1/2}$	from maleic anhydride (isomerizes mainly to fumaric acid residues)
Polycarbonate	H	O-*p*Ph-$C(CH_3)_2$-*p*Ph-O	CO	OC_6H_5	
	Na	O-*p*Ph-$C(CH_3)_2$-*p*Ph-O	CO	Cl	
Polyarylates	Na	O-Ar-O	CO-Ar-CO	Cl	different Ar
Poly(phenylene sulfide)	Na	S	*p*Ph	Cl	
Polysulfides	Na	S	Y	Cl	different Y
Polysulfones	K	O-*p*Ph-O	SO_2-*p*Ph-O-*p*Ph-SO_2	Cl	different V and W
Polyamides	H	$NH(CH_2)_iNH$	$CO(CH_2)_jCO$	OH	PA 6.6 (i=6; j=4) PA 6.10 (i=6; j=8) PA 4.6 (i=4; j=4)
Polyether-etherketone	K	O-*p*Ph-CO-*p*Ph-O-	*p*Ph-CO-*p*Ph	F	
AB polycondensations					
Polyamide 11	H	$NH(CH_2)_{10}CO$	-	OH	
Polycarbo-diimides	O	C=N-R-N	-	CO	
Polysulfones	H	C_6H_4-O-(*p*-C_6H_4)-SO_2	-	Cl	

3.4.3. Multifunctional Reactions

Multifunctional polycondensations are condensation reactions with the participation of monomers with higher functionalities ($f_0 > 2$). At low conversions, branched molecules with increasing higher number of functional endgroups are formed, i.e., with increased functionality f per molecule (Eq. (3-2)). Only one functional group is however needed to link two molecules; the probability of such linkages thus increases dramatically with increasing degree of polymerization, X. At a certain (critical) degree of conversion of functional groups, molecules are linked throughout the volume of the reactor. The viscosity increases strongly, and the mass-average molar mass of the cross-linked polymer approaches infinity at this "gel point". Not all reactants are linked at the gel point; some are still soluble up to high conversions. The number-average degree of polymerization of reactants (includes monomers!) is thus fairly small at the gel point, usually in the range $10 < \overline{X}_n < 50$.

The knowledge of the gel point is very important. Polymers cannot, or can only with difficulty, be processed beyond the gel point and must thus be shaped before or during the cross-linking reaction. The gel point can be calculated from the functionality and proportion of reactive groups if intramolecular cyclizations are assumed absent. In practice, such cyclizations are often present and cross-linking occurs at higher extents of reaction than calculated. Theoretical calculations thus provide a safety margin.

Chemical cross-linking reactions are utilized for "hardening" or "curing" of thermosetting resins, the "tanning" of certain leathers, and the "vulcanization" of elastomers from rubbers. The hardening of phenolic resins with formaldehyde is caused by the formation of methylene, formal, and ether bridges between the *ortho* and *para* positions of the phenol groups; the hardening of novolacs with hexamethylenetetramine (urotropin) leads to imine structures. Amino plastics result from the reaction of urea or melamine with formaldehyde, and alkyd resins from multifunctional alcohols with bifunctional organic acids or their anhydrides (Chapter 12). Cross-linking also occurs inadvertently during cyclopolycondensations, e.g., to polyimides.

3.4.4. Polyaddition

No leaving molecules have to be removed during polyadditions, an important advantage for process engineering. Practically all industrially important polyadditions are thus performed with thermosetting resins where entrapped low molar-mass compounds would have deleterious effects on properties.

Polyurethanes are formed from diisocyanates (or triisocyanates) and polyols. The polyurethane formation may be so rapid that cross-linking and shaping can be performed in one step through reaction injection molding (RIM) with short cycle times. Epoxides (epoxy resins) contain two or more epoxide groups per prepolymer molecule and are cured (cross-linked) with either difunctional or multifunctional amines or with anhydrides of organic acids (see Chapter 12).

3.5. Polymer Reactions

3.5.1. Polymer Transformations

Several polymers are chemically transformed into industrially useful polymers by polymer-analogous reactions of their substituents (Table 3-8). These reactions are per group chemically identical to those of their low molar mass analogues. They differ in the effect of side reactions, which do not lead to removable byproducts as in micromolecules but to "wrong" structures in the polymer chain. In most cases, such structures decrease the desired polymer properties. Polymer transformations that are prone to side reactions are thus in general avoided. This is one reason why only five types of polymer-analogous reactions are utilized commercially: transesterification/saponification, chlorination/sulfochlorination, etherification, hydrogenation, and cyclization.

Tab. 3-8 Industrial polymer transformations (for ring formations see text). cell = Cellulose, EO = ethylene oxide, PO = propylene oxide, R" = fluorinated hydrocarbon residue, [a] = chlorine content in %, [b] ether groups per glucose unit, [c] additional reactions if OR' = OH.

	Monomeric units in primary polymer	Reagent	Conversion in %	Monomeric units in final polymer	Application
1	$CH_2CH(OOCCH_3)$	ROH	98	CH_2-CHOH	Thickener
2	ditto + CH_2-CH_2	CH_3OH	99	CH_2-CHOH/CH_2-CH_2	Engng. plastics
3	cell–$NHCOCH_3$	H_2O	?	cell-NH_2	Paper additive
4	cell-OH	CH_3COOH	83-100	cell-$OOCCH_3$	Fibers, films
5	cell-OH	HNO_3	67-97	cell-ONO_2	Plastics, fibers
6	cell-ONa	EO	53-87	cell-OCH_2CH_2OH	Thickener
7	cell-OH	PO	100-400[c]	cell-$OCH_2CH(OR')CH_3$	Thickener
8	CH_2–CH_2	Cl_2	25-40[a]	CH_2-CHCl	Elastomer
			> 40[a]	CH_2-CHCl/CHCl-CHCl	Impact improver for PVC
9	CH_2-CHCl	Cl_2	64[a]	CH_2-CHCl/CHCl-CHCl	Adhesives, paints
10	CH_2-CH_2	Cl_2/SO_2	<42[a]; <2[b]	ditto, SO_2Cl	Coatings
11	CH_2CH=$CHCH_2$- co-CH_2-CHCN	H_2	?	$(CH_2)_4$/CH_2-CHCN	Elastomer
12	CH_2-$CH(C_6H_5)$	SO_3	few %	CH_2-$CH(C_6H_4SO_3H)$	Ion exchange
13	N=PCl_2	R"ONa	?	N=P(OR")$_2$	Elastomer

All reactions listed in Table 3-8 may lead, in principle, to 100 % substitution of functional groups. A limit of reaction exists, however, in kinetically controlled reactions if a bifunctional reagent attacks adjacent functional groups at the polymer. An example is the reaction of poly(vinyl alcohol) with butyraldehyde which leads to acetals. For statistical reasons, not all hydroxyl groups can be transformed. Some OH groups remain unreacted in this ring-forming reaction, for example, a fraction $f = \exp(-2q)$ in head-to-tail vinyl homopolymers with a proportion q of reactive groups (i.e., $f = 0.135$ at $q = 1$; only 86.5 % of all groups have reacted):

(3-21)

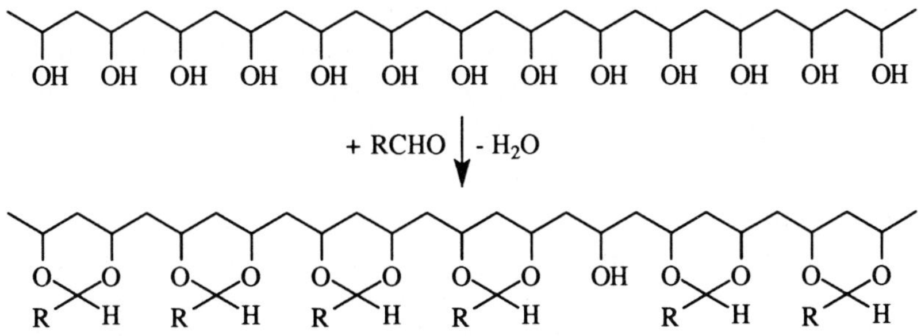

OH OH OH OH OH OH OH OH OH OH OH OH OH

$+ RCHO \downarrow - H_2O$

O O O O O O O O OH O O O O
R H R H R H R H R H R H

At high concentrations of poly(vinyl alcohol), some acetals are formed intermolecularly, leading to cross-links between chains. The resulting poly(vinyl butyral) is used for wash primers and as an intermediate layer in safety glass.

Heating of methacrylic acid/methacrylonitrile copolymers with ammonia produces hard foams with poly(methacrylimide) structures:

(3-22)

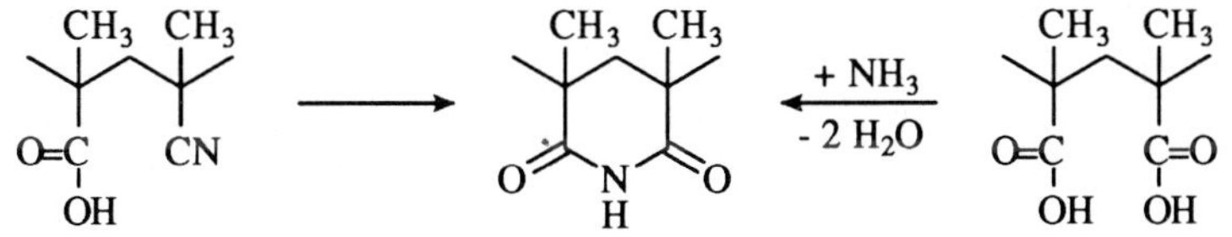

CH_3 CH_3 — O=C(OH), CN $\longrightarrow$ CH_3 CH_3 — O, N–H, O $\xleftarrow[- 2\ H_2O]{+ NH_3}$ CH_3 CH_3 — O=C(OH), C=O(OH)

3.5.2. Formation of Block and Graft Copolymers

Reactive chain ends can either initiate the polymerization of other monomers

(3-23) $\sim A_m{}^* + n\,B \rightarrow \sim A_mB_n{}^*$

or couple with other preformed macromolecules

(3-24) $\sim A_m{}^* + {}^*B_n\sim \rightarrow \sim A_mB_n\sim$

to give block polymers poly(A)-*block*-poly(B).

Both methods are used industrially with a variety of strategies for the synthesis of diblock copolymers A_nB_m as compatiblizing agents for polymer blends, triblock copolymers $A_nB_mA_n$ as thermoplastic elastomers, and various multiblock copolymers for different purposes, especially for thermoplastic elastomers. Anionic living polymerizations are used for the syntheses of poly(styrene)-*block*-poly(butadiene)-*block*-poly(styrene) and poly(styrene)-*block*-poly(isoprene)-*block*-poly(styrene). Polycondensations serve for the preparation of multi[{poly(butylene terephthalate)}-*block*-poly{(polytetrahydrofuran)-terephthalate}], and various polyesteramide and polyetheramide segmented block polymers.

Graft copolymers result from the grafting of monomers on chemically different polymer chains. Grafts of isobutene/isoprene on poly(ethylene)s, vinyl chloride on poly(ethylene-*stat*-vinyl acetate), ethylene/propylene on poly(vinyl chloride), or styrene/acrylonitrile on saturated acryl rubber also yield thermoplastic elastomers.

3.5.3. Cross-Linking Reactions

Cross-linking of preformed polymers is widely used for the manufacture of elastomeric materials such as the vulcanization of unsaturated rubbers with sulfur, vulcanization of saturated rubbers with peroxides, or tanning of hides to leather by cross-linking the collagen with certain reagents.

In the plastics industry, it is utilized for the hardening of unsaturated polyester chains. This reaction is a free radical copolymerization of styrene or methyl methacrylate and the carbon-carbon double bonds of the unsaturated polyester molecules ~[O–CH_2–CH_2–O–OC–CH=CH–CO]$_n$~; the polyester is the cross-linking agent, not the monomers styrene or methyl methacrylate (see Chapter 12).

Most thermosets are however prepared by cross-linking of prepolymers (thermosetting resins), which are oligomers with molar masses between ca. 500-2000 g/mol.

3.5.4. Degradation Reactions

Degradation of polymers is mostly undesirable for their application but often desirable for their disposal after use. The term degradation includes the decrease of the degree of polymerization with preservation of the chemical structure of the monomeric units (depolymerization), the unwanted chemical transformation of some monomeric units with preservation of the degree of polymerization (decomposition), and a combination of both. The terms degradation, depolymerization, and decomposition are sometimes used with the same meaning as in this book, but sometimes with broader or more restricted meanings; no accepted usage exists.

Depolymerizations involve chain scissions. They are retro-reactions (i.e., the reverse of chain polymerizations, polyeliminations, polycondensations, or polyadditions). Depolymerizations may be chain scissions, unzipping reactions, or a combination of both. Retro-polycondensations and retro-polyadditions involve chain scissions, i.e., break-ups of chains at any link between monomeric units in the chain (e.g., amide groups of polyamides). Living polymers from chain polymerizations can unzip one monomeric unit after the other from the living ends of the chains. Dead (terminated) polymer chains from chain polymerization cannot unzip, however. They first have to undergo chain scissions which are usually bond cleavages with formation of macroradicals; the macroradicals then unzip.

Unzipping can occur only with activated chains above their ceiling temperatures, that is with living macroions, polymer molecules with chain ends comprising certain functional groups, or with radical ends generated by a prior homolytic chain scission.

Unzipping at higher temperatures results in gaseous monomers that lead to voids in molded parts. This can be prevented or reduced by shorter sequence lengths of the unzippable polymer blocks. This strategy is utilized, for example, in the formation of acetal copolymers by copolymerization of trioxane (the cyclic trimer of formaldehyde) with ethylene oxide. Chain scission of oxymethylene sequences $\sim[OCH_2]_n\sim$ leads to unzipping and formation of formaldehyde; the unzipping stops at the unzippable oxyethylene sequences $\sim(OCH_2)_m{-}OCH_2CH_2^{\bullet}$.

Decomposition changes the chemical structure of monomeric units; the degree of polymerization of polymer molecules may be maintained or lowered or even increased through branching and cross-linking reactions. Decomposition can be caused by heat, light, oxygen, water, or other environmental agents as well as mechanical forces such as extensional flow, ultrasound, or a combination of these. It may lead to undesired changes of mechanical, electrical, or optical properties. Unstabilized poly(vinyl chloride) $\sim[CH_2CHCl]_n\sim$ discolors on decomposition because HCl is eliminated and colored polyene sequences $\sim[CH{=}CH]_i\sim$ are formed. The polymer ultimately becomes brittle.

Decomposition is the primary degradation process in the pyrolysis of polymers, which may be desirable (rocket fuels, waste disposal) or not (fires), often causing toxic fumes and/or smoke at lower flame temperatures. The risk of pyrolysis is enhanced for polymers with branching points, electron-accepting groups, long methylene sequences, and all groups capable of forming five- or six-membered rings. Thermostability is enhanced by aromatic rings, ladder structures, fluorine as substituent, and low hydrogen contents.

3.6. Polymerization Reactors

3.6.1. Reactor Types

The architecture, molar mass, and molar mass distribution of polymers is influenced not only by the constitution of the monomers, the types of initiators or catalysts for polymerization processes, and the physical state of the reacting system but also by the polymerization reactors. The choice of a reactor depends on

- the desired amount of polymer (small amounts are usually prepared batchwise, large amounts preferentially by continuous polymerizations);
- the properties of monomers and polymers (state of matter, solubility, thermal stability, etc.);
- the properties of the reaction system (viscosities, heats of polymerization, etc.); and
- the desired product (solution, powder, beads, bales, etc.).

Stirred tank reactors (STR) are used for batchwise operations (Fig. 3-3). Discontinuous stirred tank reactors (batch reactors, BR) allow a conservation of mass; in semibatch reactors (SBR), material may be added (but not removed) during a polymerization. Well-stirred batch reactors lead to very narrow distributions of residence times. The largest batch reactors have volumes of ca. 200 m^3. Smaller reactors may be externally cooled via the mantle, larger reactors require built-in heat-exchangers or the removal of heat by, for example, boiling off the solvent.

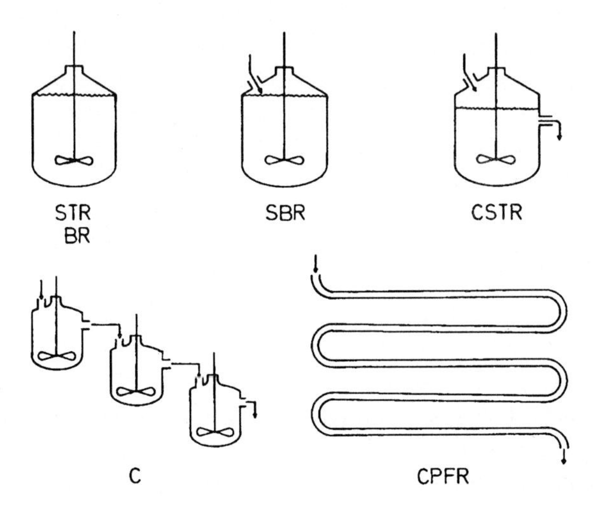

Fig. 3-3 Types of polymerization reactors:
STR = Stirred tank reactor
BR = batch reactor
SBR = semi-batch reactor
CSTR = continuous stirred tank reactor
C = cascade of continuous stirred tank reactors
CPFR = continuous plug-flow reactor

Stirred tank reactors are used for all types of polymerizations (Table 3-8): polycondensations in bulk (PA 6.6) or at interfaces in suspension; living Ziegler-Natta polymerizations as slurry processes (PE, PP, EPDM); radical polymerizations in bulk (PVC), in solution (SBR), or under precipitation (SAN). Suspension polymerizations are almost exclusively performed in STRs, emulsion polymerizations predominantly.

Material is continuously added and removed in *continuous stirred tank reactors* (CSTR). These reactors may be arranged in series to form a *cascade* (C). CSTRs lead to a broad distribution of residence times because the mixing by stirrers (agitators, turbines, etc.) is always a so-called macromixing that creates fluid elements with different residence times at different zones. Macromixing may not necessarily lead to molecular mixing of reaction components because of the often high viscosity of polymerizing systems. The continuous flow in CSTRs may also remove polymer particles from the reactor, that is, pearls can have different residence times in suspension polymerizations. Systems without any molecular mixing are called "segregated" (S) whereas those with complete molecular mixing are said to be "homogeneous" (H).

Single continuous stirred tank reactors are used for the Ziegler-Natta polymerization of propylene, the cationic copolymerization of isobutene + isoprene in a slurry, and free radical polymerizations in solution, precipitation, or emulsion. Fluid-bed reactors are special types of CSTRs; they serve for gas-phase polymerizations of ethylene and propylene, respectively.

Cascades are the second most important types of reactors. They serve in processes for the continuous melt polymerization of ε-caprolactam or styrene, solution polymerizations of ethylene and propylene with Ziegler catalysts, and free radical emulsion polymerizations to various rubbers. Polycondensations include those to amino, phenol, and unsaturated polyester resins.

Tube-type reactors generate a plug-flow and are therefore called *continuous plug-flow reactors* (CPFR). Special CPFRs include tube reactors used in the high-pressure free radical polymerization of ethylene, extruders in the final step of continuous polycondensations to polyamide 6.6, or band reactors in the cationic polymerization of isobutene in ethylene as solvent.

Table 3-9 Reactors for industrial polymerizations. STR = Stirred tank reactor, batch-type (BR) or semibatch reactor (SB); CSTR = continuous stirred tank reactor, single (S) or as cascade (C); CPFR = continuous plug-flow reactor. After a compilation by H.Gerrens.

Polymerization	STR BR	STR SB	CSTR S	CSTR C	CPFR
Polycondensation, polyaddition					
-, in melt or solution	+	+	+	+	+
-, at interfaces	+			+	+
-, as solids	+				+
Living chain polymerizations					
-, in solution	+	+	+	+	
-, under precipitation	+		+		+
Chain polymerizations with terminations					
-, in bulk or solution	+	+	+	+	+
-, under precipitation	+		+		+
-, in suspension	+				
-, in emulsion	+	+	+	+	

3.6.2. Effects on Molar Mass Distributions

The effect of reactor type on the resulting molar mass distribution of the polymer depends on the type of the polymerization reaction. It is different for polymer/monomer reactions (chain polymerization, polyelimination) and polymer/polymer reactions (polycondensations, polyadditions) (Eq. (3-1)). Polymer/monomer reactions can be further subdivided into those with termination and those without.

Polycondensations and Polyadditions. The number-average degree of polymerization of reactants depends on the fractional conversion p of functional groups according to $\overline{X}_n = 1/(1 - p)$ and the mass-average according to $\overline{X}_w = (1 + p)/(1 - p)$ in equilibrium reactions to linear polymers. Such equilibria can be established in batch reactors STR and in continuous plug-flow reactors CPFR. Polymers with high molar masses and Schulz-Flory distributions of molar masses are obtained at high conversions of functional groups (Section 3.4).

Molecules grow during the whole residence time in homogeneous continuous stirred tank reactors HCSTR. Monomer molecules have here a high probability to leave the reactor before they have reacted. On the other hand, a larger molecule is more likely to be formed from the reaction of two smaller ones at higher functional group conversions. The first effect leads to higher molar masses and the second one to relatively smaller ones: The molar mass distribution broadens considerably. The mass-average degree of polymerization is given by $\overline{X}_w = (1 + p^2)/(1 - p)^2$ instead of $\overline{X}_w = (1 + p)/(1 - p)$ whereas the number-average remains $\overline{X}_n = 1/(1 - p)$. The polymolecularity ratio becomes $\overline{X}_w / \overline{X}_n \approx 198$ instead of 1.99 for $p = 0.99$ (corresponds to $\overline{X}_n = 100$)! HCSTRs are thus unlikely choices for polycondensation and polyaddition reactions.

Segregation reduces the proportions of very high and very low molar mass compounds. The molar mass distribution of polymer molecules produced in an SCSTR is more narrow than that from a HCSTR but still broader than that from STR or CPFR; at $p = 0.99$, the ratio $\overline{X}_w / \overline{X}_n$ is 198 (HCSTR) vs. 11 (SCSTR) vs. ca. 2 (STR, CPFR).

Living Chain Polymerizations. Growing polymer chains are not terminated in living polymerizations, except by impurities or deliberately added chain stoppers. If initiation is much faster than propagation, all polymer chains are started "at once" and each chain competes for the same proportion of monomer molecules. The molar mass distribution becomes very narrow. If the polymerization is conducted in batch reactors BR or continuous plug-flow reactors CPFR, life times of chains equal residence times. If furthermore no diffusion effects are present upon addition of initiator or chain stoppers, the molar mass distribution will be a Poisson distribution.

Residence times also equal life times of polymer chains if homogeneous continuous stirred tank reactors HCSTR are used and all growing chains are immediately terminated at the exit of the reactor. Molar mass distributions will be broader, however, because there is a distribution of residence times in HCSTRs; they will be of the Schulz-Zimm type according to calculations. In segregated CSTRs, molar mass distributions shift from the Schulz-Zimm type at low monomer conversions to the Poisson type at high conversions because residence time variations have no effect if the monomer is completely converted into a polymer.

Chain Polymerizations with Termination. Monomer concentrations decrease according to second order microkinetics, e.g., $-\mathrm{d}[\mathrm{M}]/\mathrm{d}t = k[\mathrm{M}][\mathrm{I}]$ for the chain polymerization of a monomer M by an initiator I. Monomer concentrations are however practically constant for small conversion intervals. If initiators have half live times much greater than residence times in a batch reactor BR, then the initiation rate is constant and independent of the monomer concentration. The molar mass distribution is of the Schulz-Zimm type for such incremental conversions ($\overline{X}_w / \overline{X}_n = 2$ for termination by disproportionation, 1.5 for combination of polymer radicals).

Molar mass distributions after monomer conversions from 0 to p are obtained by integration over all distributions from incremental monomer conversions. Molar mass distributions of polymers from free radical polymerizations in batch reactors (and CPFRs) thus become broader than Schulz-Zimm distributions with increasing monomer conversion, especially after the setting-in of the Trommsdorff-Norrish effect. Bulk polymerizations are restricted accordingly often to 40-60 % monomer conversions. Batch reactors are, however, very well suited for suspension and emulsion polymerizations because the viscosity of the system does not vary much with the polymer concentration due to the spherical (or nearly so) shape of the polymer particles. Batch reactors are thus also useful for slurry polymerizations with Ziegler catalysts and for free radical chain polymerizations under precipitation.

Monomer concentrations are constant in the steady state of polymerization in homogeneous continuous stirred tank reactors HCSTR (Fig. 3-4) and the molar mass distributions become narrower than in BRs. The broadening of molar mass distributions due to distributions of residence times can be neglected because the average life time of the growing polymer radicals is much greater than the average residence time in the reactor. HCSTRs thus deliver Schulz-Zimm distributions of molar masses.

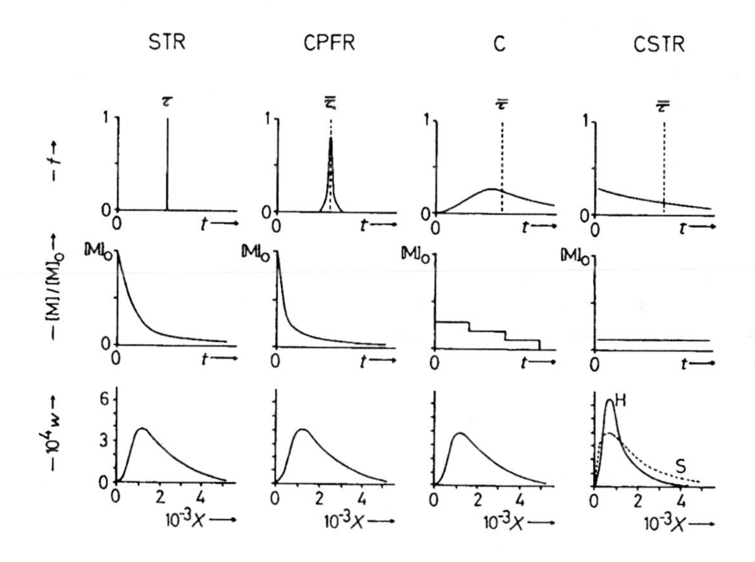

Fig. 3-4 Characteristics of stirred tank reactors STR, continuous plug-flow reactors CPFR, cascades of continuous stirred tank reactors C, and continuous stirred tank reactors CSTR.
First row: Differential distribution of proportions f of residence times τ ($\overline{\tau}$ = average residence time).
Second row: Time dependence of ratios of monomer concentrations [M] to initial monomer concentrations $[M]_0$ of a second order reaction.
Third row: Differential mass distribution of degrees of polymerization X of a polymer ($\overline{X}_n$ = 1000) from a free-radical polymerization with termination by radical combination at a monomer conversion of $p = 0.6$ (60 %) (H = homogeneous, S = segregated).

Bulk polymerizations in CSTRs are most likely to occur as segregated polymerizations, however. At small monomer conversions, viscosities are still low; the reactor operates like an HCSTR and delivers polymers with Schulz-Zimm distributions of molar masses. High viscosities will prevent molecular mixing at larger monomer conversions. Each volume element in an SCSTR can then be treated as a micro-batch reactor; molar mass distributions of polymers for the conversions interval $0 \rightarrow p$ are obtained by integration over the molar mass distributions from all the micro-batch reactors. The molar mass distributions thus broaden with increasing monomer conversions; they become much broader than those from batch reactors which are in turn broader than those from HCSTRs.

Literature

3.1. Raw Materials for Polymers

D.G.Altenpohl, Materials in World Perspective, Springer, Berlin 1980
H.A.Wittcoff, B.G.Reuben, Industrial Organic Chemicals in Perspective, Wiley, New York 1980
H.H.Szmant, Industrial Utilization of Renewable Resources, Technomic, Lancaster (PA) 1986
K.Weissermel, H.J.Arpe, Industrial Organic Chemistry, VCH, Weinheim, 2nd ed. 1993

3.2.-3.4. Polymer Syntheses: Chain Polymerization, Polycondensation, Polyaddition

Houben-Weyl, Methoden der Organischen Chemie, Thieme, Stuttgart, 5th ed. 1987, vol. XX (in three parts)
G.E.Ham, ed., Copolymerization, Interscience, New York 1964
G.E.Ham, ed., Vinyl Polymerization, Dekker, New York 1969 (2 vols.)
D.H.Solomon, ed., Step-Growth Polymerizations, Dekker, New York 1972
S.R.Sandler, W.Karo, Polymer Syntheses, Academic Press, New York 1974-1980 (3 vols.)
H.Sawada, Thermodynamics of Polymerization, Dekker, New York 1976
J.Boor, Jr., Ziegler-Natta Catalysts and Polymerizations, Academic Press, New York 1979
R.W.Lenz, F.Ciardelli, eds., Preparation and Properties of Stereoregular Polymers, Reidel, Dordrecht 1980
P.Rempp, E.W.Merrill, Polymer Syntheses, Hüthig and Wepf, Basle, 2nd ed. 1981
M.Morton, Anionic Polymerization: Principles and Practice, Academic Press, New York 1983
R.P.Quirk, ed., Transition Metal Catalyzed Polymerizations. Alkenes and Dienes, Harwood Academic Publ., Chur 1983 (2 vols.)
K.J.Ivin, T.Saegusa, eds., Ring-Opening Polymerization, Elsevier, New York 1984
Y.V.Kissin, Isospecific Polymerizations of Olefins, Springer, Berlin 1986
F.Rodriguez, Principles of Polymer Systems, Hemisphere Publ., New York, 3rd ed. 1989
H.-G.Elias, Makromoleküle, Hüthig and Wepf, Basle, 5th ed.; vol. I (1990), Grundlagen, Chapters 6-15; vol. II, Technologie (1992), chapters 2, 4-11 (in German)
G.Odian, Principles of Polymerization, Wiley, New York, 3rd edition 1991
J.P.Kennedy, B.Iván, Carbocationic Macromolecular Engineering, Hanser, Munich 1992
G.B.Butler, Cyclopolymerization and Cyclocopolymerization, Dekker, New York 1992
D.J.Brunelle, ed., Ring-Opening Polymerization, Hanser, Munich 1992
C.M.Paleos, ed., Polymerization in Organized Media, Gordon and Breach, New York 1992
M.Szwarc, M.Van Beylen, Ionic Polymerization, Chapman and Hall, London 1992
M.Kucera, Mechanism and Kinetics of Addition Polymerizations, Elsevier, Amsterdam 1992
R.K.Sadhir, R.M.Luck, Expanding Monomers: Synthesis, Characterization, and Applications, CRC Press, Boca Raton (FL) 1992

3.5. Polymer Reactionss

E.M.Fetters, ed., Chemical Reactions of Polymers, Interscience, New York 1964
J.A.Moore, ed., Reactions on Polymers, Dordrecht 1973

3.6. Polymerization Reactors

D.C.Blackley, Emulsion Polymerization: Theory and Pratice, Halsted, New York 1975
K.E.J.Barrett, ed., Dispersion Polymerization in Organic Media, Wiley, New York 1975
H.Gerrens, Polymerization Reactors and Polyreactions. Proceedings of the 4th International/6th European Symposium on Chemical Reaction Engineering, Dechema, Frankfurt/Main 1976, Vol. 2, p. 585
J.L.Throne, Plastics Process Engineering, Dekker, New York 1979
J.A.Biesenberger, D.H.Sebastian, Principles of Polymerization Engineering, Wiley, New York 1983
K.H.Reichert, W.Geiseler, eds., Polymer Reaction Engineering, Hanser, Munich 1983
F.J.Schork, P.B.Deshpande, K.W.Leffew, Control of Polymerization Reactors, Dekker, New York 1993
C.McGreavy, ed., Polymer Reaction Engineering, Blackie (Chapman and Hall), New York 1993

References

[1] H.-G. Elias, Makromoleküle, vol. II: Technologie, Hüthig and Wepf, Basle, 5th ed. 1992, Fig. 4-4.

4. Supermolecular Structure

Polymer properties are influenced not only by the chemical structure of polymers (constitution, molar mass, configuration, microconformation) but also by their physical structure. These structures may range from totally irregular arrangements of chain segments and shorter or longer parallelizations of chains to voids and other defects in otherwise highly organized assemblies of polymer molecules.

Two possible ideal structures exist in the solid state: perfect crystals and totally amorphous polymers. Polymer molecules are perfectly ordered in ideal crystals. They convert at the thermodynamic melting temperature into melts, which ideally are totally disordered. Amorphous polymers can be viewed as frozen-in polymer melts. They are polymer glasses that convert to melts at the glass temperature.

4.1. Noncrystalline State

4.1.1. Structure

Isolated polymer coils possess approximately a Gaussian distribution of chain segments; their segment density (segment concentration) decreases with increasing chain length (Section 2.3.3.). However, the macroscopic densities of polymer melts do not change with chain lengths if endgroup effects on small molar mass molecules are neglected. Coils must thus overlap considerably in polymer melts. At the glass temperature of noncrystallizable polymers, cooperative segmental movements freeze in and the physical structure of the melt and its constituent polymer molecules is conserved. Small-angle neutron scattering has shown that the radii of gyration of amorphous polymers are practically identical for their melts, glasses, and theta solutions.

The absence of long-range order in melts and amorphous polymers does not exclude the presence of short-range order in these states. Because of the persistence of polymer chains, a parallelization of short segments seems probable, as is found, for example, for melts of alkanes according to X-ray investigations. This local order does not exceed 1 nm in each direction; it is similar to that of simple organic liquids.

The random packing of chain segments cannot be perfect. Amorphous polymers thus possess "free volumes", which are regions of approximately atomic diameter. The volume fraction $f_{free} = (v_{am} - v_L)/v_{am}$ of the free volume can be calculated from the specific volumes v of the amorphous (glassy) polymers (am) and their melts (liquids L). The fractional free volume is ca. 2.5 % at the glass temperature; it is approximately independent of the polymer constitution (Section 4.2.5.). Polymer segments in melts can move more freely than in the glassy state. They pack better and the densities of melts are thus greater than the densities of glasses at the same temperature.

4.1.2. Orientation

Polymer segments, polymer molecules, and crystalline domains may be oriented along the machine direction by drawing or other mechanical processes. The orientation of chain segments need not necessarily lead to crystallization, however. An example is injection-molded poly(styrene), which shows optical birefringence due to the orientation of segments but no X ray crystallinity.

The degree of orientation of segments and/or crystallites can be measured by wide angle X-ray or small-angle light scattering, infrared dichroism, optical birefringence, polarized fluorescence, and ultrasound velocity. The orientation is usually characterized for each of the three directions *a*, *b* and *c* by a *Hermans orientation factor* f_i (averaged second Legendre polynom), where β is the angle between the draw direction and the principal axis of segments:

(4-1) $$f_i = (1/2)[3 \langle \cos^2 \beta \rangle - 1]$$

The orientation factor becomes 1 for a complete orientation of the principal axis in the draw direction (β = 0°), -1/2 for a complete orientation perpendicular to the draw direction (β = 90°), and 0 for a random orientation of segments.

Since methods to determine orientation factors are often expensive, time consuming, and/or difficult to perform, the easy-to-calculate *draw ratio* (= length after drawing/length before drawing) is often used to characterize orientations. The draw ratio is, however, not a good measure of the degree of orientation because it depends on the history of the specimen. Drawing may also lead to shear flow of the polymer without any orientation of segments and crystallites.

4.2. Crystalline Polymers

4.2.1. Introduction

Spheres and other geometrically simple entities can be arranged in *crystal lattices* irrespective of whether they are atoms, small molecules, large molecules such as enzymes, or latex particles. The centers of gravity of these entities occupy lattice sites that are regularly spaced in three dimensions. The existence of three-dimensional order on the atomic level is called *crystallinity*. It can be detected by the sharp diffraction patterns generated by X rays because the wave length of X rays is comparable to the distances between atoms in crystal lattices.

Lattices are called *superlattices* if the distances between lattice points significantly exceed atomic dimensions. Superlattices are formed, for example, by the spherical domains of certain block-copolymers (Section 4.3.6.).

Chain molecules can also be arranged in crystal lattices. An all-trans or helical chain may be completely parallel to other chains as in extended-chain crystals (Fig. 4-1). In most cases, only parts of chains may be parallel to each other, either sections of a single chain as in folded chain crystals or those of neighboring ones as in fringed micelles. Folded chains form lamellae that are connected to each other by tie molecules (crystal bridges). The ordered segments of fringed micelles form crystalline regions in otherwise less ordered (but not always completely disordered) polymers.

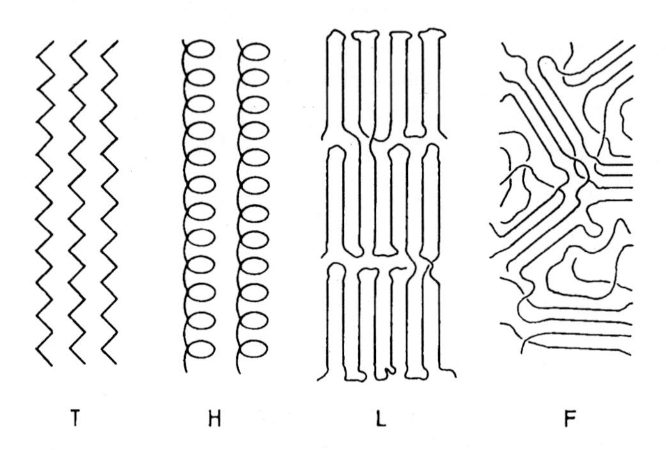

Fig. 4-1 Structure of crystalline polymers (no bond angles are shown for H, L, and F).
T = extended-chain crystals with chains in all-*trans* conformations (zig-zag chains);
C = crystals with chains in helical conformations;
L = lamellae of folded chains with amorphous interlayers interconnected by tie molecules;
F = fringed micelles.

Polymers showing crystallinity are called *crystalline polymers*. Since the fractional amount of crystallinity (the *degree of crystallinity*) may be small, the term *semi-crystalline polymer* is also used. Crystalline polymers with low degrees of crystallinity obviously cannot not form *polymer crystals* with well-defined boundaries but rather *polymer crystallites* with irregular boundaries. Parts of the constituent macromolecules of a crystallite may extend beyond the crystallite boundary (Fig. 4-1).

The types and degrees of order/disorder depend on external conditions such as temperature, pressure, cooling and heating rates, and presence of solvents. A variety of morphologies may thus exist for any crystallizable polymer, for example, spherical, platelet-like, fibrillar, and sheaflike structures of a polyamide 6 (Fig. 4-2).

4.2.2. Crystal Structures

The unit cell describes the smallest, regularly repeating structure of a parallelepiped that generates a crystal by parallel displacements in three directions. The parallelepipeds are characterized by three lattice constants (distances between lattice

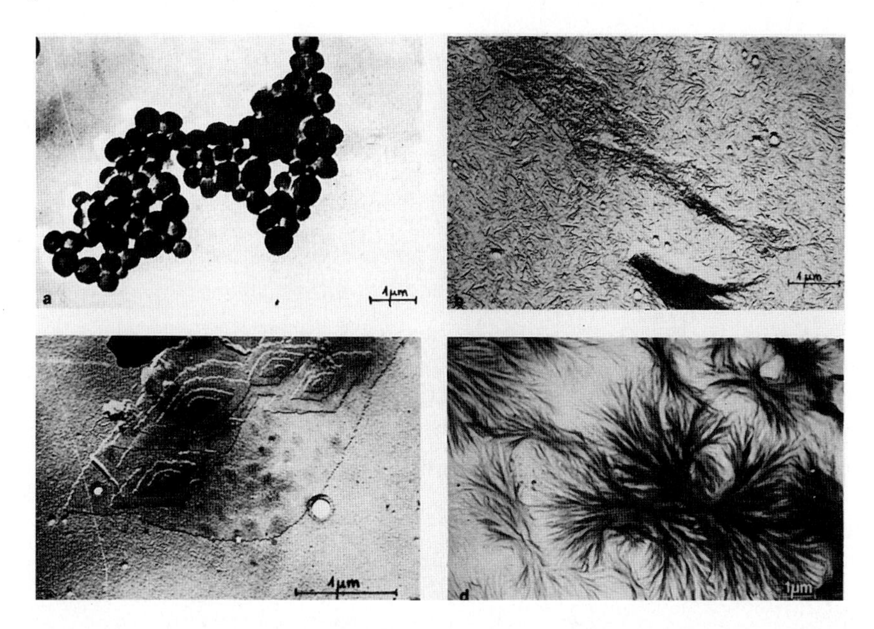

Fig. 4-2 Electron micrographs of different morphologies of a polyamide 6. Upper left (a): From a 260°C solution in glycerol quenched into 20°C glycerol; Upper right (b): Same solution fast cooled (40 K/min); Lower left (c): Same solution slowly cooled (1-2 K/min); Lower right (d): Slow evaporation of formic acid solution at room temperature. Reprinted with permission by the copyright holders, Ch.Ruscher and E.Schulz [1].

points in three directions) and three lattice angles (Table 4-1). Lattice constants reflect the microconformations of chains in crystal lattices.

The c-direction is normally assigned to the chain direction; a value of c = 0.254 nm thus gives the distance between repeating units $-CH_2-CH_2-$ in all-*trans* conformation (bond distance 0.154 nm, bond angle 112°). The c-values in carbon chains that are nonmultiples of 0.254 indicate the presence of non-all-*trans* conformations (e.g., helix structures in it-poly(propylene) and it-poly(butylene-1)). The a and b values reflect physical bonds (such as van der Waals or hydrogen bonds) that are longer than the chemical bonds in the c-direction.

The anisotropy of bond lengths in the three spatial directions prevents the existence of cubic lattices for chain molecules. The other six lattice types are however found (hexagonal, tetragonal, trigonal, (ortho)rhombic, monoclinic, triclinic). Superlattices may be cubic, however.

Most polymers form polymorphs, that is, polymers of the same constitution and configuration possess various energetically different crystal modifications. *Polymorphism* may be caused by different microconformations of polymer chains (e.g., poly(butylene-1)) or by different packing of chains with identical chain conformations (e.g., it-poly(propylene)). Such differences can be generated by small changes in processing conditions, e.g., by stretching, different crystallization temperatures, overgrowth on nucleating agents (incl. pigments and fillers), etc. The slow transformation of one polymorph into another causes a physical aging of the plastic.

Copolymers can furthermore show *isomorphism* if different monomeric units can replace each other without change of lattice structure. Isomorphism of chains is also possible if the two corresponding homopolymers have analogous crystal modifications, similar lattice constants, and the same helix types. An example is the cocrystallization of a mixture of the γ-modification of it-poly(propylene) and the modification I of it-poly(butylene-1) (Table 4-1). This phenomenon is sometimes also called *allomerism*. One company calls its crystalline copolymers from two or more olefinic monomers *polyallomers* because the monomeric units are isomorphic.

Table 4-1 Crystal structures of poly(ethylene) PE, poly(propylene) PP, and poly(butene-1) PB. N_u = Number of monomeric units per unit cell; a, b, c = lattice constants; α, β, γ = angles of unit cell; IV = tetragonal, VI = hexagonal, M = monoclinic, R = (ortho)rhombic, T = triclinic.

Polymer Type	Crystal modification	N_u	a/nm	b/nm	c/nm	α/°	β/°	γ/°	Helix type	Crystal system
PE	I	2	0.742	0.495	0.254	90	90	90	2_1	R
	II	2	0.809	0.479	0.253	90	90	107.9	2_1	M
PP, st		8	1.450	0.560	0.74	90	90	90	2_1	R
it	α	12	0.665	2.09	0.495	90	99.6	90	3_1	M
it	β	3	0.638	0.638	0.633	90	90	120	3_1	VI
it	γ	12	0.647	2.140	6.50	89	100	99	3_1	T
PB, it	I	18	1.769	1.769	0.650	90	90	90	3_1	VI
it	II	44	1.485	1.485	2.060	90	90	90	11_3	IV
it	III	8	1.238	0.892	0.745	90	90	90	4_1	R

4.2.3. Crystallinity

Perfect crystal lattices should give sharp X-ray reflections and sharp melting points. Polymers do not exhibit these features, even if they show a crystallike appearance in electron micrographs. In addition, X-ray patterns of polymers frequently exhibit continuous, diffuse scattering. These findings can be interpreted by either a 1-phase model (voids in crystals) or a 2-phase model (coexistence of completely crystalline and completely amorphous regions). For example, an 83 % crystallinity of poly(ethylene) according to the 2-phase model corresponds to 2.9 % lattice defects in the 1-phase model.

In reality, segments are ordered to various degrees. Since most experimental methods are sensitive to different types and degrees of order, various degrees of crystallinities are found depending on the experiment. The degree of crystallinity may be a mass fraction (weight fraction) $w_c = \phi_c\rho_c/\rho$ or a volume fraction ϕ_c, where ρ_c is the density of the crystalline fraction and ρ the density of the entire sample. Stretched poly(ethylene terephthalate) has, for example, a degree of crystallinity of 59 % by infrared spectroscopy, 20 % by density determination, and 2 % by X ray diffraction.

Semicrystalline polymers are furthermore not in true equilibrium states. Slow recrystallizations and transformations of polymorphs may lead to physical aging.

4.2.4. Morphology

Early experimental findings were interpreted exclusively in terms of a 2-phase model of partially crystalline polymers: the coexistence of crystalline reflexes and amorphous halos in X-ray diagrams; the independence of X-ray short periodicities on molar masses of chain polymers; the disappearance of X-ray long periodicities with increasing molar masses; lower macroscopic densities than predicted by X-ray densities of unit cells; broad melting ranges instead of sharp melting points; optical birefringence of oriented polymers; and heterogeneity of partially crystalline polymers against chemical reactions.

In 1957, however, three different groups found that rhombic platelets are obtained if very dilute solutions of poly(ethylene) are cooled. Similar platelets of 5-20 nm thickness are now known from many other crystallizable polymers (e.g., see Fig. 4-2). Electron diffraction of these platelets showed the sharp reflections typical of *single crystals* and chain directions perpendicular to the surface of the platelets. Since the contour lengths of the chains are greater than the thicknesses of the platelets, chains must be folded back as shown in Figure 4-1 L.

Crystallization from poly(ethylene) melts similarly shows stacked single crystals. These lamellae are 75-80 % X-ray crystalline. After oxidation of the surface, a 100 % crystalline material remains. The folds and/or the interface between the lamellae must thus be "amorphous".

Various models have been suggested for the surface layers: loose folds with adjacent reentry of the chains, a switchboard-like structure, adsorbed chains not involved in chain folding, loose or entangled cilia (tangling chain ends), and crystal bridges composed of tie molecules (see also Fig. 4-1).

Electron microscopy has shown that crystal bridges exist in crystallized mixtures of high and low molar mass poly(ethylene)s. Such interlamellar connections are responsible for some remarkable properties of partially crystalline polymers (see below). Highly crystalline polymers result if tie molecules and surface layers are removed (e.g., by grinding); these fairly low molar mass materials are called microcrystalline polymers.

Crystallization of melts sometimes delivers nearly spherical, polycrystalline entities, called spherulites, that exhibit radial symmetry and show a Maltese cross pattern in the polarization microscope (Fig. 4-3). They consist of lamellar, fibrous, or lath crystallites that grew from the center and have chain axes tangential to the circumference of the spherulite. Spherulites with radial chain axes seem to be not known. The formation of spherulites can be controlled by the crystallization temperature, the cooling rate, and the addition of nucleating agents.

Spherulites lead to opaque plastics if their diameters are greater than one-half the wavelength of incident light and their crystalline and noncrystalline regions have different densities and refractive indices. The number and size of such spherulites affects the fracture behavior of polymers.

Spherulites develop in a viscous environment if crystallization nuclei are present and if equal crystallization rates prevail in all spatial directions. In strongly stirred dilute solutions, however, polymer chains try to orient themselves along the flow gradients where they crystallize in an extended (nonfolded) conformation. The resulting fibrils are organized in bundles with chains parallel to the fibril axes. These bundles act as nuclei/substrates for the epitaxial growth of the remaining chains. Since the shear rate is strongly reduced between bundles, crystallization leads to lamellae with folded chains whose axes are parallel to the fibril axes. The resulting extended-chain structures are responsible for the remarkable properties of some ultraoriented polymers, e.g., high Young's moduli and high tensile strengths in chain direction.

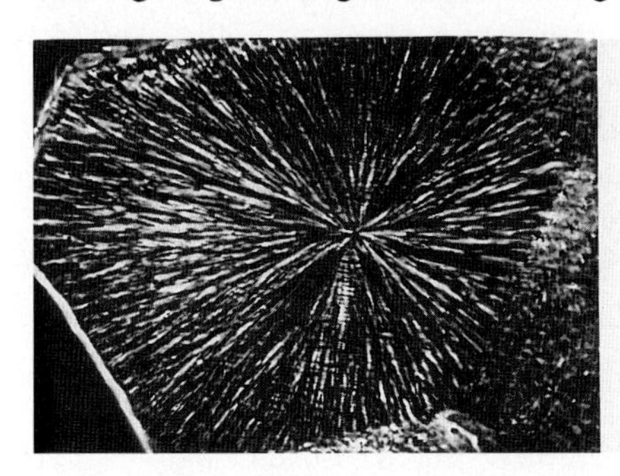

Fig. 4-3 Spherulites of it-poly(propylene) under the phase contrast microscope (left) and the polarization microscope (right). Reproduced with permission by the copyright holder, R.J.Samuels [2].

4.3. Mesophases

4.3.1. Introduction

Mesomorphic chemical compounds exhibit microscopic structures (mesophases) that are intermediate between crystals with long-range three-dimensional order and liquids without long-range order (Greek: *mesos* = middle, *morphe* = shape, form). Three types of mesophases are distinguished: liquid crystals, plastic crystals, and condis crystals. Since liquid crystals were discovered first and are the most common of the three classes, the term "mesophase" is often used as a synonym for "liquid crystal".

Liquid crystallinity arises if the molecules contain *mesogens*, that is short rod-like sections that comprise the whole molecule in low molar mass liquid crystals (LCs), or segments thereof in polymer liquid crystals (LCPs). Mesogens are anisotropic; they are oriented with respect to their long axes but not with respect to their positions.

Liquid crystals possess certain ordered structures like crystals, but they flow like liquids. *Thermotropic liquid crystals* exhibit this phenomenon in the neat state above a certain solid/mesophase transition temperature, where the solid may be a crystal or a glass. *Lyotropic liquid crystals* display mesophase behavior in certain solvents above critical concentrations.

Plastic crystals feature order of positions but disorder of orientations of molecules. They are found for certain spherical low molar mass molecules packed in cubic lattices (i.e., they are isotropic). The absence of strong attractive forces between molecules and the presence of slip planes lead to easy deformability of plastic crystals, sometimes even under their own weight. No polymers are known that form plastic crystals.

Condis crystals are *con*formationally *dis*ordered crystals containing several conformational isomers in more or less ideal crystalline positions with respect to position and orientation of segments (conformational isomorphs). An example is the hexagonal high-pressure phase of poly(ethylene) extended-chain crystals. It is often difficult to determine whether polymer crystals are normal or condis crystals.

On cooling, thermotropic liquid crystals either crystallize or solidify to "mesomorphic glasses" with retainment of their liquid crystal structures. These glasses do not carry special names, but since "liquid crystalline glass" and similar names are oxymorons, the terms "LC glass", "PC glass", and "CD glass" have been suggested for the three classes of mesomorphic materials. The LCPs are mainly used as LC glasses, not as liquid crystals per se, except in processing.

Condensed matter can thus exist in several of the five possible classes of pure phases: completely ordered crystal oC, mesomorphic glass mG, amorphous glass aG, ordered mesophase oM, and isotropic melt (liquid) iL. Seven types of biphasic materials exist: oC+mG, oC+aG, oC+oM, oC+iL, mG+aG, mG+iL, oM+iL, and three types of triphasic materials: oC+mG+aG, oC+mG+iL, oC+oM+iL. Since each class can be subdivided further, many physical structures (and thus properties) are possible for the same polymer.

Mesophasic polymers are characterized by domains of microscopic size (see below). "Microdomains" of similar size are also exhibited by certain block copolymers and, in much smaller sizes, by ionomers (see Section 4.3.7.).

4.3.2. Types of Liquid Crystals

Liquid crystals and LC glasses are generated by rod- or disc-like mesogens, which either comprise the whole molecule (as in low molar mass LCs) or sections of it (LCPs). Examples are

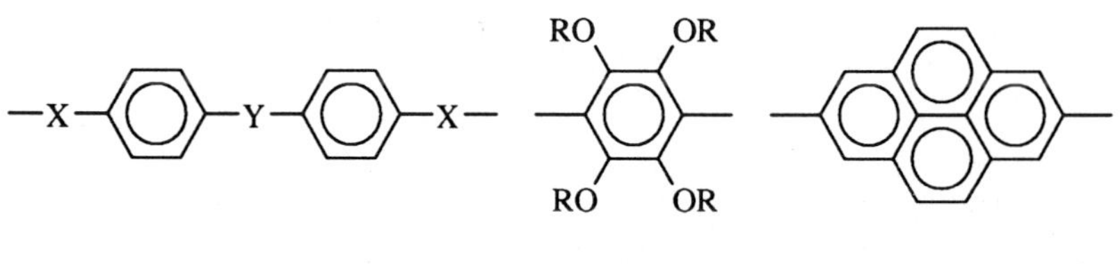

Rod-like Disc-like

—X—⟨⟩—Y—⟨⟩—X— RO OR RO OR

where X = O, OOC, COO, etc; Y = COO, p-C_6H_4, CH=CH, N=N, etc; R = $CO(CH_2)_nCH_3$, etc; and phenylene residues may be replaced by 1,4-cyclohexane rings. The lengths of polymer mesogens are often comparable to those of repeating units; they seem to be identical with persistence lengths.

Rod-like mesogens are also called calamitic (Greek: *calamos* = reed) and disc-like ones discotic (Latin: *discus*; Greek: *diskos*, from *diskein*: to throw). Calamitic liquid crystals are further subdivided into smectic, nematic, and cholesteric mesophases. Smectic LCs show fan-shaped structures under the polarization microscope; they feel like soaps (Greek: *smegma* = soap). Nematic LCs form thread-like schlieren (Greek: *nema* = thread; German: *Schlieren*, plural of *Schliere* = streak). Cholesteric LCs exhibit beautiful reflective colors, which were first discovered with cholesterol.

The birefringence of LCs originates from the anisotropy of mesogens, which are more or less aligned in microscopic domains with diameters in the micrometer range. These alignments lead to different refractive indices parallel and perpendicular to the direction of polarization of incident light. The domains themselves are arranged randomly. The presence of birefringence is no proof for the existence of mesogens, however. It may also be due higher-melting crystallites in partially molten polymers or from crystallites in gels or concentrated solutions.

Liquid crystals are turbid because their domains are anisotropic structures whose dimensions are greater than the wavelength of incident light. At a certain "clearing temperature", these domains melt to a transparent (isotropic) liquid.

The soap-like character of *smectic LCs* is due to two-dimensional, layer-like arrangements of mesogens (Fig. 4-4). At present, twelve different smectic types are known: four with the average direction of the long axes perpendicular to layer planes,

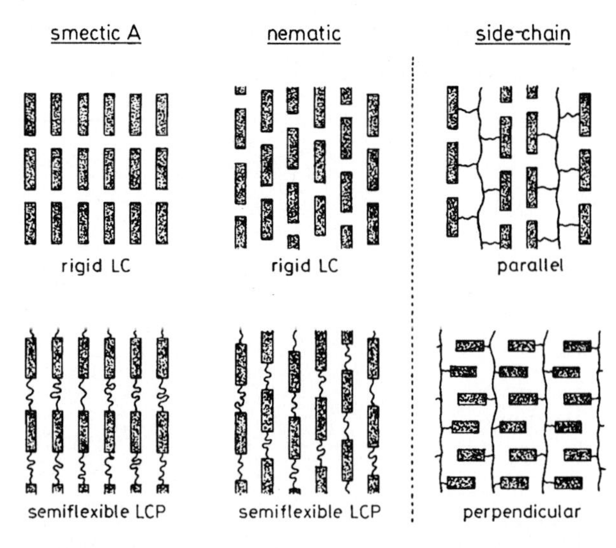

Fig. 4-4 Schematic representation of mesogens in different types of mesophases.
Left: Smectic A arrangements of rigid (low molar mass) liquid crystal molecules LC (above) and mesogens of liquid main-chain crystal polymers LCP (below).
Center: Nematic arrangements of rigid LCs (above) and mesogens of semiflexible main-chain LCPs (below).
Right: Nematic side-chain liquid crystalline polymers with parallel (above) and perpendicular (below) arrangements of mesogens relative to the main chains.

seven with tilted axes, and one with unknown structure. Examples are type S-A with mesogens perpendicular to the layer planes; type S-B, the same but with perfect hexagonal packing of mesogens; type S-C, with mesogen axes at angle to the planes, etc.

Smectic liquid crystalline polymers LCP are relatively rare. In contrast to low molar mass LCs, mesogens do not comprise the whole molecule but may be interconnected by flexible segments (Fig. 4-4).

Nematic mesophases (Fig. 4-4) are one-dimensionally ordered. The principal axes of mesogens are more or less parallel to each other; the centers of mass are however distributed at random. In nematic LCPs, nematically ordered mesogens are interconnected by flexible segments. Axes of mesogens in side-chain LCPs may be parallel or perpendicular to chain axes.

Cholesteric mesophases are observed only with chiral mesogens. They consist of layers of nematically ordered mesogens. These layers have a sense of rotation relative to each other due to the presence of chirality centers in mesogens. Cholesteric mesophases are thus screw-like nematics.

Discotic mesophases may exist in many different structures. The disc-like mesogens are randomly arranged in nematic-discotic types but stacked like coins in the columnar-discotic ones.

4.3.3. Mesogens

Mesogens of polymers can be contained in main chains (MC-LCPs) or in side-chains (SC-LCPs). Thermotropic MC-LCP glasses are used as engineering plastics; lyotropic MC-LCP glasses, for fibers. Thermotropic SC-LCPs are not used commercially at present but may find use as optical recording devices.

Mesogens usually constitute the whole molecule in low molar mass LC molecules. In macromolecules, this is true only for dissolved helices up to certain molar masses. Such helical macromolecules behave like rigid rods at medium molar masses but become coil-like at very high degrees of polymerization (see Fig. 2-5). Examples are certain poly(α-amino acid)s in helicogenic solvents, the helical macromolecules of poly(isocyanate)s, the double helices of deoxyribonucleic acids DNA, and the tobacco mosaic virus TMV with its protein studded helical nucleic acids.

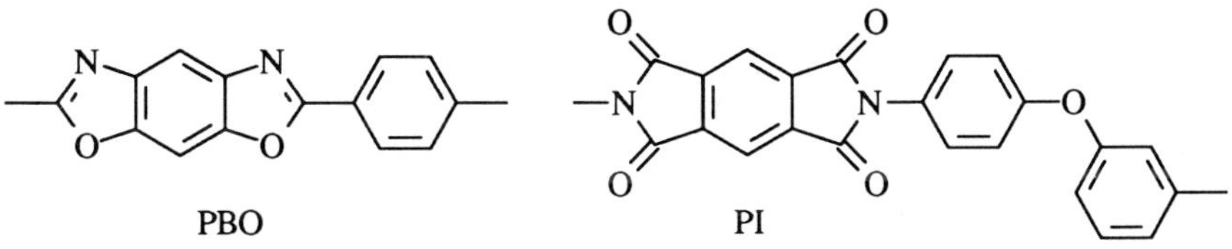

PBO PI

All other polymer chains form either flexible or semiflexible coils in in their isotropic melts or in dilute solution. Mesogens are produced from segments with linear chain axes; examples are poly(*p*-phenylenebisbenzoxazole) PBO, poly(*p*-hydroxybenzoic acid) PoHB, and poly(*p*-phenylene ethylene) PPE. In such LCPs, the linearity of the mesogen axis is either not disturbed (PBO) or only slightly (PoHB) by a rotation around the long axis of a conjugated mesogen. The molecules of PBO and PoHB are semiflexible.

The "stiffness" is removed if the potential mesogen axes are bent by chain groups in *ortho* or *meta* positions such as in the flexible polymer molecules of polyimides PI from pyromellithic acid and 4,4'-diaminidiphenylether. The liquid crystalline structure is also removed if a 180° rotation around a chain bond leads to a significant deviation from the linearity of the mesogen axis. An example is the flexible molecule of poly(*m*-hydroxybenzoic acid) PmHB, whereas as PoHB is semiflexible.

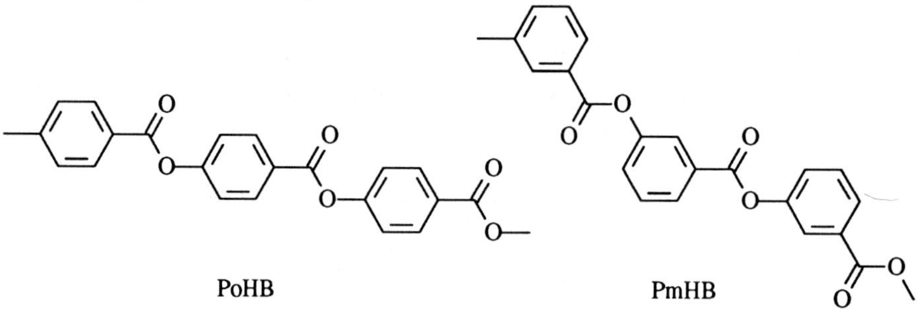

PoHB PmHB

Essential for a rod-like character of mesogens is the retainment of linearity upon rotation around the mesogen long axis. Conjugation helps but is not required. Even nonconjugated mesogens with $-CH_2-CH_2-$ groups remain rod-like in melts and/or concentrated solutions because of packing requirements. The molecules of poly(*p*-phenylene ethylene) P*p*PE are semiflexible, whereas those of poly(*p*-phenylene trimethylene) PPTM are flexible; PpPE is an LCP but PPTM is not.

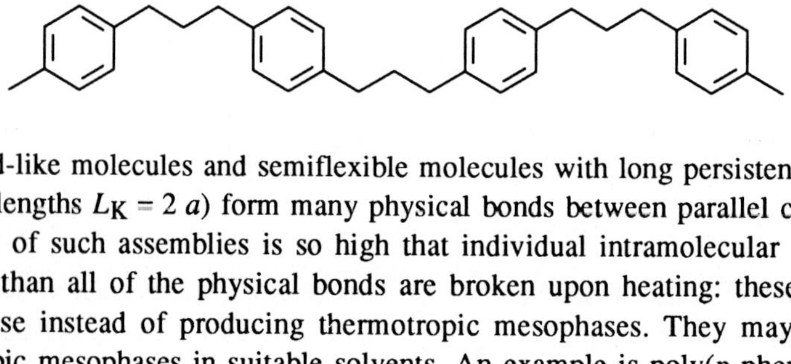

P*p*PE

PPTM

Rod-like molecules and semiflexible molecules with long persistence lengths a (or Kuhn lengths $L_K = 2\ a$) form many physical bonds between parallel chains. The total energy of such assemblies is so high that individual intramolecular chemical bonds rather than all of the physical bonds are broken upon heating: these molecules decompose instead of producing thermotropic mesophases. They may however form lyotropic mesophases in suitable solvents. An example is poly(*p*-phenylene bisbenzoxazole) PBO.

The "stiffness" of the chain segments can be removed or reduced by several means (Fig. 4-5). Bulky substituents prevent the parallelization of chains (FC); they "frustrate" the crystallization. Nonlinear chain elements (such as *ortho* and *meta* substituents) work in the same way (NL). Flexible chain segments (FL) reduce the persistence lengths of the chains.

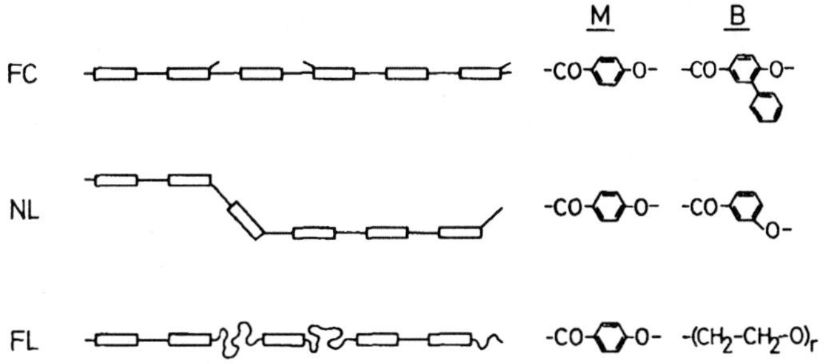

Fig. 4-5 Left: Flexibilization of rigid chain segments by frustrated crystallization (FC), nonlinear chains (NL), or flexible chain elements (FL).
Right: Examples for FC, NL, and F; M = mesogen-forming monomeric units (examples), B = stiffness-breaking units (examples).
Reprinted with permission by Hüthig and Wepf Publ., Basle, Switzerland [3].

4.3.4. Lyotropic Liquid Crystalline Polymers

Molecule axes of low molar mass rod-like molecules are disordered in isotropic melts, whereas mesogen axes are more or less parallel to each other in nematic and smectic mesophases. Very short rods resemble spheroids. They can arrange themselves in various stable ways, but the parallel ordering need not be much more stable than other arrangements. Therefore, a critical axial ratio Λ_{crit} = length/diameter (also called critical aspect ratio) exists above which the simple geometric anisotropy is sufficient to stabilize a mesophase; no attractive forces between mesogens are needed as postulated by the Maier-Saupe theory.

This critical axial ratio has been calculated by Flory with the lattice theory of polymer solutions as $\Lambda_{crit} \approx 6.42$. A geometric stabilization by repulsion is insufficient for an axial ratio $\Lambda < \Lambda_{crit}$ and the mesogens must exert additional orientation-dependent attraction forces.

In solution, repulsive forces between mesogens can act only at sufficiently high mesogen concentrations. At a critical volume fraction ϕ_p^*, phase separation occurs into a polymer-rich mesophase and a dilute isotropic phase. The dependence of ϕ_p^* on Λ can be described to a first approximation by

$$(4\text{-}2) \qquad \phi_p^* \approx 8(1 - 2\,\Lambda^{-1})/\Lambda$$

The axial ratio Λ of this equation is the molar mass-dependent true axial ratio L/d of rod-like molecules (i.e., helices) and the molar mass-independent Kuhn length L_K

of semiflexible molecules. Equation (4-2) is remarkably well fulfilled for both helical and semiflexible molecules (Fig. 4-6). An exception is the tobacco mosaic virus, probably because of additional charge effects that are not considered by the theory.

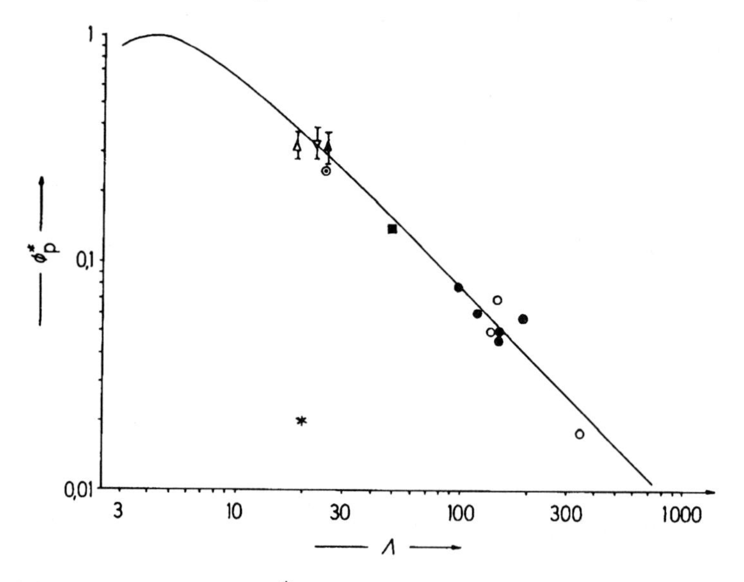

Fig. 4-6 Critical volume fractions ϕ_p^* for phase separations of isotropic solution → nematic mesophase as function of Λ = the axial ratio of various helical (rigid) molecules (○,⊙,●) or the Kuhn length of semiflexible molecules (△,▽,▲,■); * Tobacco mosaic virus. Solid line: Prediction of Equation (4-2). Reprinted with permission by Hüthig and Wepf Publ., Basle, Switzerland [4].

Lyotropic mesophases are formed by the helices of tobacco mosaic virus ($c > 2$ %) and deoxyribonucleic acids ($c > 6$ %) in dilute aqueous salt solutions as well as by poly(α-amino acid)s in helicogenic solvents, e.g., poly(γ-methyl-L-glutamate) in dichloromethane/ethyl acetate (12:5) ($c > 15$ %). Hydroxypropyl cellulose HPC forms a cholesteric mesophase at room temperature in water at $w_{HPC} > 0.41$.

Liquid crystalline polymers consist of domains. In each domain, all mesogen axes are oriented; the resulting domain directions are however distributed at random. On shearing, domains orient themselves with their mesogen axes in flow direction. This alignment becomes frozen-in on rapid quenching of thermotropic LCs or by precipitation of lyotropic ones into baths (half-lives of orientation between a few seconds and a few hundredths of seconds). The resulting LC glasses possess good orientations of mesogen axes, high tensile strengths, and low extensibilities in shear direction (e.g., fiber direction) (see Chapter 12).

Several types of LCP fibers are produced commercially. Poly(*p*-phenylene terephthalamide) PPB-T (Du Pont)

~NH–C_6H_4–NH–CO–C_6H_4–CO~ Kevlar™; PPB—T

was originally spun from concentrated sulfuric acid solutions, whereas chlorinated N-methyl pyrrolidine is now used. A similar copolyamide with the units

—NH—C₆H₄—NH—CO—C₆H₄—CO— / —NH—C₆H₄—O—C₆H₄—NH—CO—C₆H₄—CO— Technora™

is also on the market. Even better LCP fibers are produced from poly(*p*-phenylene-2,6-benzoxazole)s PBO and poly(*p*-phenylene-2,6-benzthiazole)s PBT; these are spun from poly(phosphoric acid) solutions, at present on pilot scale. The semicommercial products have "cis" structures (PBO) and "trans" structures (PBT), respectively:

PBO PBT

4.3.5. Thermotropic Liquid Crystalline Polymers

Non-Newtonian shear viscosities of liquid crystalline polymers drop drastically with increasing shear rates due to the orientation of mesogen axes. Less energy is thus needed for the processing of thermotropic LCPs than for isotropic thermoplastics. The high orientation times permit a freezing-in of the mesogen orientation by, for example, injection molding. Because this leads to higher than usual moduli of elasticity and tensile strengths, such LCPs are also called *self-reinforcing polymers.*

The first self-reinforcing polymer was the copolyester X7G with 60 mol-% *p*-hydroxybenzoyl and 40 mol-% terephthaloyl glycol units. It is no longer produced because the less expensive glass fiber-reinforced saturated polyesters possess similar properties. Xydar™ and Vectra™ are, however, industrially produced:

—O—C₆H₄—CO— —O—CH_2—CH_2—O— —OC—C₆H₄—CO— X7G

—O—C₆H₄—CO— —O—C₆H₄—C₆H₄— —OC—C₆H₄—CO— Xydar™

—O—C₆H₄—CO— —O—C₁₀H₆—CO— Vectra™

Polymeric side-chain LCs can serve for the thermooptical storage of information. A guided laser beam increases the temperature locally, whereupon phase transformations occur. The change of order can then be frozen-in on cooling below the glass transition temperature. The resolution is ca. 0.3 μm.

4.3.6. Block Copolymers

Block copolymers consist of two or more blocks of constitutionally and/or configurationally different monomeric units (e.g., A_m-B_n, A_m-B_n-A_m, etc.). In most cases, these blocks are thermodynamically immiscible since their homopolymers A_m, B_n,, etc., are not miscible. The demixing of the blocks cannot proceed, however, to macroscopic phase separations because the blocks are chemically coupled. Similar blocks of different block copolymer molecules can thus only aggregate and form domains in a matrix of the other blocks. This phenomenon is called *microphase separation*; it controls the morphology of neat block copolymers.

The morphology of amorphous diblock copolymers is mainly governed by the relative spatial requirements of the blocks. Both blocks, if independent, would form unperturbed random coils. If the unperturbed volumes of both blocks are of equal size, all A_m blocks would line up in one layer and all B_n blocks in another because the blocks are (1) coupled to each other and (2) inmiscible. The A_m layer faces another A_m layer, where the A_m blocks are coupled to B_n blocks etc. (Fig. 4-7, L). Such diblock polymers thus form lamellae with the thickness of two coil diameters. Similar structures result for triblock polymers A_r-B_n-A_r, where $r = m/2$.

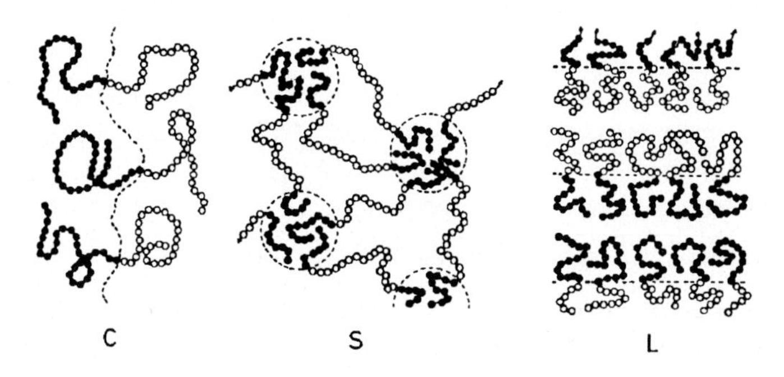

Fig. 4-7 Arrangement of A_m blocks (with A units ●) and B_n blocks (with B units O) in diblock copolymers A_m-B_n (C, L) and triblock (S) copolymers A_r-B_n-A_r.
C: Compatibilizer molecules at an interface - - - between A-polymer and B-polymer;
S: Formation of spherical domains of A_r-blocks in a continuous matrix of B_n-blocks by triblock copolymers A_r-B_n-A_r (with $r = m/2$ and $r < n$);
L: Lamellae of A_m and B_n of diblock copolymers A_mB_n with equal volumes of blocks.

If the volume of the A_m blocks is much smaller than the volume of the B_n blocks, then the A_m blocks can no longer be packed into layers without violating the demand for tightest packing or deviating from the shape of unperturbed coils. Both possibilities are energetically unfavorable, and the smaller A_m blocks thus cluster together and form spherical domains in a continuous matrix of the larger B_n blocks (Fig. 4-7 S). If the A_m blocks are somewhat larger (but not large enough to form lamellae), then cylinders would result. The domain morphology thus depends on the relative unperturbed dimensions of the blocks, which are in turn proportional to the molar masses. Theoretical predictions of the concentration regions of the various morphologies have been confirmed experimentally (Fig. 4-8).

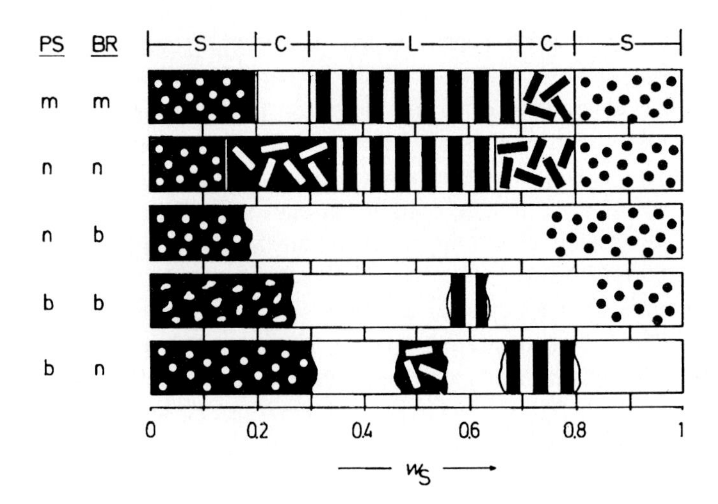

Fig. 4-8 Morphology of styrene-butadiene diblock copolymers PS-BR as a function of the mass fraction w_S of the styrene units.
Letters on top of graph: ranges of domain structures according to Meier theory;
S = Spherical domains, C = cylinders (rods), L = lamellae (layers).
Rows 2-5: experimentally-found ranges (schematic after results of many authors)
White: poly(styrene) blocks PS; black: poly(butadiene) blocks BR;
Block size distribution: (m) molecularly uniform (theory), (n) narrow, or (b) broad.
Reprinted with permission by Hüthig and Wepf Publ., Basle, Switzerland [5].

Three different types of applications of block copolymers are known: polymeric detergents, thermoplastic elastomers, and compatibilizers for polymer blends. Molecules of *polymeric detergents* consist of hydrophilic and hydrophobic blocks. The hydrophobic blocks associate in water. Intramolecular association between blocks of the same molecule leads to unimolecular "micelles"; intermolecular association between blocks of different molecules produces high viscosities. The polymers also accumulate at the interface water/air which lowers the surface tension of water. Examples are multiblock copolymers with water-soluble ethylene oxide blocks $-[OCH_2CH_2]_n-$ and water-insoluble propylene oxide blocks $-[OCH_2CH(CH_3)]_m-$.

A similar action is exerted by *compatibilizers* for immiscible blends of A_n and B_m polymers. Such compatibilizers may be D_p-E_q diblock copolymers, linear triblock copolymers $D_pE_qD_p$, or multiblock copolymers $D_pE_qD_pE_qD_p$, etc. The D_p blocks reside in the A_n phases and the E_q blocks in the B_m ones; compatibilizers thus anchor the two types of phases (Fig. 4-7 C). Not all compatibilizer molecules accumulate at the interface, however, some are wasted as micelles or mesophases.

D units have usually the same chemical structure as A units and E units the same as B units. Chemical identity is not required, however, only miscibility of D_p blocks with A_n polymers and of E_q blocks with B_m polymers. The blocks of the compatibilizer molecules should be longer than the molecules of the corresponding blend component in order to assure good anchoring. Multiblock compatibilizers are suspected to be more effective than diblock ones. Other compatibilizers act through cocrystallization of their blocks with the polymer molecules of the blend.

Thermoplastic elastomers consist of tri- or multiblock copolymers. Triblock copolymers S_m–B_n–S_m possess a soft center poly(butadiene) block –B_n– (glass temperature below use temperature) and two hard end poly(styrene) blocks –S_m– (glass temperature above use temperature). Below certain m/n ratios, styrene blocks form spherical microdomains in a continuous matrix of butadiene blocks. These spherical domains thus act as cross-linking units in an elastomer (Fig. 4-7 S). The hard domains "melt" above their glass temperatures, and the block copolymers can then be processed like thermoplastics.

A great number of thermoplastic elastomers of either the triblock or multiblock type is known. The chemical structures range from the (styrene)$_m$–(butadiene)$_n$–(styrene)$_m$ triblock copolymers to polyether-urethane multiblock polymers, ethylene-propylene copolymer/poly(olefin) blends, polyether-esters, and graft copolymers of butyl rubber on poly(ethylene)s.

4.3.7. Ionomers and Acidic Copolymers

Acidic copolymers are defined as copolymers of a hydrophobic parent monomer (or several thereof) and a small proportion of an acidic comonomer. Examples are copolymers of ethylene E and up to 15 wt-% acrylic acid AA (EAA copolymers) or less than 15 mol-% methacrylic acid MAA (EMAA copolymers), and a 1:1 copolymer of ethylene and methyl acrylate with a small content of (meth)acrylic acid units. The association of the COOH groups of these copolymers leads to physical cross-links; moduli and yield strengths of these thermoplastics are higher than those of their parent polymers.

Ionomers are partially or completely neutralized acidic copolymers. Examples are thermoplastic copolymers of ethylene and methacrylic acid that are partially neutralized with zinc or sodium ions and membranes from copolymers of tetrafluoroethylene with perfluoro vinylethers.

Rubbers may also carry acid groups or partially neutralized acid groups. Those with COOH groups are known as carboxylated rubbers or, after vulcanization, as carboxylated elastomers. Acidic rubbers behave as thermoplastic elastomers if they are partially neutralized by zinc ions; no chemical cross-linking is needed. Examples are post-chlorosulfonated poly(ethylene)s, post-sulfonated EPDM polymers from the terpolymerization of ethylene and propylene with small proportions of nonconjugated dienes, and terpolymers of butadiene and acrylonitrile with small proportions of acrylic acid.

The hydrophobic environments of these ionomers prevent the salt groups from dissociation; ionomers do not contain free ions. The salt groups are rather present as contact ion pairs, multiplets consisting of associated ion pairs, and ion clusters. These ion/ion interactions are controlled by coordination numbers and not by valences; not only bivalent ions such as Zn^{2+} but also monovalent ions such as Na^+ are effective.

Ion multiplets of two, three and possibly up to eight ion pairs are small entities of less than 0.6 nm diameter that do not contain polymer chain segments. Because of their small size, they are not separate phases in solid polymers; they thus affect directly matrix properties such as segment mobilities.

Multiplets can aggregate into considerably larger ion clusters which contain significant amounts of polymer segments. The domain sizes of ion clusters are limited to 2-10 nm by the elastic forces generated by the backbone chains. Clusters thus do not affect markedly matrix properties except that they act as cross-links with reinforcing action (Fig. 4-9).

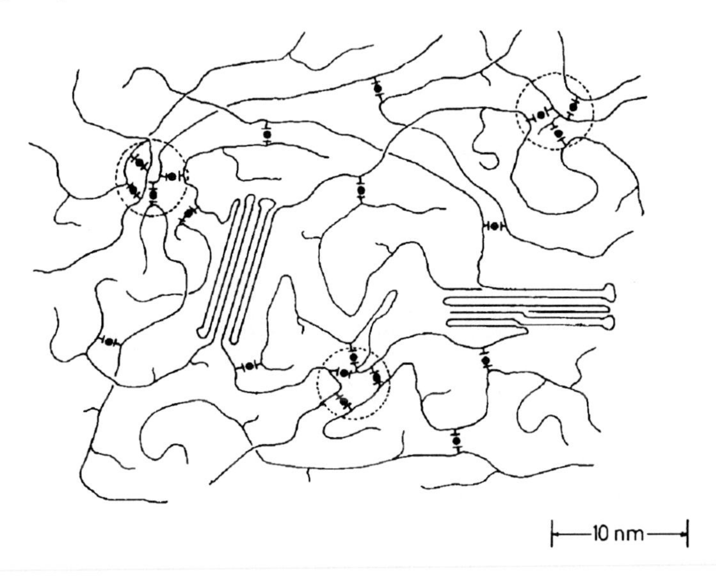

Abb. 4-9 Schematic representation of a semicrystalline ionomer with crystalline lamellae, triple ions ~$COO^-/Mt^{2+}/^-OOC$~ (-I●I-), and ion clusters (broken circles). The scale is approximate.

Ion pairs, multiplets, and clusters dissociate at higher temperatures; ionomers can then be processed like thermoplastics. At lower temperature, they behave as either thermoplastic elastomers (T_G of polymer segments between cross-linking sites lower than use temperature) or as "reversible" thermosets (T_G of segments greater than use temperature).

4.4. Gels

Chemically lightly cross-linked polymers swell upon addition of solvent and form gels. The maximum degree of swelling results from the attempt of segments located between cross-linking sites to attain their energetically most favored coil dimensions which in turn is counteracted by the elastic retraction due to cross-links. Physically cross-linked polymers behave in the same way if the physical cross-links (crystallites, ion clusters, etc.) survive the dissolution process.

The latter phenomenon is utilized in poly(vinyl chloride) pastes. The strong dipole-dipole interactions between C-Cl bonds leads to associations of chain segments, which at room temperature resist dissolution by plasticizers such as phthalates, adipates, sebacates, or citrates. Heating of PVC with these plasticizers and subsequent cooling leads to gelation by formation of a lightly physically cross-linked network. Plasticized PVC thus behaves like an elastomeric material. The physical cross-links dissociate at higher temperatures and the resulting PVC paste can be processed with extruders or roll mills.

4.5. Polymer Surfaces

4.5.1. Structure

Outer layers of solid or liquid polymers do not exhibit against air, water, metal surfaces, fillers, etc., the same average composition as the interiors of these polymers. Groups or segments with the lowest Gibbs interfacial energy will reside preferentially at the surface or interface.

Surface structures may also be altered by chemical attack during processing or on prolonged use (e.g., by oxidation) and/or by physical processes such as transcrystallization. In addition, coil molecules are flattened out and adopt a more pancake-like macroconformation because of their broken translational symmetry. These phenomena effect surface tensions and wetting of polymers, phase interfaces in heterogeneous blends, adhesion between polymers and fillers, etc.

Surface structures can be studied by a wide variety of new spectroscopic methods:

surface groups:	Fourier transform infrared spectroscopy (FT-IR)
upper 0.3-0.5 nm:	ion scattering spectroscopy (ISS, LEIS)
upper 1-10 nm:	photoelectron spectroscopy (UPS, PE(S))
	electron spectroscopy for chemical analysis (ESCA, XPES, IEE)
	Auger spectroscopy (AES)

These methods have shown that poly(2-vinyl pyridine) and poly(4-vinyl pyridine) have much more hydrophobic surfaces than their chemical structures indicate: CH_2 and CH groups face the surface, pyridine residues $-C_5H_4N$ the interior. Imide-carbonyl residues of aromatic polyimides, on the other hand, are preferentially found at the surface and aromatic rings in the interior. The surface is enriched by siloxane residues in polycarbonate-*block*-poly(dimethyl siloxane)s and by styrene units in poly(styrene)-*block*-poly(ethylene oxide). Since the surface structure depends on the contact during processing and use (air, water, metals, etc.) and also on kinetic effects (thermal history, solvents, etc.), one and the same polymer may possess various surface compositions and thus also surface properties.

4.5.2. Interfacial Tension

Interfacial tension is the force that acts at the interface between two phases; it is called surface tension for the interface condensed phase-gas phase.

Surface tensions γ_{lv} of liquid polymers against air decrease with the two-third power of the molar mass according to

$$\gamma_{lv} = \gamma_{lv}^{\infty} + K_{lv}M^{-2/3} \quad (4\text{-}3)$$

Surface tensions do not vary markedly with temperature. Typical values (at 150°C) range from 13.6 mN/m [poly(dimethylsiloxane)], 22.1 mN/m [it-poly(propylene)], 28.1 mN/m [poly(ethylene)], and 33.0 mN/m [poly(ethylene oxide)]. At present, no correlation of these surface tensions with chemical constitution is possible because of the unknown surface structures of liquid polymers.

Interfacial tensions between two liquid polymers are always lower than surface tensions. They are low for two apolar polymers but higher if one polymer is polar (or both). For example, the interfacial tension poly(ethylene)-it-poly(propylene) is only 1.1 mN/m but that of poly(ethylene)-poly(ethylene oxide) 9.5 mN/m and poly-(ethylene oxide)-poly(dimethylsiloxane) 9.8 mN/m.

The interfacial tensions between solid polymers and air are mostly unknown. They do not differ markedly, however, from the *critical surface tensions* determined by the *Zisman method.* In this method, the liquid-air surface tensions of various liquids are plotted against the cosine of contact angles ϑ of these liquids against the polymer and extrapolated to cos $\vartheta \to 0$. The resulting "critical" surface tensions of polymers are always lower than the surface tension of water (72 mN/m): poly(tetrafluoroethylene), 18.5 mN/m; poly(ethylene), 33 mN/m; poly(vinyl alcohol), 37 mN/m; polyamide 6.6, 46 mN/m; and urea-formaldehyde resins, 61 mN/m. Fluorinated polymers have especially low critical surface tensions. They are not wetted by water or oils and fats (20-30 mN/m) and are therefore used as surface coatings to prevent sticking.

4.5.3. Adsorption

Macromolecules possess many adsorbable groups and segments. The type of adsorption depends on the adsorption energy per segment (group), the concentration of adsorbable macromolecules, and the duration of the experiment.

"Adsorption equilibria" (based on adsorbed amounts) are established in minutes to hours at smooth surfaces but may take days to attain at rough surfaces and powders. The adsorption time increases with the concentration and molar mass of the polymer.

Polymer coils tend to overlap even in fairly dilute solutions: the adsorbed polymer layers are almost always multilayers, except for adsorptions of oligomers ($10 < X < 100$) and/or from very dilute solutions ($10^{-4} < \phi < 10^{-3}$). At higher volume fractions ϕ and/or molar masses M, adsorbed amounts and structures of polymer layers are mainly kinetically controlled. Reorganizations may occur with time, for example, adsorption of polar polymers at polar surfaces can change from an initial loose loop structure to a more compact, flat covering.

Physical structures of adsorbed polymers vary widely. The layer of poly(ethylene oxide) adsorbed on chromium surfaces is only ca. 2 nm thick and so tightly packed that the refractive index of the adsorbed layer is identical with that of the crystalline polymer. The thickness of the surface layer of poly(styrene) of M = 176 000 g/mol adsorbed from a 5 mg/mL solution in the thermodynamically bad solvent cyclohexane is however about 27 nm (i.e., identical with the end-to-end distance in the unperturbed state). The physical structures of adsorbed polymers obviously play an important, yet widely unknown, role in many technological applications of polymers.

Literature

4.1. Noncrystalline State

G.S.Y.Yeh, Morphology of Amorphous Polymers, Crit.Revs.Macromol.Sci. **1** (1972) 173
R.N.Haward, The Physics of the Glassy State, Interscience, New York 1973
G.Allen, S.E.B.Petrie, eds., Physical Structure of the Amorphous State, Dekker, New York 1977
S.E.Keinath, R.L.Miller, J.K.Rieke, eds., Order in the Amorphous "State" of Polymers, Plenum, New York 1987

4.2. Crystalline Polymers

L.E.Alexander, X-Ray Diffraction Methods in Polymer Science, Wiley, New York 1969
K.Kakudo, N.Kasai, X-Ray Diffraction by Polymers, Elsevier, Amsterdam 1972
B.Wunderlich, Macromolecular Physics, Academic Press, New York, 3 vols., 1973 ff.
R.J.Samuels, Structured Polymers Properties, Wiley, New York 1974
I.M.Ward, ed., Structure and Properties of Oriented Polymers, Halsted Press, New York 1975
H.Tadokoro, Structure of Crystalline Polymers, Wiley, New York 1979
L.Marton, C.Marton, eds., Methods of Experimental Physics, Bd. **16 B,** R.A.Fava, ed., Crystal Structure and Morphology, Academic Pres, New York 1980
D.C.Bassett, Principles of Polymer Morphology, Cambridge Univ.Press, Cambridge 1981
F.A.Bovey, Chain Structure and Conformation of Macromolecules, Academic Press, New York 1982
O.Glatter, O.Kratky, Small Angle X-Ray Scattering, Academic Press, New York 1982
U.D.Standt, Crystalline Atactic Polymers, J.Macromol.Sci.-Revs.Macromol.Chem.Phys. **C 23** (1983) 317
I.H.Hall, ed., Structure of Crystalline Polymers, Elsevier Appl. Sci. Publ., London 1984
N.March, M.Tosi, eds., Polymers, Liquid Crystals, and Low-Dimensional Solids, Plenum, New York 1984
D.A.Hemsley, The Light Microscopy of Synthetic Polymers, Oxford Univ.Press, New York 1985
R.A.Komoroski, ed., High Resolution NMR Spectroscopy of Synthetic Polymers in Bulk, VCH, Weinheim 1986
L.C.Sawyer, D.T.Grubb, Polymer Microscopy, Chapman and Hall, New York, 1987
A.E.Woodward, Atlas of Polymer Morphology, Hanser/Oxford Univ.Publ., Munich/Oxford 1988

4.3. Mesophases

Liquid Crystalline Polymers

A.Ciferri, W.R.Krigbaum, R.B.Meyer, eds., Polymer Liquid Crystals, Academic Press, New York 1982

G.W.Gray, ed., Thermotropic Liquid Crystals, Wiley, New York 1987

A.E.Zachariades, R.S.Porter, eds., Structure and Properties of Oriented Thermotropic Liquid Crystalline Polymers in the Solid State, Dekker, New York 1988

A.Ciferri, ed., Liquid Crystallinity in Polymers, VCH, Weinheim 1991

N.A.Platé, ed., Liquid-Crystal Polymers, Plenum, New York 1992

Block Copolymers

A.Nohay, J.E.McGrath, Block Copolymers: Overview and Critical Survey, Academic Press, New York 1976

M.J.Folkes, ed., Processing, Structure and Properties of Block Copolymers, Elsevier, New York 1985

N.R.Legge, G.Holden, H.E.Schroeder, ed., Thermoplastic Elastomers, Munich 1987

Ionomers

L.Holliday, ed., Ionic Polymers, Halstead Press, New York 1975

A.Eisenberg, M.King, eds., Ion Containing Polymers, Academic Press, New York 1977

A.Eisenberg, M.Pineri, eds., Structure and Properties of Ionomers, Reidel, Den Haag 1987

4.4. Gels

W.Burchard, S.B.Ross-Murphy, eds., Physical Networks, Elsevier, London 1990

4.5. Polymer Surfaces

B.W.Cherry, Polymer Surfaces, Cambridge University Press, Cambridge (UK) 1981

W.J.Feast, H.S.Munro, Polymer Surfaces and Interfaces, Wiley, New York 1987

I.C.Sanchez, ed., Physics of Polymer Surfaces and Interfaces, Butterworth-Heineman, Stoneham (MA) 1992

References

[1] Private communication by Professor C. Ruscher and Dr. E. Schulz, formerly at Deutsche Akademie der Wissenschaften (former German Democratic Republic).

[2] Private communication by Professor R.J.Samuels, Georgia Institute of Technology, Atlanta (GA), USA.

[3] H.-G. Elias, Makromoleküle, vol. I: Grundlagen, Hüthig and Wepf Publ., Basle, 5th ed. 1990, Fig. 20-6.

[4] H.-G. Elias, Makromoleküle, vol. I: Grundlagen, Hüthig and Wepf Publ., Basle, 5th ed. 1990, Fig. 20-13.

[5] H.-G. Elias, Makromoleküle, vol. I: Grundlagen, Hüthig and Wepf Publ., Basle, 5th ed. 1990, Fig. 20-23.

5. Thermal Properties

5.1. Overview

Polymers pass through various physical states upon heating: crystalline polymers change from solids to viscoelastic melts at the melting temperature T_M; amorphous polymers convert from glassy to rubber-like materials at the glass temperature T_G (Fig. 5-1). On slow heating, semicrystalline polymers exhibit a glass temperature of the amorphous domains, followed by a partial crystallization of the amorphous regions, and finally a melting temperature. The response on cooling may be different because of low thermal conductivities and high viscosities. Crystalline domains may overheat and polymers may remain solid above the melting temperature; melts may supercool and stay liquid below the melting temperature.

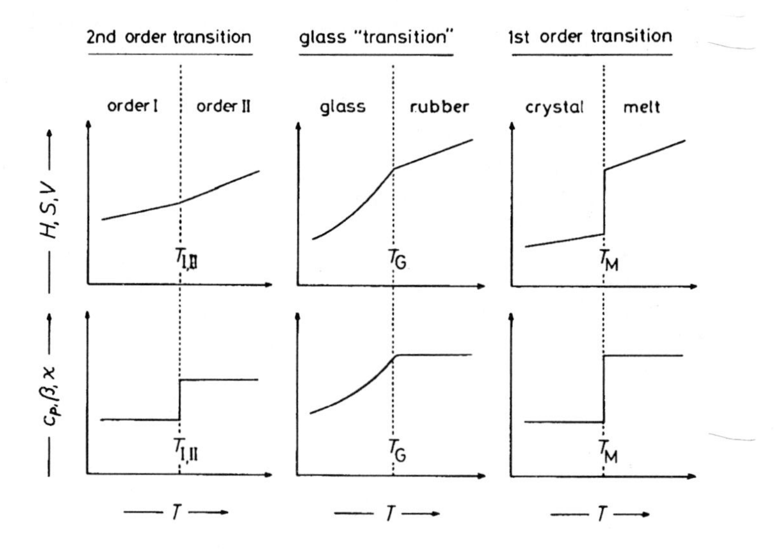

Fig. 5-1 Schematic representation of the temperature dependence of first derivatives H, S, and V (top) and second derivatives c_p, β, and κ (bottom) of the Gibbs energy G.

Left: Second order transition (example: smectic to nematic).
Center: Glass transition from glass to rubber.
Right: First order transition (example: melting of crystals).

H = enthalpy $\quad S$ = entropy $\quad V$ = volume
c_p = specific heat capacity $\quad \beta$ = cubic expansion coefficient $\quad \kappa$ = compressibility
$T_{I,II}$ = transition temperature I→II $\quad T_G$ = glass temperature $\quad T_M$ = melting temperature

At the melting temperature, crystal lattices loose their long-range three-dimensional order. The melting temperature T_M is a first order thermodynamic transition: the first derivatives of the Gibbs energy G all rise stepwise at T_M, that is, the volume $V \equiv (\partial G/\partial p)_T$ and the entropy $S \equiv (\partial G/\partial T)_p$, and thus also the enthalpy $H \equiv G - (\partial G/\partial T)_p$.

The second derivatives of G also change stepwise at the melting temperature: the heat capacity $C_p \equiv (\partial H/\partial T)_p$, the cubic expansion coefficient $\beta = V^{-1}(\partial V/\partial T)_p$, and the isothermal cubic compressibility $\kappa = - V^{-1}(\partial V/\partial p)_T$.

Polymer segments begin cooperative thermal motions at the glass temperature T_G, which causes the volume to raise more strongly with T above T_G. The cubic expansion coefficient $\beta = V^{-1}(\partial V/\partial p)_T$ of amorphous polymers changes stepwise at this temperature, which resembles in some respect a second order thermodynamic transition. The increased mobility of segments leads to a successive melting of crystalline domains; no stepwise change of volume is found for semicrystalline polymers.

5.2. Molecular Motions

5.2.1. Thermal Expansion

Isotropic bodies expand upon heating equally in all three spatial directions because of the increasing thermal motions of atoms, groups, and molecules. The expansion is characterized by the cubic expansion coefficient $\beta = V^{-1}(\partial V/\partial T)_p$, which is usually converted by $\beta = 3\,\alpha$ into the linear expansion coefficient $\alpha = L^{-1}(\partial L/\partial T)_p$. Such isotropic materials are, for example, diamond (α = $1.06 \cdot 10^{-6}$ K^{-1}), iron ($12 \cdot 10^{-6}$ K^{-1}), and decane ($353 \cdot 10^{-6}$ K^{-1}) (all data at 25°C).

Table 5-1 X-ray crystallinities w_X, linear expansion coefficients α, isobaric specific heat capacities c_p, and heat conductivities λ of various materials at 25°C. Bonding may be by covalent (c), metallic (m), or hydrogen bonds (h), dipole-dipole interactions (p), or dispersion forces (d).

Densities of poly(ethylene)s PE are ρ = 0.85 g/cm³ (amorphous; data by extrapolation), ρ = 0.92 g/cm³ (low density; branched), ρ = 0.96 g/cm³ (high density, slightly branched), or (ρ = 1.00 g/cm³ (100 % crystalline; data extrapolated).

Material	Structure	$\frac{w_X}{\%}$	Bonds	$\frac{10^6\,\alpha}{K^{-1}}$	$\frac{c_p}{J\,g^{-1}\,K^{-1}}$	$\frac{\lambda}{W\,m^{-1}\,K^{-1}}$
Quartz glass	$[SiO_2]_n$	0	c c c	0.54	1.33	1.4
Quartz crystal	$[SiO_2]_n$	100	c c c	1	1.33	10.5
Diamond, type 1	$[C]_n$	100	c c c	0.8	0.52	900
Iron	Fe	100	m m m	12	0.54	58
Aluminum	Al	100	m m m	23	0.88	260
PA 6	$\sim[NH(CH_2)_5CO]_n\sim$	50	c h d	83	1.47	0.31
PVC	$\sim[CH_2\text{-}CHCl]_n\sim$	5	c p d	80	0.96	0.18
PS	$\sim[CH_2\text{-}CHC_6H_5]_n\sim$	0	c p d	70	1.22	0.15
PE, 100 % cryst.	$\sim[CH_2\text{-}CH_2]_n\sim$	100	c d d	104	1.52	0.75
PE, high density	$\sim[CH_2\text{-}CH_2]_n\sim$	80	c d d	120	1.86	0.52
PE, low density	$\sim[CH_2\text{-}CH_2]_n\sim$	48	c d d	250	2.24	0.36
PE, amorphous	$\sim[CH_2\text{-}CH_2]_n\sim$	0	c d d	287	2.28	0.16
Decane	$H[CH_2]_{10}H$	0	d d d	353	2.26	0.14

Each of these three materials has the same type of bonds in the three spatial directions: covalent bonds between carbon atoms in diamond, metallic bonds between iron atoms, and dispersion forces between decane molecules.

Polymer chains are however anisotropic; the bonds between chain atoms are almost always covalent; the intermolecular bonds perpendicular to the chain, physical (dispersion forces, dipole-dipole interactions). On thermal expansion of polymer crystals, chains contract in the direction of their axes because of the increasing amplitude of lateral motions: the linear thermal expansion coefficient in chain direction is zero to negative. The crystals expand, however, in the two directions perpendicular to the chain axes and the resulting cubic expansion β coefficient is positive. Reported linear thermal expansion coefficients $\alpha = \beta/3$ of polymers are thus averages over the three spatial directions; they lie between those of metals and liquids (Table 5-1).

Significant problems may thus arise due to different expansion coefficients for polymer-metal composites upon thermal stress. Another problem is the low dimensional stability of polymers on temperature change. This problem may be aggravated by a concomitant change of the water content of polymers on conditioning and/or by recrystallization phenomena, both of which can lead to warping.

In fillers such as minerals, glass fibers, etc., atoms are joint via chemical bonds; fillers have thus small thermal expansion coefficients. Their addition to polymers reduces the thermal expansion coefficients of plastics considerably.

5.2.2. Heat Capacity

The molar heat capacity of an atom should be 3 R according to the law of equal distribution of energy. In reality, degrees of freedom are always frozen-in and the molar heat capacity is lowered. Empirically, values of ca. 1 R have been found for solid polymers at room temperature. For example, poly(oxy-1,4-(2,6-dimethyl)phenylene), $(C_8H_8O)_n$, has a specific heat capacity of 1.22 J K^{-1} g^{-1} and a molar heat capacity (per monomeric unit) of 146.4 J K^{-1} mol^{-1} at 25°C. The molar heat capacity per average atom of this polymer is thus (146.4 J K^{-1} mol^{-1}) /17 = 8.61 J K^{-1} mol^{-1}, i.e., approximately 1 R = 8.314 J K^{-1} mol^{-1}.

The segment mobility increases strongly in a very narrow temperature interval at the melting temperature T_M of perfect crystals and at the glass-rubber transition T_G of completely amorphous materials, causing steps in the specific heat capacity temperature curve (Fig. 5-1). Below the glass temperature T_G, heat capacities are practically not influenced by the degree of crystallinity of the polymer.

Specific heat capacities c_p of polymers increase continuously with temperature; at 25°C, the temperature function is $(1/c_{p,25°C})(dc_p/dT) = (3.0 \pm 0.21)\cdot 10^{-3}$ K^{-1} for many solid polymers. Specific heat capacities of amorphous polymers do not change very abruptly at the glass temperature; the static glass temperature of amorphous polymers is near the end of the step zone (SBR, PS, PVC in Fig. 5-2). Specific heat capacities of semicrystalline polymers usually do not reflect glass temperatures (PPO, PTFE) or show only a change of slope of the $c_p = f(T)$-curve (PA 6).

Most semicrystalline polymers have crystallinities of less than 50 %. They do not show the stepwise increase of specific heat capacities of perfect crystals because recrystallization sets in at approximately the glass temperature and maxima of c_p are observed for conventional (fast) heating (see PPO in Fig. 5-2). Such maxima are also caused by changes in crystal modifications (see PTFE in Fig. 5-2).

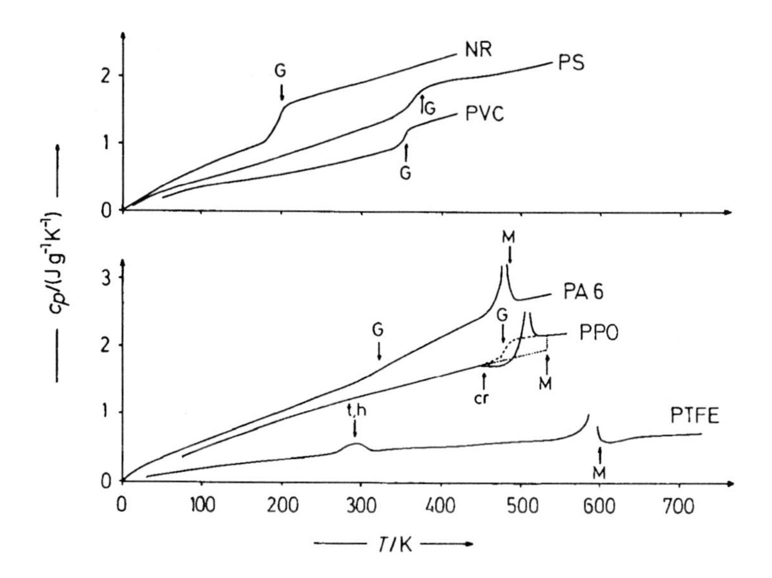

Fig. 5-2 Temperature dependence of specific heat capacities c_p at constant pressure.

Top: Amorphous polymers: styrene-butadiene rubber SBR, poly(styrene) PS, and poly(vinyl chloride) PVC; static glass temperatures are indicated by G.

Bottom: Semicrystalline polymers: polyamide PA 6, 65 % crystalline poly(oxy-2,6-dimethyl-1,4-phenylene) PPO®, and poly(tetrafluoroethylene) PTFE. Glass temperatures are indicated by G, melting temperatures by M, beginning of recrystallization by cr, and a change of crystal modifications from triclinic to hexagonal by t,h. The broken line at PPO® indicates a completely amorphous polymer with glass temperature T_G; the dotted line, a perfect crystal with melting at T_M.

5.2.3. Heat Conductivity

Conventional polymers have much lower heat conductivities than metals because they are electrical insulators and not conductors (Table 5-1). In polymers, heat is thus not transported by electrons but by elastic waves (phonons in the corpuscular model). The free path length of phonons is defined as the distance at which the intensity of elastic waves has decreased to 1/e. This free path length is about 0.7 nm for glasses, amorphous polymers, and liquids; it is practically independent of temperature.

Heat conductivities λ (thermal conductivities) of amorphous plastics and elastomers are not independent of polymer type and temperature, however; they rather increase with temperature T and pass then through a weak maximum near the glass temperature (Fig. 5-3). The variation of λ with T must thus be caused by changes in the heat capacity (see Fig. 5-2).

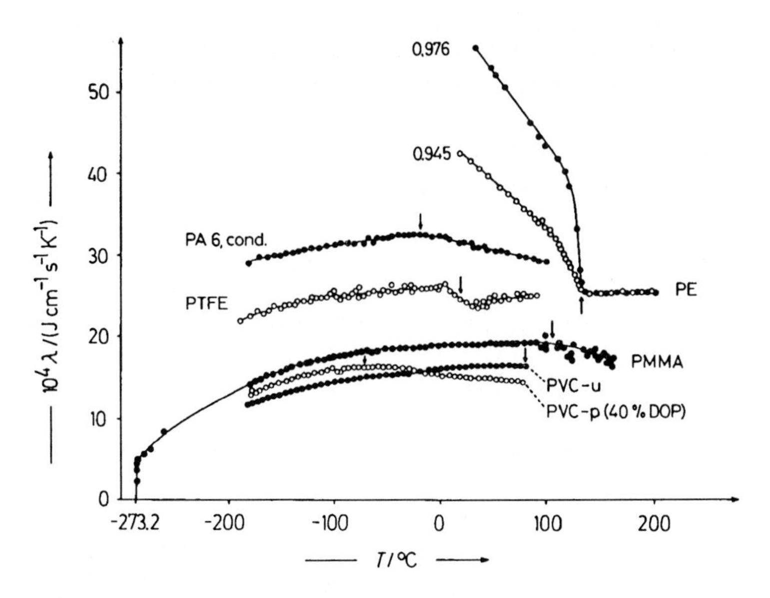

Fig. 5-3 Temperature dependence of heat conductivities λ of polymers.
PE: poly(ethylene)s PE (ρ = 0.976 g/cm³ and 0.945 g/cm³; T_M = 132°C at ↑);
PA 6: polyamide 6 (conditioned at 20°C and 65 % relative humidity; T_G = -18°C at ↓);
PTFE: poly(tetrafluoroethylene) with transition tricline-hexagonal ↓ at 18°C;
PMMA: poly(methyl methacrylate) with glass transition ↓ at 105°C;
PVC-u: unplasticized (hard) poly(vinyl chloride) with glass transition ↓ at 82°C;
PVC-p: poly(vinyl chloride) PVC-p, plasticized with 40 wt% di(2-ethylhexyl)phthalate DOP and glass temperature ↓.

Heat capacities of crystalline polymers decrease strongly at the melting points because packing densities decline drastically at these temperatures (see PE in Fig. 5-3). A weaker decline is found for crystal-crystal transformations (PTFE, Fig. 5-3).

5.3. Thermal Transitions and Relaxations

5.3.1. Overview

Micro- and macroconformations of polymer molecules and thus supermolecular structures change at melting and glass temperatures. Thermal transitions are distinguished from thermal relaxations:

In a true *thermal transition*, chemical compounds are in equilibrium on both sides of the transition temperature. An example is the melting transition, which is a true thermodynamic first order transition because the first and the second temperature derivatives of the Gibbs energy G show jumps at the melting temperature (Fig. 5-1). A second order thermodynamic transition has only a jump of the second temperature derivative of G; an example is the smectic → nematic transition of liquid crystals.

Thermal relaxations, on the other hand, are kinetic effects. They depend on the frequency of the experimental method and thus on the time scale. Typical thermal relaxations are caused by the onset of translations and rotations of charges, dipoles, and chemical groups (i.e., by atomic motions).

Some experimental methods work at frequencies that such relaxations appear to be thermal transitions. The best-known example is the glass transition temperature T_G at which hard, glassy polymers convert into soft, rubbery materials ($T > T_G$) and vice versa ($T < T_G$). In many cases, a thermal effect cannot be unambiguously classified as either transition or relaxation; for this reason, the glass transition temperature is better called a glass temperature.

Thermal transitions and relaxations can be detected and determined by many different experimental methods. The most commonly applied methods for *thermal transitions* are differential thermoanalysis DTA (measures temperature differences between specimen and standard on heating or cooling with constant rate), differential scanning calorimetry DSC (does the same for enthalpy differences), thermomechanical analysis TMA (deformation of specimen under load), dynamic-mechanical analysis DMA (either free or forced vibration of specimen), and torsional braid analysis TBA (specimen on vibrating support). Figure 5-4 shows a typical thermogram and the accompanying volume changes.

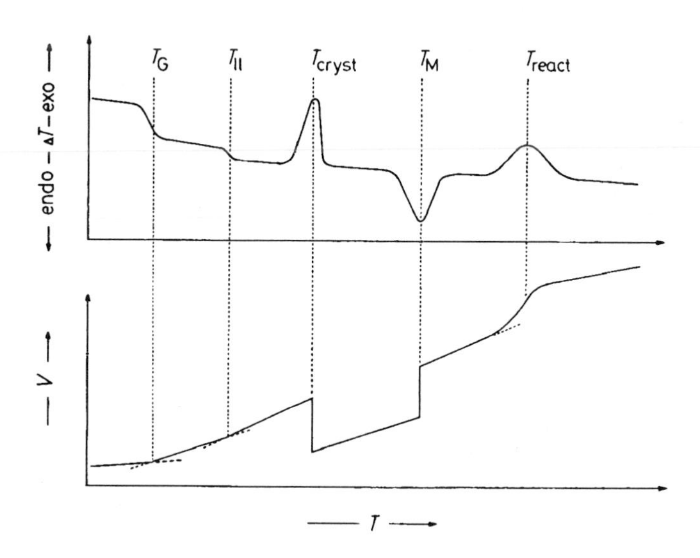

Fig. 5-4 Idealized thermogram with exothermic and endothermic temperature changes ΔT (above) and volumes V (below) of a partially crystalline polymer.
T_G = glass temperature T_G | T_{ll} = liquid-liquid transition
T_{cryst} = maximum crystallization temperature | T_M = melting temperature
T_{react} = maximum temperature of a chemical transformation with gas formation

Many methods are available for the study of molecular motions and thus *thermal relaxations*. These methods work with frequencies ν that correspond to correlation times t_c between 10^{-12} s $< 1/\nu < 10^6$ s (= 11.5 days). Typical methods include quasielastic neutron scattering ($10^{-12} < t_c/s < 10^{-8}$), NMR spin lattice relaxation ($10^{-12} < t_c/s < 10^{-5}$), dielectric relaxation ($10^{-10} < t_c < 10^{-5}$), and photon correlation spectroscopy ($10^{-4} < t_c/s < 10^2$). Thermal relaxations furthermore manifest themselves in sudden changes of mechanical properties, such as rebound elasticity ($t_c/s \approx 10^{-5}$), penetrometry ($t_c/s \approx 10^2$), mechanical loss ($10^3 < t_c/s < 10^7$), and thermal expansion ($t_c/s \approx 10^4$). Slow methods (large t_c) are called "static" methods; fast ones, "dynamic". Transition/relaxation phenomena are also detected by several empirical, standardized methods that measure the resistance of specimens against flow under various loads (Vicat temperature, heat distortion temperature, Martens temperature, etc.).

Various characteristic signals are observed at a fixed temperature for a given frequency (see insert in Fig. 5-5). They often cannot be correlated with molecular processes and are commonly indicated with descending temperature by letters in the sequence of the Greek alphabet, starting with the melting temperature (crystalline polymers, subscript c) or glass transition temperature (amorphous polymers, subscript a).

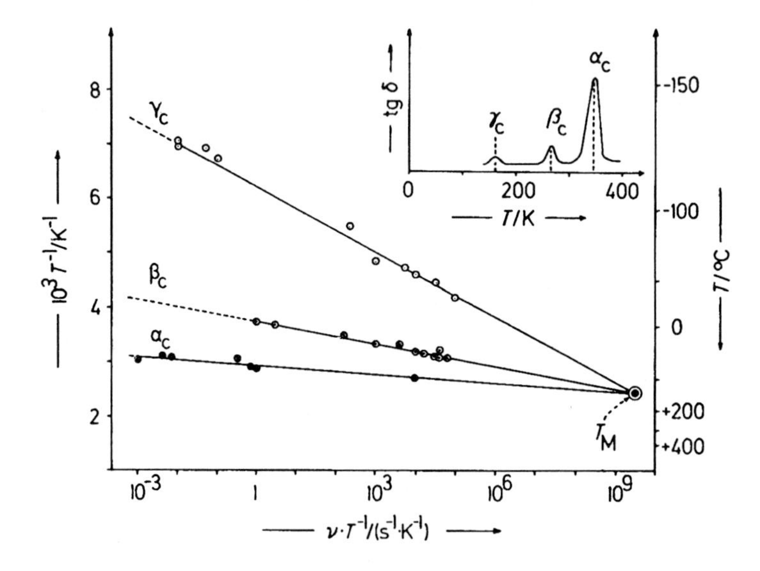

Fig. 5-5 Dependence of inverse relaxation temperature $1/T$ on the logarithm of reduced frequence ν/T for various relaxation processes of a low density poly(ethylene). The insert shows the mechanical relaxation spectrum at $\nu = 1000$ Hz.

Data of H.W.Starkweather, Jr. [1]. Reprinted with permission by Hüthig and Wepf Publ., Basle, Switzerland [2].

The various methods work at different frequencies and thus give different relaxation temperatures for the same molecular process. The frequency dependence of relaxation temperatures can be described by the Eyring equation for rate processes:

(5-1) $$\nu = (k_BT/2\pi h)\cdot \exp(-\Delta H^{\ddagger}/RT)\cdot \exp(\Delta S^{\ddagger}/R)$$

where k_B is the Boltzmann constant, h the Planck constant, $\Delta H^{\ddagger}$ the activation enthalpy, and $\Delta S^{\ddagger}$ the activation entropy. Transformation of Equation (5-1) leads to

$$\frac{1}{T} = \frac{R}{\Delta H^{\ddagger}} \left[\left(\frac{\Delta S^{\ddagger}}{R} + \ln \frac{k_B}{2\pi h} \right) - \ln \frac{v}{T} \right] \quad (5\text{-}2)$$

In an $(1/T) = f[\ln(v/T)]$ plot, lines for various processes intersect at the melting temperature T_M = 131°C (Fig. 5-5). Such common intersects seem to be general for nonhelical polymers.

5.3.2. Crystallization

The crystallization of coil-like polymers from dilute solutions leads to platelets (Fig. 4-2) in which folded polymer chains are arranged with their stems perpendicular to the fold surface (Fig. 4-1). From concentrated solutions and melts at rest, lamellae may grow into spherulites (Figs. 4-2 and 4-3). Spherulitic structures may be transformed into row structures by strain such as shearing or drawing (Fig. 5-6). Topological constraints and kinetically hindred diffusion of segments control the crystallizability; crystallization is never complete and may even change slowly with time. "Crystalline" polymers are thus semicrystalline materials composed of crystalline and noncrystalline ("amorphous") regions.

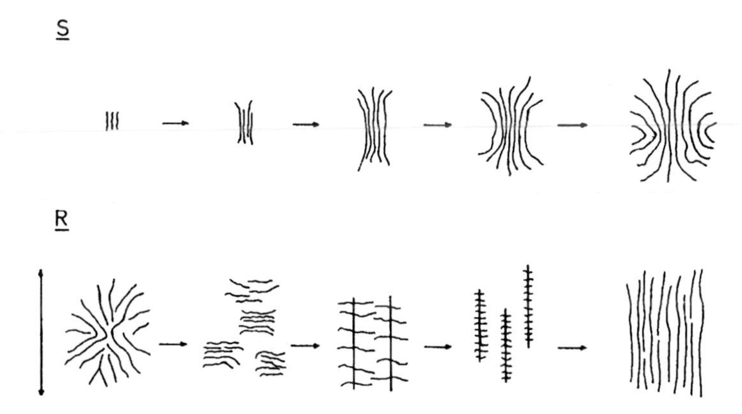

Fig. 5-6 Schematic representation of formation and transformation of spherulites.
Top: Formation of spherulites from lamellae on crystallization of melts at rest.
Bottom: Transformation of spherulites into row structures by strain in ↕ direction.
Reprinted with permission by Hüthig and Wepf Publ., Basle, Switzerland [3].

Crystallization is subdivided into two elementary processes: primary nucleation and crystal growth (secondary nucleation). Both processes determine crystallization rates, which depend strongly on temperature and polymer structure. Symmetrically structured polymers usually crystallize rapidly; polymers with bulky groups and/or low tacticities, only slowly. At a temperature of 30 K below the melting temperature, the linear crystallization rate may, for example, range between 5000 μm/min for poly(ethylene) and 0.01 μm/min for poly(vinyl chloride).

High crystallization rates may make it impossible to obtain amorphous polymers. Quenching of poly(ethylene) melts never gives amorphous polymers, even if liquid nitrogen is used. Quenching of poly(ethylene terephthalate) melts, on the other hand, leads to amorphous polymers under certain conditions (see Chapter 12).

Primary nucleation may be homogeneneous (spontaneous, sporadic, thermal) or heterogeneous (simultaneous, athermal). Homogeneous nuclei are formed from segments of the crystallizing polymer molecules; they are very rare. Heterogeneous nuclei result from extraneous materials such as additives, dust particles, container walls, or specially added nucleation agents. Such nuclei must have minimum sizes of 2-10 nm. Their concentrations can range from ca. 1 nucleus per cubic centimeter in poly(oxyethylene) to ca. 10^{12} nuclei per cubic centimeter in poly(ethylene).

Above the melting temperature T_M, fragments of crystallites may survive for certain time periods. These fragments act as athermal primary nucleation agents on subsequent cooling. They are responsible for the "memory effect" (i.e., the reappearance of spherulites at the same locations before the melting and after melting/cooling), which takes place because of low diffusion rates at very high melt viscosities.

Growth of primary nuclei occurs by addition of chain segments in a *secondary nucleation process*. The growth rate is low at temperatures just below the melting temperature because secondary nuclei are formed and dissolved rapidly. About 50 K below the glass temperature, on the other hand, motions of molecule segments are practically zero and the crystal growth rate is therefore low as well. Crystallization rates must thus exhibit a maximum between the melting and glass temperatures; the maximum crystallization rates are usually at temperatures of (0.80-0.87) T_M (in kelvin).

In addition, the entire crystallization process can be subdivided into a primary and a secondary stage. The *primary stage* comprises the conversion of the melt to a solid. It ends when the crystallized entities (crystallites, spherulites, etc.) start touching each other. This primary crystallization is described by the Avrami equation:

$$\text{(5-3)} \qquad \phi/\phi_\infty = 1 - \exp(-zt^n)$$

where ϕ = fraction of crystallized volume, ϕ_∞ = fraction of maximal attainable crystallinity for a given entity (e.g., spherulite), and z and n are constants that depend on both the nature of the nucleation process (homogeneous, heterogeneous) and the type of the growing entity (rod, disc, sphere, sheaflet, etc.). Theoretical exponents n are whole numbers between 1 and 7; empirically, they may assume fractional values.

Segments between the contacting crystallites or between the lamellae of spherulites may not have crystallized during the primary stage: the polymer has not attained its maximal crystallinity (the crystallizability). The crystallization may thus continue during a *second stage*. In this after-crystallization, lamellae may thicken, lattices become more perfect, etc. The polymer may undergo a shrinkage. It may even warp if the crystallization rates differ in the different spatial directions such as in the extrusion of belts. These effects are the greater, the bigger the crystallized entities of the primary stage. They can be reduced by the addition of nucleating agents that rapidly form large concentrations of small primary crystallites.

5.3.3. Melting

Melting is defined as the thermal transition from a crystal to an isotropic melt. The word "melting" is, however, used also more loosely for a variety of phenomena, for example, the softening of an amorphous glass at the glass temperature or the helix-coil transition of DNA double-helices in aqueous solutions.

The melting temperature T_M (fusion temperature) is defined as the temperature at which crystallites are in equilibrium with the melt. Melting starts at the corners and edges of crystal surfaces because these are less ordered than the interior of a lattice. In contrast to crystallization, no nuclei are needed for the initiation of melting.

Segments of about 60-100 chain atoms participate in the melting process. During heating, segments are redistributed continuously between crystalline and noncrystalline regions; a melting range exists and no sharp melting point is observed. The theoretical melting temperature is defined as the upper end of the melting range because the biggest and most perfect crystals melt there. Published melting temperatures often refer to the maxima of $\Delta T = f(T)$ curves, however. Observed melting temperatures are in general lower than the thermodynamic melting temperatures of perfect crystals but may be occasionally higher because of overheating effects.

Melting temperatures increase with molar masses and become practically constant at molar masses of ca. 50 000-150 000 g/mol (ca. 10 times the molar mass of the participating segments). It is these melting temperatures of "infinite" molar mass polymers that are usually given as *the* melting temperature of a polymer although they are not necessarily those of perfect crystals.

The melt can be considered as a dilute solution of endgroups in monomeric units. The decrease of melting temperatures with decreasing degree of polymerization (i.e., increasing concentration of endgroups) can be described by the thermodynamic law for the lowering of freezing temperatures:

$$\text{(5-4)} \quad \frac{1}{T_M} = \frac{1}{T_M^o} + \left(\frac{2R}{\Delta H_{M,u}^m} \right) \cdot \frac{1}{\overline{X}_n}$$

where $\Delta H_{M,u}^m$ is the molar melt enthalpy per monomeric unit, $\overline{X}_n$ the number-average degree of polymerization, and T_M and T_M^o are the thermodynamic melting temperatures at finite and infinite molar masses.

Similar depressions of the melting temperature are caused by addition of low molar mass solvents 1 or amorphous polymers A to the parent polymer 2 or by non-crystallizable comonomer units in statistical copolymers; in all these cases, the term $2/\overline{X}_n$ of Equation (5-4) must be replaced by

$$\text{(5-5)} \quad \left({}^*V_u^m / {}^*V_1^m \right) \left[(1-\phi_2) - \chi (1-\phi_2)^2 \right] \quad \text{(added solvent)}$$

$$\text{(5-6)} \quad \left({}^*V_u^m / {}^*V_A^m \right) \left[-\chi (1-\phi_2)^2 \right] \quad \text{(added amorphous polymer)}$$

(5-7) $-\ln x_u - \chi(1-\phi_2)^2$ (dissolved statistical copolymer with long sequences of crystallizable units u)

where $^*V^m$ is the partial molar volume of added solvent 1, added amorphous polymer A, or crystallizable monomeric units u of statistical copolymer; x_u the mole fraction of units u; ϕ_2 the volume fraction of crystallizable parent polymer 2 (copolymer in Equation (5-7)); and χ an interaction parameter.

Melting temperatures $T_M = \Delta H_M^m / \Delta S_M^m$ are determined by the changes in molar melting enthalpies ΔH_M^m and molar melting entropies ΔS_M^m. The *melting entropy* results from changes in conformation and volume upon melting. Conformational changes contribute theoretically $R \ln 3 = 9.12$ J K^{-1} mol^{-1} for the formation of three conformers with equal energy or 7.41 J K^{-1} mol^{-1} for one *trans* and two *gauche* conformers in the case of poly(methylene) ~$[CH_2]_n$~. The volume change adds another 10.9 J K^{-1} mol^{-1}. The theoretical melt entropy of poly(methylene) should thus be ca. (18.3-20.0) J K^{-1} mol^{-1}. Experimentally, only 9.9 J K^{-1} mol^{-1} is observed, which indicates either the existence of local order in melts or a high segment mobility below the melting temperature. The latter was found experimentally by broad-line NMR for *cis*-1,4-poly(isoprene) ($\Delta S_M^m = 4.8$ J K^{-1} mol^{-1}).

Melting enthalpies are usually between 1 and 5 kJ per mole of chain atom. Low values are expected for polymers with high chain mobilities below the melting temperature (*cis*-1,4-poly(isoprene), aliphatic polyesters and polyethers). High values are found for polymers with strong interactions between chains and tight packing of chains in crystals [poly(oxymethylene), it-poly(styrene)]. Some of these strong interactions may survive the melting process; for example, most of the hydrogen bonds of polyamides are still detected by IR spectroscopy above the melting temperature; the high order in the melt leads to low melting entropies (PA 6: 7.0 J K^{-1} mol^{-1}).

Thus, the primary factors for high melting temperatures are not intermolecular interactions but reduced chain flexibilities. High melting temperatures are therefore found for tightly packed helices such as poly(oxymethylene) and it-poly(3-methylbutene) and for ladder-like polymers such as poly(*p*-phenylene) and poly(benzimidazole). Low melting temperatures are observed for polymers with low rotational barriers (ester, oxygen, sulfide groups in chains). These factors are responsible for the variation of melting temperatures with the number i of methylene units in polymer chains ~$[X\text{–}(CH_2)_i]_n$~ and isotactic poly(α-olefin)s ~$[CH_2\text{–}CH(CH_2)_iH]_n$~ (Fig. 5-7). At large i, the melting temperature of poly(ethylene) ~$[CH_2\text{–}CH_2]_n$~ is approached.

5.3.4. Liquid Crystal Transitions

Thermal transitions of thermotropic LC polymers from their crystals to smectic (T_{cs}) or nematic phases (T_{cn}), and from nematic phases to isotropic melts (clearing temperature T_{ni}), are thermodynamic first order transitions (discontinuity of first derivative of Gibbs energy with temperature). They exhibit step-like changes in volume,

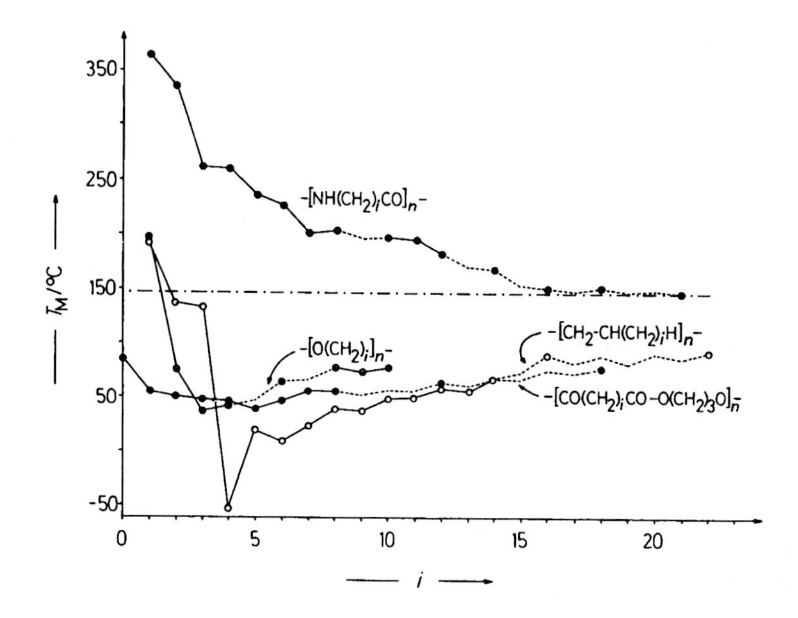

Fig. 5-7 Melting temperatures T_M of isotactic poly(olefin)s ~[CH$_2$-CH{(CH$_2$)$_i$H}]$_n$~ (O), aliphatic polyamides ~[NH-(CH$_2$)$_i$-CO]$_n$~ (⊕), aliphatic polyoxides ~[O-(CH$_2$)$_i$]$_n$~ (ʘ), and aliphatic polyesters ~[OC-(CH$_2$)$_i$-CO-O-(CH$_2$)$_i$-O]$_n$~ of trimethylene glycol (●), as function of the length i of methylene sequences. - · · · - Theoretical melting temperature of poly(ethylene).

enthalpy, and entropy, just like the melting of three-dimensional crystals to isotropic melts. Thermal transitions smectic phase → another smectic phase or smectic phase → nematic phase (T_{sn}) are second order thermodynamic transitions, however; changes are not step-like at these transition temperatures. Thermodynamically stable phases exist only between melting and clearing temperatures (T_{cs}, $T_{cn} < T_{ni}$); they are called *enantiotropic phases.*

Mesophases form dispersions in supercooled isotropic melts if clearing temperatures are lower than melting temperatures (T_{cs}, $T_{cn} > T_{ni}$). Such phases are thermodynamically unstable compared to the crystalline one; they are called *monotropic.*

Mesophases may be supercooled below T_{cs} to smectic liquids sL* and below T_{sn} to nematic liquids nL* if crystallization can be suppressed. At even lower temperatures T_{gn} and T_{gs}, these supercooled liquids may yield anisotropic glasses nG and sG, respectively. Some of these transition temperatures cannot be measured directly; their existence can be deduced from extrapolations of transition temperatures of copolymers to 100 % of the pure mesogenic compound (virtual transition temperatures).

Thermal transition temperatures T_{trans} (= T_{cs}, T_{gs}, T_{sn}, T_{cn}, T_{gn}, T_{ni}) depend on the degree of polymerization X of polymers in the same way as the melting temperature: $1/T_{trans} = f(1/X)$. The transition enthalpy ΔH_{ni} is always smaller than the transition enthalpies ΔH_{gs} and ΔH_{sn} because the n→i transition is from order to disorder, whereas the g→s and s→n transitions are from order to less order. Transition entropies of mesophases are lower than melt entropies; they have values of ca. (0.5-1.5) J K^{-1} mol^{-1}, with $\Delta S_{sn} < \Delta S_{ni}$.

5.3.5. Glass Temperatures

Glass temperatures are phenomenologically characterized by a change from a "hard", noncrystalline, glass-like material to a rubbery to highly viscous "melt". The viscosities at glass temperatures are ca. 10^{12} Pa s, independent of chemical structures. Glass "transitions" were thus thought to be "iso-viscous" phenomena. Today, glass transitions are considered to happen at that physical state where all materials exhibit the same "fractional free volume" (see also Section 4.1.1.).

Various free-volume fractions are discussed in the literature. The empirical Boyer-Simha rule relates a free-volume fraction, f_{exp}, to the cubic expansion coefficients, β, of liquid (L) and amorphous, glass-like (G) polymers and their glass temperatures T_G (see Table 5-2):

(5-8) $$f_{exp} \approx (\beta_L - \beta_G)T_G \approx 0.11 \pm 0.02$$

Table 5-2 Free-volume fractions of selected amorphous polymers at their static glass temperatures T_G; fractional free volumes are calculated from crystalline densities at 0°C (not 0 K).

Polymer	T_G/°C	f_{exp}	f_o	f_{fluc}
Poly(ethylene)	- 80	0.098	0.025	
Poly(isobutylene)	- 73	0.079	0.026	0.0017
Poly(butyl methacrylate)	20	0.13	0.026	0.0010
Poly(vinyl acetate)	27	0.128	0.028	0.0023
Poly(styrene)	100	0.133	0.025	0.0035
Poly(methyl methacrylate)	105	0.118	0.025	0.0015

The Williams-Landel-Ferry (WLF) approach assumes a linear dependence of the fractional free volume $f = V_f/V$ on temperature, i.e., $f = f_o + \beta_f(T - T_o)$. This linear free volume expansion

(5-9) $$\beta_f = (f - f_o)/(T - T_o)]$$

approximates the exponential expansion of volume given by the true cubic expansion coefficient $\beta = (1/V)(dV/dT)$; the WLF approach is thus restricted to temperatures in the range $T_o < T < (T_o + 100 \text{ K})$. The reference temperature T_o is usually the glass temperature T_G. The fractional volumes are expressed in terms of a shift factor

(5-10) $$\lg a_T = \frac{B}{2.303}\left(\frac{1}{f} - \frac{1}{f_o}\right) + \lg\left(\frac{T_o \rho_o}{T\rho}\right) \approx \frac{B}{2.303}\left(\frac{1}{f} - \frac{1}{f_o}\right)$$

which is deduced from the reduced variable $a_T = (\eta T_o \rho_o/(\eta_o T \rho)$ and the Doolittle equation $\ln \eta = \ln A + B(V - V_f)/V_f$ for the dependence of the viscosity η on the free volume V_f. The shift factor a_T corresponds to a ratio of relaxation times at two different temperatures. Combination of Equations (5-9) and (5-10) delivers the Williams-Landel-Ferry equation

$$(5\text{-}11) \quad \lg a_T = -\frac{\{B/(2.303 f_o)\}[T - T_o]}{\{f_o/\beta_f\} + [T - T_o]} = \frac{K[T - T_o]}{K' + [T - T_o]}$$

The WLF equation applies to all relaxation processes. Values of K = 17.44, K' = 51.6, and $f_o \approx 0.025 \pm 0.005$ were found empirically for reference temperatures $T_o = T_G$.

The WLF equation allows calculation of the static glass temperature from the various dynamic glass temperatures if the deformation times (inverse effective frequencies) of the methods are known. For example, the glass temperatures of poly(methyl methacrylate) PMMA are given as 105°C (thermal expansion; "static"), 120°C (penetrometry), and 160°C (rebound elasticity). The same polymer may thus exhibit very different mechanical properties if subjected to different stresses; at 130°C, PMMA behaves as either a glass (rebound elasticity) or an elastomer (penetrometry).

Another free-volume fraction f_{fluc} is obtained by measuring the speed of sound in polymers. It describes the movements of the center of gravity of a molecule that are caused by thermal motions.

The glass temperature indicates the onset of cooperative movements of chain segments of 25-50 chain atoms, which can be deduced from the ratio of molar activation energies and melting energies. These cooperative movements very probably involve *trans/gauche* transitions that proceed cooperatively along greater distances since only small changes of the direction of chain axes are involved according to deuterium NMR. The participation of segments of 25-50 chain atoms is also indicated by cross-linking experiments: no change of glass temperatures is observed for average chain segments N_{seg} > 25-50 between two cross-linking points. In cross-linked polymers with N_{seg} < 25-50, T_G increases with the inverse molar mass of segments.

Since glass and melting temperatures depend both on segmental motions, close relationships between these two temperatures can be expected. The empirical Beaman-Boyer rule states that $T_G \approx (2/3)\, T_M$, which holds reasonably well for many polymers except for, e.g., poly(ethylene) and poly(oxyethylene) for which $T_G/T_M \approx 1/2$.

A vast literature exists about effects of constitution on T_G. Cyclic macromolecules possess no endgroups, and thus no free-volume effects from these. Small rings are furthermore strained (less possible microconformations). The greater the molar mass, the more microconformations can be adopted, the greater is the chain flexibility and the lower is T_G. The same is true for segment flexibilities of star-branched polymers and long side chains in comb-like molecules (side-chain "crystallization").

Linear relationships are found between the logarithms of glass temperatures T_G and the logarithms of cross-sectional areas A of either carbon, carbon-oxygen, or carbon-nitrogen chains (another measure of segment flexibilities). The three lines intersect at A = 0.17 nm^2 and T_G = 141 K, which should be the lowest glass temperature possible. The lowest experimentally found glass temperature (134 K ≙ -139°C) is that of poly(diethylsiloxane), ~[O-Si(C_2H_5)$_2$]$_n$~.

Glass temperatures of polymers can be lowered by copolymerization with suitable monomers (internal plasticization) and by addition of external plasticizers (see Section 10.2.5); certain comonomers may also increase T_G. The dependence of glass temperature on copolymer composition is predicted by various theories. The Couchman theory treats the glass transition T_G as a true second order thermodynamic transition caused exclusively by entropy changes (T_Gs are relaxation phenomena and not thermodynamic transitions, however). It delivers

$$(5\text{-}12)\qquad \frac{\ln T_{G,2} - \ln T_G}{\ln T_G - \ln T_{G,1}} = k\left(\frac{w_1}{w_2}\right); \quad k = \Delta c_{p,1}/\Delta c_{p,2}$$

where w_i is the mass fraction of monomeric units (i = 1, 2) and Δc_p the specific heat capacity. The Couchman equation reduces for $k = 1$ to the Pochan equation $\ln T_G = w_1 \cdot \ln T_{G,1} + w_2 \cdot \ln T_{G,2}$ for $k = 1$. If furthermore $\Delta c_{p,1}/\Delta c_{p,2} = T_{G,2}/T_{G,1} \approx 1$, then the Fox equation $1/T_G = [w_1/T_{G,1}] + [w_2/T_{G,2}]$ is obtained.

The Gordon-Taylor theory assumes that the specific volume of the copolymer is a mass average of the (additive) specific volumes of monomeric units and arrives at

$$(5\text{-}13)\qquad \frac{T_G - T_{G,2}}{T_{G,1} - T_G} = \frac{1}{k'}\left(\frac{w_1}{w_2}\right) \quad ; \quad k' = \frac{\alpha_{R,2} - \alpha_{G,2}}{\alpha_{R,1} - \alpha_{G,1}}$$

where α is the linear expansion coefficient in the glassy (G) or rubbery (R) state. For $k' = 1$, T_G increases linearly with the mass fraction $w_1 \equiv 1 - w_2$ of one component because $T_G = w_1 T_{G,1} + w_2 T_{G,2}$.

Interactions between adjacent chain units contribute enthalpic effects to the glass temperature. Copolymers with such effects exhibit maxima or minima in the $T_G = f(w_1)$ function; that is, the glass temperature of the copolymer may be higher or lower than the glass temperatures of the homopolymers at a certain mass fraction w_1 of monomeric unit 1.

External plasticizations can be treated by analogy (see Section 13.3.2). Water is an effective external plasticizer for polar polymers: The glass temperature of dry poly(ε-caprolactam) decreases from 70°C (dry) to 20°C (50 % relative humidity) and -22°C (100 % relative humidity). Amide group-rich polyamides must be therefore "conditioned" to the required relative humidity before processing.

5.3.6. Other Transitions and Relaxations

Experimentally, a number of other transition/relaxation temperatures is observed, mostly of unknown origin. Amorphous polymers exhibit weak "liquid-liquid" transitions at ca. $T_{LL} \approx 1.2\ T_G$. Below the critical molar mass for entanglements, transition temperatures T_{LL} equal flow temperatures T_F at which polymers start to flow under their own weight. At higher molar masses, $T_F > T_{LL}$; probably due to entanglements.

Another transition temperature $T_U \approx 1.2\ T_M$ seems to exist for crystalline polymers. This transition has been interpreted as the dissolution of smectic structures.

Few β-relaxations have been correlated with molecular phenomena. An example is the frequency-dependent boat-chair transition of cyclohexane rings which occurs, e.g., at -125°C (10^{-4} Hz) and +80°C (10^5 Hz).

5.2.7. Technical Methods

Technical testings on thermal transitions and relaxations of plastics are usually performed with simple methods under standardized conditions and always under load.

Martens numbers measure temperatures at which the specimen has experienced a certain bend.

Vicat softening temperatures give the temperatures at which a rod penetrates 1 mm into the plastics under a force of 10 N at a heating rate of 50 K/h (Vicat A) or 50 N at 120 K/h (Vicat B).

Heat distortion temperatures (heat deflection temperatures HDT) indicate the temperatures for a certain bending with a three-point method if the specimen is heated with 120 K/h: 0.21 mm at a load of 1.85 MPa (HDT A), 0.33 mm at 0.45 MPa (HDT B), or at 5.0 MPa (HDT C) (Table 5-3).

Table 5-3 Thermal properties reported by CAMPUS® data bases. For dimensions of specimens see ISO and DIN standards and CAMPUS literature; $T_{processing}$ = processing temperature.

Property	Conditions	Unit	ISO	DIN	Remarks
Heat deflection temperature HDT/A	1.80 MPa	°C	75	53461	rigid and soft plastics
Heat deflection temperature HDT/B	0.45 MPa	°C	75	53461	soft plastics
Heat deflection temperature HDT/C	5.00 MPa	°C	-	53461	rigid plastics
Vicat VST/A/50	10 N	°C	306	306	
Vicat VST/B/50	50 N	°C	306	306	
Thermal expansion coefficient (long.)	(23-80)°C	K^{-1}	-	53752	mean value in range
Thermal expansion coefficient (vert.)	(23-80)°C	K^{-1}	-	53752	mean value in range
Thermal conductivity of melt	$T_{processing}$	W/(m K)			mean value at $T_{proc.}$
Specific heat capacity of melt	$T_{processing}$	J/(kg K)			mean value at $T_{proc.}$
Effective thermal diffusivity	$T_{processing}$	m^2/s			mean value at $T_{proc.}$
No-flow temperature		°C			
Freeze temperature		°C			≈ Vicat A/50

The temperatures from these three methods do not only depend on transitions or relaxations but also on the elasticity of the specimen; Vicat temperatures correlate for example with certain values of shear moduli. Vicat and heat distortion temperatures are in addition affected by the surface hardness.

The resulting softening temperatures are neither identical with glass transition nor with melting temperatures (see also Section 13.2.1). Methods with the smallest load (HDT B) or force (Vicat A) usually give softening temperatures close to the highest major transition temperature of the specimen, i.e., near the glass temperature of amorphous polymers and somewhat below the melting temperature of crystalline ones (Table 5-4). Both Vicat and heat distortion temperatures decrease with increasing load or force: a few degrees with amorphous polymers and much more strongly with crystalline ones because the latter have usually glass temperatures below or near the test temperature and are thus prone to flow under higher loads/forces. For the same reason, Vicat and heat distortion temperatures are often no good measures of the continuous service temperature of a plastic.

Table 5-4 Melting temperature T_M, glass temperatures T_G, Vicat temperatures VT, and heat distortion temperatures HDT of some plastics. HDT and Vicat temperatures from CAMPUS data, melting and glass temperatures from the scientific literature.

Polymer Type	Grade	Company	T_M/°C	VT in °C A	VT in °C B	HDT in °C B	HDT in °C A	HDT in °C C	T_G/°C
Amorphous plastics									
PESU	Ultrason E 1000	BASF	-	226	222	-	215	-	230
PC	Calibre 600-3	Dow Chem.	-	152	146	146	143	140	150
PMMA	Plexiglas 8 H	Röhm	-	-	106	103	98	-	105
SAN	Lustran SAN 32	Monsanto	-	108	103	-	96	-	?
SAN	Tyril	Dow Chem.	-	-	102	101	98	95	?
PS	Styron 648	Dow Chem.	-	-	100	99	97	97	100
Crystalline plastics									
PA 6.12	Vestamid D 22 nf	Hüls	217	208	178	140	59	42	46
PP	Moplen X 94 J	Himont	176	153	89	109	57	-	- 15
PP	VC 12 33 B	Neste	176	155	85	95	61	-	-15
PP-CoPM	SB 15 10 LU	Neste	?	150	80	85	55	-	?
PE-HD	NCPE 7003	Neste	135	130	85	84	50	-	-80
PE-HD	Lupolen 4261 A	BASF	135	125	77	70	41	-	-80
PE-LLD	NCPE 8065	Neste	116	86	45	50	37	-	?

5.4. Transport

5.4.1. Self-Diffusion

Brownian movements cause molecules and their segments to interchange positions in fluid phases. If all entities are of the same type, such interchanges lead to a "self-diffusion", which involves no net transport of polymers. Self-diffusion can be measured by pulsed field gradient spin-echo NMR (segments) and radioactive tracers (molecules). Self-diffusion coefficients in melts range from approximately 10^{-6} cm^2/s for oligomers to 10^{-18} cm^2/s for ultrahigh molar mass polymers; they are very low below the glass temperature and in polymer crystallites.

Coiled molecules behave in their melts as unperturbed coils that are filled with segments of other coils. Self-diffusion of a segment must therefore occur by interchanging positions with a segment of another molecule. Segments of the same chain do not move independently of each other, however; their movements are more difficult if the polymer chains are entangled. Self-diffusion coefficients thus depend differently on molar masses M above and below the critical molar mass for entanglements.

Diffusion coefficients are inverse proportional to the friction coefficients f of the diffusing particles according to the Einstein-Sutherland equation $D = RT/(N_A f)$. Polymer coils can be thought of as consisting of N_{seg} segments with segmental molar masses $M_{seg} = M/N_{seg}$ and friction coefficients f_{seg} that move in the "solvent" of like matrix molecules. The friction coefficient of the total molecule is $f = N_{seg} f_{seg}$. Self-diffusion coefficients $D = (RTM_{seg})/(N_A f_{seg} M)$ should thus be inversely proportional to molar masses M of nonentangled polymers.

Polymer coils become entangled at higher molar masses; such entanglements are temporary but fairly long-lived. The resulting topological restraints make the polymer chain move ("reptate") through the maze of other segments like a reptile through brush. According to the Doi-Edwards theory of *reptation*, chains reptate in tubes of ca. 5 nm diameter, which are formed by other segments (Fig. 5-8). The theory predicts $D = k_B T/(6 N_{seg}^2 f_{seg})$, that is, a dependence of D on the inverse square of the molar mass, $D = f(M^{-2})$, at molar masses above a critical molar mass required for the onset of entanglements. Experimentally, such a functionality is also found below the entanglement molar mass (see Fig. 6-1), probably, because endgroups provide successively more free volume at lower molar masses.

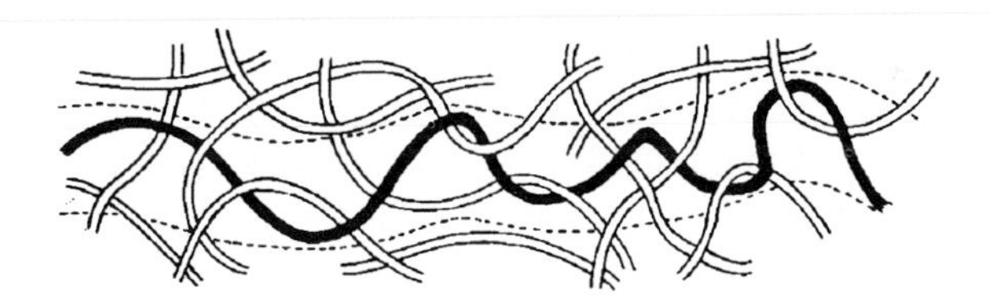

Fig. 5-8 Reptation of a test chain (black) through the segments of a matrix polymer (white). The "walls" of the tube are indicated by dotted lines.

5.4.2. Permeation

The transport of extraneous material through polymers is called permeation. The resulting net flow of mass is caused by differences in chemical potentials, that is, concentration differences (at constant temperature) or thermal gradients (at constant concentration). Permeation may be desirable as in the permeation of dyestuffs into textile fibers during dyeing and the controlled transdermal delivery of pharmaceuticals, or undesirable such as the migration of plasticizers from plastics or the loss of carbon dioxide from plastic bottles for carbonated soft drinks.

Permeation of chemical compounds through amorphous polymers below their glass temperatures or through semicrystalline polymers below their melting temperatures can occur either by flow through pores or by molecular transport. Pores have diameters much greater then the diameters of permeating substances (diameter of spheres, crosssection of coil-forming chains); the interactions of permeants and pore walls are negligible. Molecular transport, on the other hand, depends on such interactions between permeant and matrix (i.e., on the solubility of the former in the latter). Both types of transport can be distinguished by their temperature dependence: permeation coefficients of gases decrease with temperature at pore membranes; they increase with temperature at solubility membranes.

The counteraction of the two types of permeation can be utilized in lamination. Oxygen permeates through pores in aluminum films but by molecular transport through poly(ethylene) films. At 1 bar, permeation rates decrease from $5 \cdot 10^{-5}$ cm^3/s for 0.025 mm-thick aluminum films to $5 \cdot 10^{-13}$ cm^3/s for the same films laminated with 0.025 mm-thick PE films.

The permeation is characterized by a permeation coefficient P, which is the product of the diffusion coefficient D and the solubility coefficient S of the permeant in the matrix. In the steady state, P is given by the amount Q of a gas that permeates in the time t through an area A of a film (membrane, etc.) with a thickness L under a pressure difference Δp:

$$P = DS = \frac{QL}{A \Delta p t} \tag{5-14}$$

The technical literature often uses different "practical units" for the various quantities of Equation (5-14), for example, Q in cm^3, L in mm, A in m^2, Δp in atm and t in 24 h; P is then given in (cm^3 mm)/(24 h m^2 atm). If like units are used, then the unit of P is of course length^2 time^{-1} pressure^{-1} and the unit of S is pressure^{-1}. Permeation coefficients P^* of liquids are usually measured without a pressure differential; their unit of P is cm^2 s^{-1} and their unit of S is 1.

Permeation coefficients of gases in polymers vary widely, for example, oxygen permeates 10 million times faster through poly(dimethylsiloxane) than through poly(acrylonitrile) (Table 5-5). Polymers with very low permeation coefficients of gases are called *barrier polymers*. Gases, in general, have lower permeation coefficients in thermoplastics than in elastomers because segmental movements of the former are frozen-in below the glass transition temperature. Bulky substituents, orientation of polymer segments, crystalline domains, and added fillers all increase the pathway for a gas molecule through a polymer matrix; these tortuosity factors decrease permeation coefficients.

The *permeation of nondissolving liquids* into a polymer is proportional to t^n. The exponent n depends on the ratio of the relaxation time of the polymer-solvent system to the diffusion time of the solvent, i.e., on the Deborah number De (Section 7.4.1.).

Three different regimes are normally considered. In the regime denoted as *Case I*, the mobility of the permeant is much smaller than the relaxation of the polymer seg-

Table 5-5 Permeation coefficients of gases (P) and water vapor (P^*) through polymers at 30°C. Literature values vary widely because polymers are often not identical with respect to crystallinity, orientation, water absorption, etc.
$P = 1 \cdot 10^{-14}$ cm² s⁻¹ Pa⁻¹ corresponds to $P^* = 1 \cdot 10^{-9}$ cm² s⁻¹ at normal pressure ($p = 1 \cdot 10^5$ Pa).

Polymer	10^{14} P/(cm² s⁻¹ Pa⁻¹)		10^9 P^*/(cm² s⁻¹)
	O_2	CO_2	H_2O
Poly(dimethylsiloxane)	25000	85000	40
cis-1,4-Poly(isoprene)	2000	10000	0.3
Butyl rubber	100	500	0.1
Poly(styrene), regular	200	1000	1
, biaxially oriented	0.1	100	0.5
Poly(ethylene terephthalate), regular	4	20	0.2
, biaxially oriented	0.2	1	0.2
Poly(vinylidene chloride)	0.05	0.15	0.02
Cellulose	0.03	0.1	10
Poly(acrylonitrile)	0.002	0.02	0.02
Required for bottles			
Cola	1	0.5	0.14
Beer	0.05	0.5	0.14

ments ($DE < 0.1$). The movement of the permeant causes "instantaneous" conformational changes of the polymer segments. Both permeant and polymer behave as viscous liquids: the system can be described by Fickian diffusion laws: the diffusing amount is proportional to the square root of time (i.e., $n = 1/2$).

In the *Case II* regime, the mobility of the permeant is much higher than the relaxation of the polymer segments ($DE > 10$). The physical structure of the polymer does not change during the permeation; the polymer appears to the permeant as an elastic body. Case II is characterized by a sharp demarcation line between the glassy inner polymer core and the swollen zone advancing with constant speed. The permeating amount is directly proportional to time (i.e., $n = 1$).

Relaxation and permeation become comparable for $0.1 < DE < 10$. This third regime is usually called *anomalous diffusion* or *viscoelastic diffusion* ($1/2 < n < 1$).

Literature

5.1. Overview: General

G.M.Bartenev, Yu.V.Zenlenev, Ed., Relaxation Phenomena in Polymers, Halsted, New York 1974

D.J.Meier, Ed., Molecular Basis of Relaxations and Transitions of Polymers (= Midland Macromolecular Monographs **4**), Gordon and Breach, New York 1978

R.T.Bailey, A.M.North, R.A.Pethrick, Molecular Motions in High Polymers, Clarendon Press, Oxford 1981

5.1. Overview: Experimental Methods

M.Dole, Calorimetric Studies and Transitions in Solid High Polymers, Fortschr. Hochpolym. Forschg. **2** (1960) 221
W.J.Smothers, Y.Chiang, Handbook of Differential Thermal Analysis, Chem.Publ.Co., New York 1966
R.C.MacKenzie, ed., Differential Thermal Analysis, Academic Press, New York, Bd. **1** (1972)
J.K.Gillham, Torsional Braid Analysis - A Semimicro Thermomechanical Approach to Polymer Characterization, Crit.Revs.Macromol.Sci. **1** (1972) 83
A.M.Hassan, Application of Wide-Line NMR to Polymers, Crit.Revs.Macromol.Sci. **1** (1972) 399
W.Wrasidlo, Thermal Analysis of Polymers, Adv.Polym.Sci. **13** (1974) 1
J.Chiu, Dynamic Thermal Analysis of Polymers. An Overview, J.Macromol.Sci. [Chem.] **A 8** (1974) 1
J.-M.Braun, J.E.Guillet, Study of Polymers by Inverse Gas Chromatography, Adv.Polym.Sci. **21** (1976) 107
P.Hedvig, Dielectric Spectroscopy of Polymers, Hilger, Bristol 1977
T.Murayama, Dynamic Mechanical Analysis of Polymeric Materials, Elsevier, Amsterdam 1978
E.A.Turi, ed., Thermal Characterization of Polymeric Materials, Academic Press, New York, 2.Aufl. 1982

5.2. Molecular Motions

R.F.Boyer, S.E.Keinath, eds., Molecular Motion in Polymers by ESR, Harwood Academic Publ., Chur 1980
R.T.Bailey, A.M.North, R.A.Pethrick, Molecular Motion in High Polymers, Clarendon Press, Oxford 1981
J.M.O'Reilly, M.Goldstein, ed., Structure and Mobility in Molecular and Atomic Glasses, Ann.N.Y.Acad.Sci. **371** (1981)

5.2.2. Heat Capacity

B.Wunderlich, H.Baur, Heat Capacities of Linear High Polymers, Adv.Polym.Sci. **7** (1970) 151
Y.Godovsky, Thermophysical Properties of Polymers, Springer, Berlin 1991

5.2.3. Heat Conductivity

D.R.Anderson, Thermal Conductivity of Polymers, Chem.Revs. **66** (1966) 677
W.Knappe, Wärmeleitung in Polymeren, Adv.Polym.Sci. **7** (1971) 477
C.L.Choy, Thermal Conductivity of Polymers, Polymer **18** (1977) 984
D.Hands, The Thermal Transport Properties of Polymers, Rubber Chem.Technol. **50** (1977) 480
Y.Godovsky, Thermophysical Properties of Polymers, Springer, Berlin 1991

5.3.2. Crystallization

L.Mandelkern, Crystallization of Polymers, McGraw-Hill, New York 1964
A.Sharples, Introduction to Polymer Crystallisation, Arnold, London 1966

B.Wunderlich, Macromolecular Physics, Bd. **2**, Crystal Nucleation, Growth, Annealing, Academic Press, New York 1976
R.L.Miller, ed., Flow Induced Crystallization of Polymers(= Midland Macromolecular Monographs **6**), Gordon and Breach, New York 1979
G.S.Ross, L.J.Frolen, Nucleation and Crystallization (of Polymers), Methods Exp.Phys. **16 B** (1980) 339

5.3.3. Melting

B.Wunderlich, Macromolecular Physics, Bd. **3**, Crystal Melting, Academic Press, New York 1980

5.3.4. Liquid Crystal Transitions

B.Wunderlich, S.Grebowicz, Thermotropic Mesophases and Mesophase Transitions of Linear, Flexible Macromolecules, Adv.Polym.Sci. **60/61** (1984) 1

5.3.5. Glass Temperatures

R.F.Boyer, The Relation of Transition Temperatures to Chemical Structure in High Polymers, Rubber Chem.Technol. **36** (1963) 1303
R.N.Haward, The Physics of the Glassy State, Interscience, New York 1973
N.W.Johnston, Sequence Distribution - Glass Transition Effects, J.Macromol.Sci.-Revs. Macromol.Chem. **C 14** (1976) 215
M.Goldstein, R.Simha, ed., The Glass Transition and the Nature of the Glassy State, Ann.N.Y.Acad.Sci. **279** (1976) 1

5.3.6. Other Transitions and Relaxations

A.Hiltner, E.Baer, Relaxation Processes at Cryogenic Temperatures, Crit.Revs.Macromol.Sci. **1** (1972) 215
R.F.Boyer, T_{ll} and Related Liquid State Transitions-Relaxations: A Review, in A.Pethrick, ed., Polymer Yearbook **2** (1985) 233

5.4. Transport

P.G.de Gennes, Scaling Concepts in Polymer Physics, Cornell University Press, Ithaca, NY 1979
R.B.Bird, R.C.Armstrong, O.Hassager, Dynamics of Polymeric Liquids, Bd. 1; R.B.Bird, C.F.Curtiss, R.C.Armstrong, O.Hassager, ditto, Bd. 2, Wiley, New York, 2.Aufl. 1987
M.Doi, S.F.Edwards, The Theory of Polymer Dynamics, Oxford University Press, Oxford 1987
K.F.Freed, Renormalization Group Theory of Macromolecules, Wiley, New York 1987

5.4.1. Self-Diffusion

J.Klein, The Self-Diffusion of Polymers, Contemp.Phys. **20** (1979) 611
M.Tirrell, Polymer Self-Diffusion in Entangled Systems, Rubber Chem.Techn. **57** (1984) 523

5.4.2. Permeation

B.J.Hennessy, J.A.Mead, T.C.Stening, The Permeability of Plastics Films, Plastics Institute, London 1966
J.Crank, G.S.Park, eds., Diffusion in Polymers, Academic Press 1968
H.B.Hopfenberg, Ed., Permeability of Plastic Films and Coatings, Plenum Press, New York 1974
H.Yasuda, Units of Permeability Constants, J.Appl.Polym.Sci. **19** (1975) 2529
T.R.Crompton, Additive Migration from Plastics into Food, Pergamon Press, Oxford 1979
M.B.Huglin, M.B.Zakaria, Comments on Expressing the Permeability of Polymers to Gases, Angew.Makromol.Chem. **117** (1983) 1
J.Comyn, ed., Polymer Permeability, Elsevier Appl.Sci.Publ., London 1985
G.E.Zaikov, A.P.Jordanskii, V.S.Markin, Diffusion of Electrolytes in Polymers, VNU Science Press, Utrecht 1987
N.Toshima, Polymers for Gas Separation, VCH, Weinheim 1992

References

[1] H.W.Starkweather, Jr., J.Macromol.Sci.-Phys. **B 2** (1968) 781, data of Fig. 1
[2] H.-G. Elias, Makromoleküle, vol. I: Grundlagen, Hüthig and Wepf Publ., Basle, 5th ed. 1990; Fig. 22-4
[3] H.-G. Elias, Makromoleküle, vol. I: Grundlagen, Hüthig and Wepf Publ., Basle, 5th ed. 1990; Fig. 22-7

6. Rheological Properties

6.1. Introduction

Materials exhibit two limiting types of behavior against deformation. Typical liquids such as water flow under their own weight and are irreversibly deformed (viscous behavior; Latin: *viscum* = birdlime (from mistletoe)). Typical solids such as iron resist deformation; they return from small deformations to their former states after removal of loads (elastic behavior; Greek: *elastos, elatos* = beaten). Polymers commonly combine both types of behavior: they are viscoelastic materials. Their melts show viscous behavior at small deformations and elastic properties at large ones. Polymer solids respond elastically to small deformations but begin to flow at larger ones.

Melts and concentrated solutions of polymers have extremely high viscosities. Air has, for example, a viscosity of 10^{-5} Pa s; water, 10^{-3} Pa s; and glycerol, 1 Pa s, whereas polymer melts exhibit viscosities of ca. 10^{2}-10^{8} Pa s. Polymer viscosities may also depend on deformation rates and the duration of the experiment. Three types of viscosities are usually distinguished:

shear viscosity: rate of shear flow = f(shear stress);
extensional viscosity: rate of extensional (elongational) flow = f(tensile stress);
bulk viscosity: rate of volume deformation = f(applied hydrostatic pressure).

Shear viscosities are the most often studied rheological properties of polymers; they are very important for the processing of plastics by extrusion or injection molding. Much less is known about *extensional viscosities* (important for blow forming and fiber spinning), and practically nothing about *bulk viscosities*.

6.2. Zero-Shear Viscosities

6.2.1. Fundamentals

Nine different shear stresses σ may be assigned to a three-dimensional body: three parallel to the three spatial directions (σ_{11}, σ_{22}, σ_{33}) and six perpendicular to these (σ_{12}, σ_{13}, etc.). A body is by definition sheared in the 2-1 direction. The ratio of shearing force K to contact area A is called the shear(ing) stress $\sigma_{21} = \sigma = K/A$; it produces a shear strain γ. The ratio $G = \sigma_{21}/\gamma$ is the shear modulus. Between layers of distance y moving parallel to each other with different rates v, a velocity gradient (shear rate) $\dot{\gamma} = dv/dy$ thus exists. The ratio of shear stress to shear rate is the (dynamic) viscosity (Newton's Law):

(6-1) $\eta = \sigma_{21} / \dot{\gamma}$

Liquids following Newton's law are called *Newtonian liquids*. The inverse of viscosity is the fluidity $1/\eta$. Division of the viscosity by the density ρ leads to the kinematic viscosity $\nu = \eta/\rho$.

Viscosities of Newtonian liquids are independent of shear rates, i.e., $\eta = \eta_o \neq f(\dot{\gamma})$. Water is such a liquid at a wide range of shear rates and so are polymer melts at very low shear rates. Newtonian viscosities are also called zero-shear viscosities, viscosities at rest, or stationary viscosities. The shear modulus G_o of Newtonian liquids is independent of the extent of deformation. η_o and G_o are true material constants, whereas η and G depend on shear rates and sometimes also on shearing time.

Viscosities η of non-Newtonian liquids depend on shear rates; they are thus called shear-dependent viscosities or apparent viscosities. Examples are polymer melts at higher shear rates.

Shear stresses and shear gradients (and thus viscosities) can be measured with a variety of instruments that usually belong to one of three groups: capillary, rotatory, and cone-plate viscometers. A number of industrially used instruments provides viscometric indicators (but neither shear stress, shear gradient, or viscosity); this group includes Höppler viscometers, Cochius tubes, Ford beakers, and instruments that measure melt flow indices or Mooney values (Mooney viscosities).

Thermoplastics are usually characterized by their *melt flow*. The melt flow is determined as the mass of polymer extruded in 10 min by a standard load F from a standard plastometer at a temperature T. It is a measure of the (usually non-Newtonian) fluidity. The ratio of two melt flows is known as *melt flow index MFI*. Division of *MFI* by the density gives the *melt volume index MVI*.

The higher the melt flow indices and melt flow rates, the lower are the molar masses of polymers of the same chemical structure. An increase of melt flow indices thus correlates often with a decrease of viscosity coefficients (Table 6-1). The two quantities cannot be converted into each other, however, because the former depends on the mass-average molar mass and the shear rate and the latter on the viscosity-average molar mass and the thermodynamic interaction of the polymer with the solvent.

Table 6-1 Viscosity coefficients and melt volume indices *MVI 1* (2.16 kg) and *MVI 2* (5 kg) of poly(methyl methacrylate)s PMMA of Degussa, high density poly(ethylene)s PE-HD of Solvay.

Polymer		Visc. coeff. in mL/g	Melt volume indices		T/°C
			MVI 1	*MVI 2*	
PE-HD Eltex	A 1100	110	14	46	190
	A 2080	130	11	32	190
	A 1050	140	6.6	24	190
	A 3040	150	5.1	14	190
PMMA Degalan	6	53	10	-	230
	7	53	5	-	230
	8	60	1.6	-	230

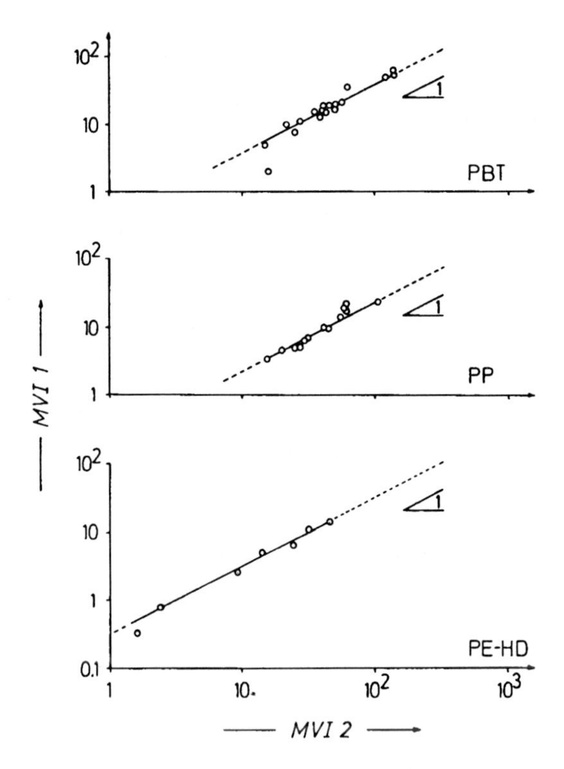

Fig. 6-1 Logarithms of *MVI 1* as function of logarithms of *MVI 2* for three families of plastics.
PBT: unfilled, glass fiber-filled, glass sphere-filled or white pigment-filled poly(butylene terephthalate)s (Pocan® grades of Bayer AG at 260°C);
PP: mineral-filled, glass fiber-filled or/and impact-modified poly(propylene)s (Moplen® grades of Himont at 230°C);
PE-HD: unfilled high density poly(ethylene)s (Eltex® grades of Solvay S.A. at 190°C).

Melt volume indices are measured at loads of 2.16 kg (*MVI 1*) and 5 kg (*MVI 2*) (ISO 1133; DIN 53735). The two values are usually directly proportional to each other for each family of plastics, except for very low *MVI* values (Fig. 6-1). The constant $K = MVI\ 1/MVI\ 2$ depends only on the chemical structure of the resin; it does not matter whether the plastics are unfilled, filled, or rubber-modified.

The *Mooney viscosity* does not indicate a viscosity but an elasticity. It is used not only for elastomers but also for polymer melts. A polymer is deformed in a standardized cone-plate viscometer at constant rotational speed and constant temperature T; after t minutes, the force to recover is read.

6.2.2. Melts

Newtonian melt viscosities depend on mass-average molar masses according to:

(6-1) $\eta_o = K\overline{M}_w^{\beta}$

Two different viscosity regimes can be distinguished at lower and higher molar masses (Fig. 6-2). The transition from the low molar-mass regime to the high molar-mass regime is thought to be caused by the onset of molecular entanglements that causes the molecules to behave as physically cross-linked networks at $\overline{M}_w > M_c$.

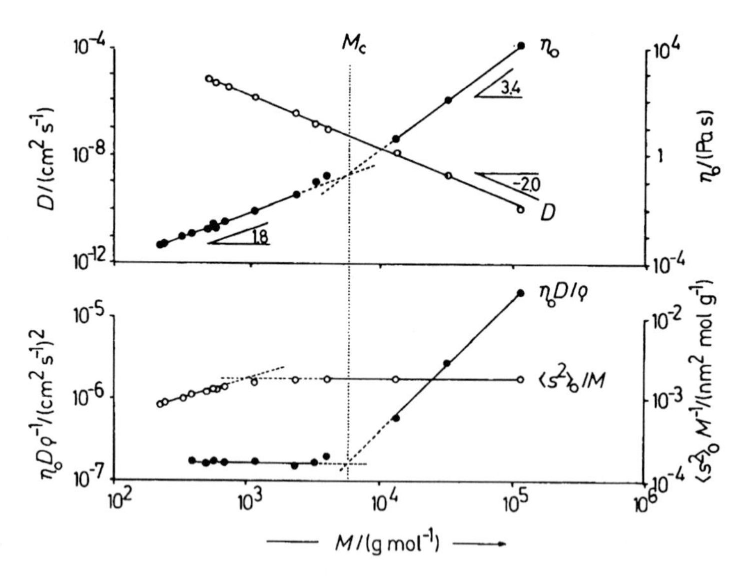

Fig. 6-2 Molar mass dependence of experimental self-diffusion coefficients D, zero-shear viscosities η, and ratios $\eta_o D/\rho$ (data of [1]) and calculated normalized mean-square radii of gyration $\langle s^2\rangle_o/M$ of alkanes and poly(ethylene)s at 175°C. M_c = critical molar mass between entanglements; ρ = density of melts.

Low molar mass. Chains are not entangled at low molar masses. They adopt a Gaussian distribution of segments if the chain has a sufficiently high number of chain atoms (Section 2.4.4).

On stress, chains deform. The relaxation of such chains can be modeled by the Rouse theory as a series of Hookean springs (see Chapter 7) of zero unstretched lengths. The springs are flexibly joined together by beads; they are immersed in a viscous medium. The theory predicts that the kinematic zero-shear viscosity η_o/ρ depends on the Avogadro constant N_A, the normalized mean-square radius of gyration, $\langle s^2\rangle_o/M$, and the number N_{seg} of segments with friction coefficients f_{seg}:

$$\text{(6-2)} \qquad \frac{\eta_o}{\rho} = \frac{N_A}{6}\left(\frac{\langle s^2\rangle_o}{M}\right)N_{seg}f_{seg}$$

The segment term $N_{seg}f_{seg}$ of Equation (6-2) can be replaced by the Rouse expression for the self-diffusion coefficient, $D = k_BT/(N_{seg}f_{seg})$ (see Section 5.3.1; $k_BN_A = R$)). The resulting Equation (6-3) shows that the term $\eta_o D/\rho$ is proportional to $\langle s^2\rangle_o/M$, which depends only on the chain parameters M_b, b, τ, and σ (see Equations (2-19) and (2-22)) and not on the molar mass M (Fig. 6-2):

(6-3) $$\frac{\eta_o D}{\rho} = \frac{RT}{6}\left[\frac{\langle s^2\rangle_o}{M}\right] = \frac{RT}{6}\left[\frac{b^2}{6M_b}\left(\frac{1-\cos\tau}{1+\cos\tau}\right)\sigma^2\right] = const.$$

The Rouse prediction $\eta_o D/\rho \neq f(M)$ is well fulfilled for alkanes and poly(ethylene)s with molar masses below the critical molar mass of $M_c \approx 6030$ g/mol (Fig. 6-2). However, η_o/ρ is not directly proportional to M for low molar masses (as predicted by Equation (6-2)) but varies with ca. $M^{1.8}$ for these polymers; this may be caused by endgroup effects. Other polymers, such as linear and cyclic poly(dimethylsilicone)s, do show the predicted direct proportionality of η and M, i.e., $\eta_o = KM$.

High molar mass. Newtonian viscosities increase more strongly with molar masses at $M_c > M$. The exponent of Equation (6-1) is often found as $\beta \approx 3.4$ for many polymers. Reptation theory explains this high exponent β as follows:

The proportion of entanglements can be assumed to be constant at low shear rates. The elasticity of the network formed by entanglements is described by its shear modulus G_o (unit of stress). Since shear viscosities have the unit stress × time, $\eta_o = G_o t$. The reptation model identifies the time t as the time required for a chain to leave the tube. This time is proportional to the third power of the number of segments per molecule (i.e., the viscosity should be proportional to the third power of the molar mass). The deviation between the theoretically predicted $\beta = 3$ and the experimental value of $\beta = 3.4$ is thought to be due to "breathing" of the tube. Breathing pushes nonentangled chain loops back into the surrounding matrix; chain ends cause additional relaxations and the tube length decreases.

The dependence of zero-shear viscosities on the 3.4th power of the mass-average molar mass means that high molar-mass tails of molar mass distributions of molecularly nonuniform polymers have an overproportional effect on viscosities.

6.2.3. Concentrated Solutions

Newtonian viscosities of concentrated solutions increase with both solute concentrations c and molar masses M, or, since $[\eta] = KM^a$ (Eq. (2-9)), with the intrinsic viscosity $[\eta]$ as well. The concentration c measures the mass of polymer per unit volume of solution; the intrinsic viscosity, the volume of polymer molecules per unit mass of polymer. The product $c[\eta]$ is thus a measure of the volume fraction of polymer molecules that would be occupied by isolated polymer coils.

Isolated polymer coils exist at very low polymer concentrations. The product $c[\eta]$ is proportional to the viscosity increment ("specific viscosity") η_i at low $c[\eta]$ values (Section 2.2.2; Fig. 6-3), where $\eta_i = (\eta/\eta_1) - 1$, η is the viscosity of the polymer solution at rest, and η_1 the viscosity of the solvent.

At higher concentrations, coils start to overlap and the total occupied volume is smaller than the one demanded by isolated coils (i.e., $c[\eta] > 1$). At higher $c[\eta]$ values, $\eta_1 \sim (c[\eta])^q$ or, since $[\eta] \sim M^a$, $\eta_1 \sim c^q M^{aq}$. Because at very high concentrations, $\eta_i \approx \eta/\eta_1$, and η approaches the Newtonian melt viscosity $\eta_o \sim M^{3.4}$ for molar masses higher than the critical molar mass M_c, aq = 3.4 is obtained. In theta solvents and melts, coils are unperturbed (α = 1/2) and thus q = 6.8 (fig. 6-3). In good solvents, $a \rightarrow 0.764$ and q = 4.55. These relationships are independent of the chemical nature of polymers and solvents.

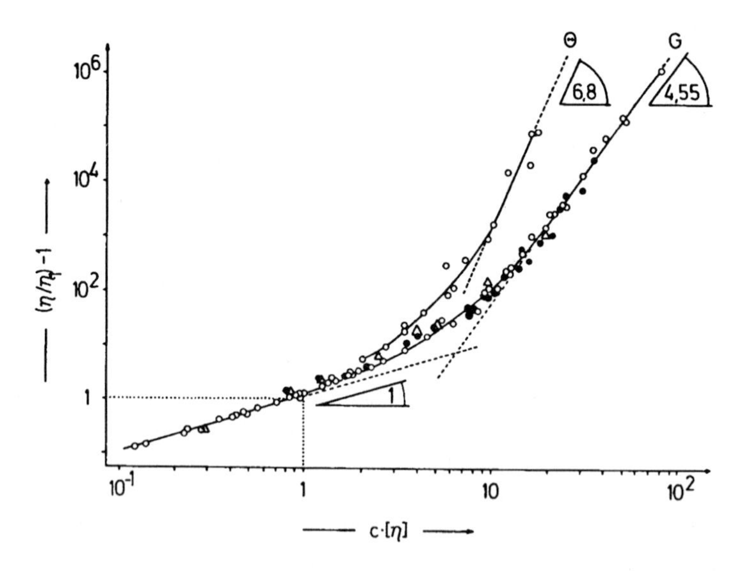

Fig. 6-3 Relative viscosity increments ("specific viscosities") (η/η_1) - 1 as function of c[η] for (o) poly(styrene)s in *trans*-decalin (theta solvent) or toluene (good solvent G) at 25°C; (●) *cis*-poly-(isoprene)s in toluene at 34°C (good solvent); and (▵) hyaluronates in water at 25°C (good solvent). Reprinted with permission by Hüthig and Wepf Publ., Basle, Switzerland [2].

6.3. Non-Newtonian Shear Viscosities

6.3.1. Overview

The shear stress σ_{21} is directly proportional to the shear gradient $\dot{\gamma}$ for Newtonian liquids (Eq.(6-1)); this relationship and thus the viscosity η is, moreover, independent of the duration of the experiment. Non-Newtonian viscosities (apparent viscosities), on the other hand, vary with shear rate and sometimes even with time. Polymers may experience high shear on processing, for example, up to 20 000 s^{-1} upon brush coating and up to 70 000 s^{-1} in the gate area of injection molding of polycarbonates for compact discs.

Various dependencies of apparent viscosities on shear rate are found for time-independent non-Newtonian liquids (Fig. 6-4). *Plastic bodies* (Bingham bodies) exhibit a yield value σ_y, i.e., shear stresses have finite values σ_o at $\dot{\gamma} \rightarrow 0$ and the bodies start to flow only above a certain critical value of the shear rate. Above σ_o, such plastic bodies

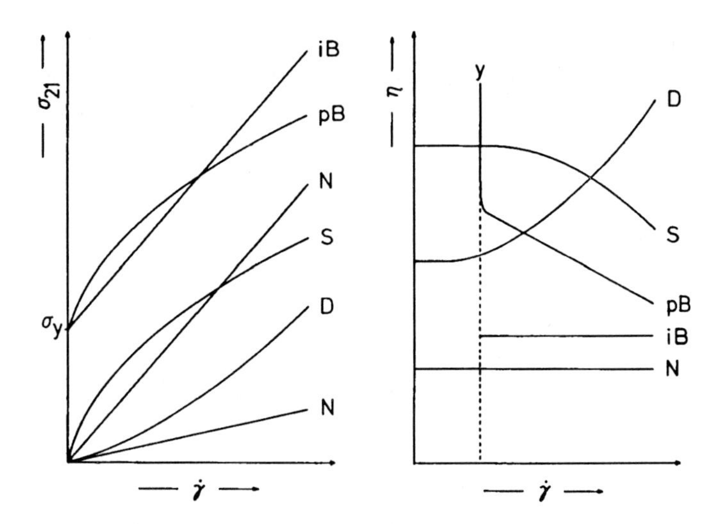

Fig. 6-4 Shear rate dependence of shear stress σ_{21} (left) and shear viscosity η (right) of Newtonian (N), dilatant (D), and pseudoplastic (P) liquids and for ideal (iB) and pseudoplastic (pB) Bingham bodies (y = yield). Reprinted with permission by Hüthig and Wepf Publ., Basle, Switzerland [3].

may behave in a Newtonian (ideal Bingham bodies) or non-Newtonian (pseudoplastic Bingham bodies) manner. This behavior seems to due to the disappearance of aggregates. An example of a Bingham body is tomato ketchup. Melts of plastics rarely behave as plastic bodies.

A decrease of apparent viscosities with shear rates is most common for polymer melts (*shear-thinning*). Because such behavior resembles that of pseudoplastic Bingham bodies (see Fig. 6-4), it is also called *pseudoplasticity* in the English literature ("structural viscosity" in German), although such pseudoplastic materials do not possess a yield value.

Pseudoplastic behavior eases polymer processing from melts and reduces the required energy: lower pressures can be applied in injection molding. It is usually caused by an orientation of chain segments that is most pronounced in the processing of liquid crystalline polymers with rigid mesogens and in the ultradrawing of flexible polymers.

In *dilatant liquids*, the shear stress increase is more than linearly proportional to shear rate; viscosities increase with shear rate (*shear-thickening*) (Latin: *dilatare* = to enlarge, extend). Shear-thickening is often found for associating charged entities such as ionomers, charged micelles, and ionically stabilized dispersions.

Dilatant and pseudoplastic liquids are characterized by an "instantaneous" adoption of shear rates on application of shear stresses (i.e., by time-independent apparent viscosities). *Thixotropic materials* exhibit a decrease of apparent viscosities with time at constant shear rate (Greek: *thixis* = touch, movement; *tropos* = to change). Examples are suspensions of bentonite and other platelet-like silicates. *Rheopectic* or *antithixotropic* materials show an increase of apparent viscosities with time at constant shear rate (Greek: *rheos* = flow; *pectous* = solidified, curdled).

The flow behavior may be furthermore complicated by wall effects. Certain dispersions and gels exude liquid on application of shear stresses. The liquid acts as an external lubricant and a plug-like flow results (example: tooth paste). An additional complication may result from the onset of turbulence, which usually occurs at much lower Reynolds numbers in non-Newtonian than in Newtonian liquids.

6.3.2. Flow Curves

A plot of log $\dot{\gamma} = f(\sigma_{21})$ or vice versa is usually called a *flow curve*. A generalized flow curve may comprise a first Newtonian region, followed by pseudoplasticity and a second Newtonian regime. Since experiments are difficult to conduct at high shear rates, many rheologists doubt the existence of true second Newtonian regions. Dilatancy may or may not be present. Finally, turbulence sets in and melt fracture occurs.

The first Newtonian region is also called the lower Newtonian region because it refers to low shear rates; the second Newtonian region, at high shear rates, is thus the upper Newtonian region. Newtonian viscosities are high at the lower Newtonian region and small at the upper Newtonian.

Several empirical functions have been suggested for the description of flow curves. Examples include

(6-4) $\dot{\gamma} = a(\sigma_{21})^m$ Ostwald-de Waele

(6-5) $\dot{\gamma} = b \cdot \sinh(\sigma_{21}/d)$ Prandtl-Eyring

(6-6) $\dot{\gamma} = f\sigma_{21} + g(\sigma_{21})^3$ Rabinowitsch-Weissenberg

where a, b, d, f, g, and m are empirical constants. The exponent m is known as the "flow exponent" or "pseudoplasticity index"; it takes values of 1 (Newtonian liquid) or < 1 (pseudoplasticity). Equation (6-4) is sometimes written as "power law" $\tau = k(\dot{\gamma})^n$ for the dependence of shear stress τ on shear rate; k is called the consistency index and n the power law index. All equations cover only limited ranges of shear rates.

The variation of viscosities with the whole range of shear rates can be described by many equations; CAMPUS® uses the Carreau equation:

(6-7) $$\eta = \eta_\infty + (\eta_o - \eta_\infty)[1 + (\lambda\dot{\gamma})^2]^{(n-1)/2}$$

where η_o is the zero-shear viscosity (viscosity at rest, first Newtonian viscosity), η_∞ is the second Newtonian viscosity at high shear rates, and λ and n are adjustable constants. The second Newtonian viscosity is usually not accessible by experiment; it is thus treated as an adjustable constant.

The model predicts Newtonian viscosities for both low and high shear rates. For small shear rates ($\dot{\gamma} \rightarrow 0$), first Newtonian viscosities (viscosities at rest) are obtained

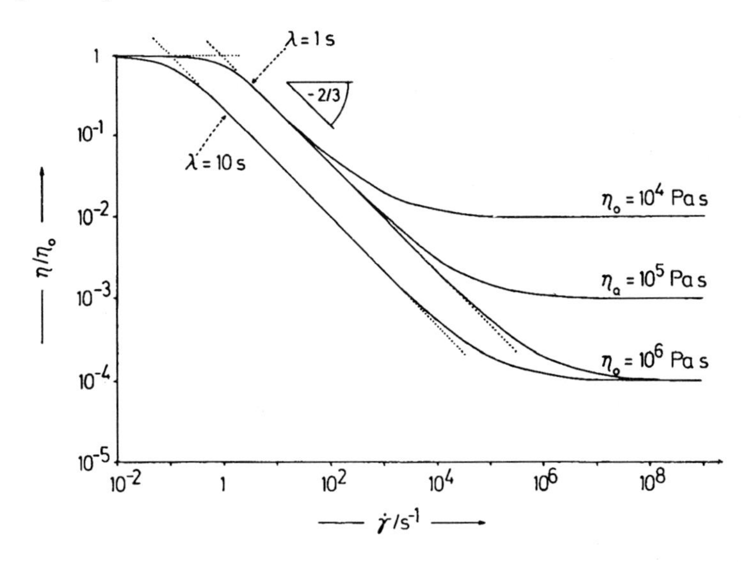

Fig. 6-5 Dependence of normalized shear viscosity η/η_o on the shear rate according to the Carreau-equation, Equation (6-7). Model parameters: η_o = 10^4 Pa s, 10^5 Pa s, or 10^6 Pa s; η_∞ = 10^2 Pa s; λ = 1 s or 10 s; n = 1/3.

($\eta \rightarrow \eta_o$). For exponential factors n < 1, second Newtonian viscosities of the upper Newtonian region should be observed for very high shear rates ($\eta \rightarrow \eta_\infty$ for $\dot{\gamma} \rightarrow \infty$).

A fairly linear dependence of lg (η/η_o) on lg $\dot{\gamma}$ is predicted for intermediate values of $\dot{\gamma}$ (Fig. 6-5); the slope is n-1. This region can thus be described by the power law. The greater the difference between η_o and η_∞, the larger is the range of $\dot{\gamma}$ over which the linear section extends. The power law line intersects with the line for η/η_o = 1 at $\dot{\gamma}$ = $1/\lambda$; the constant λ is thus the inverse of a critical shear rate for the onset of non-Newtonian viscosity.

Non-Newtonian viscosities are no longer proportional to mass-average molar masses. The broader the molar mass distribution, the more the number-average molar mass seems to be the correct corresponding quantity: Non-Newtonian viscosities are caused by entanglements; the probability of formation of entanglements depends on the number of chains, not on their masses.

6.3.3. Shear Rate Dependence

Polymers with molar masses M below the critical molar masses M_c for the onset of the effect of entanglements on melt viscosities show extended first Newtonian ranges. For polymers with molar masses greater than the entanglement molar mass ($M > M_c$), non-Newtonian behavior appears at lower shear rates as the molar masses increase (melt volume indices decrease) (Fig. 6-6). At high shear rates, a constant exponent n -1 is approached for $\eta \sim \dot{\gamma}^{n-1}$ as indicated by Equation (6-7). This exponent seems to be independent of the melt volume index (i.e., molar mass) of the polymer.

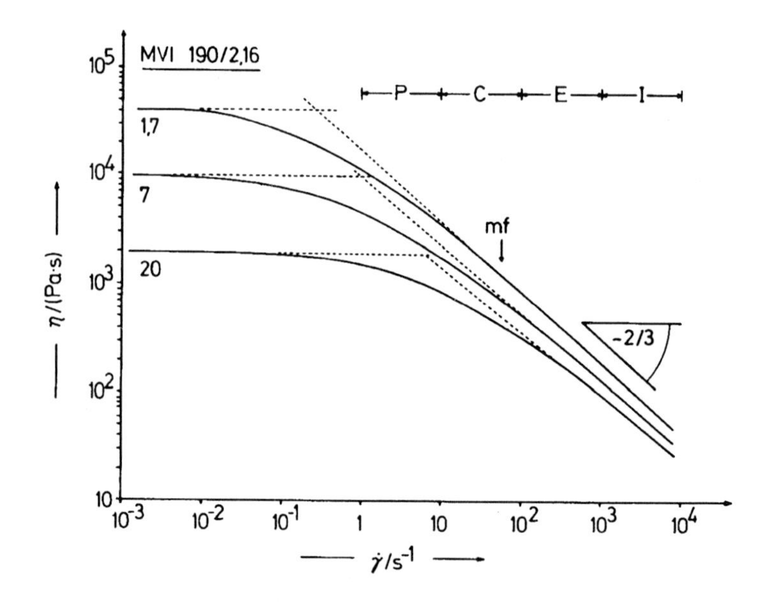

Fig. 6-6 Shear rate dependence of melt viscosities of poly(ethylene)s with different melt volume indices MVI-1 (at a load of 2.16 kg) at a melt temperature of 150°C.

Capital letters indicate the approximate ranges of various processing regimes; these ranges overlap: P = compression molding, C = calendering, E = extrusion, I = injection molding. Shear rates for processes refer to those at orifices and are much lower in the mold (tool, die). mf = Onset of melt fracture for the highest molar mass poly(ethylene) ($MVI = 1.7$).

Reprinted with permission by BASF, Ludwigshafen/Rh., Germany [4].

The lines for $\eta/\eta_o = 1$ and $\lg(\eta/\eta_o) = (n - 1) \lg \dot{\gamma}$ intersect at higher values of $\dot{\gamma} = \dot{\gamma}_c$, the higher the melt volume index and the lower the viscosity η_o. The product $\eta_o \dot{\gamma}_c$ is independent of the molar mass of the poly(ethylene)s (Fig. 6-7). It thus constitutes a critical stress of $\eta_o \dot{\gamma}_c \approx 10^4$ Pa for the onset of non-Newtonian viscosity. The transition Newtonian/non-Newtonian broadens with increasing width of the molar mass distribution of the polymer. The critical stress seems to be a universal constant for all polymers: it adopts the same value of $\eta_o \dot{\gamma}_c \approx 10^4$ Pa for poly(ethylene)s of different molar masses, glass fiber-filled poly(styrene), and plasticized poly(vinyl butyral) (Fig. 6-7).

The scaling exponent $n - 1$ varies between -2/3 for poly(ethylene) melts and -3/4 for plasticized poly(vinyl butyral) (Fig. 6-7). It must reflect the fractal dimensions d_f of the melts. Because sheared non-Newtonian liquids represent steady-state physical networks caused by entangled chains in a "solvent" of unentangled chains, the equation $n = d_f/(d_f + 2)$ derived for uniform gels might apply. It delivers a fractal dimension $d_f = 1$ for $n - 1 = -2/3$ (i.e., $n = 1/3$).

The decrease of apparent viscosities with shear rates is very important for plastics processing. Viscosities describe a frictional behavior: the higher the viscosity, the higher is the internal friction of the melt, and the greater is the proportion of energy that is converted into heat. Less added energy is needed to maintain the processing temperature; strong non-Newtonian behavior thus saves energy. Polymer melts are

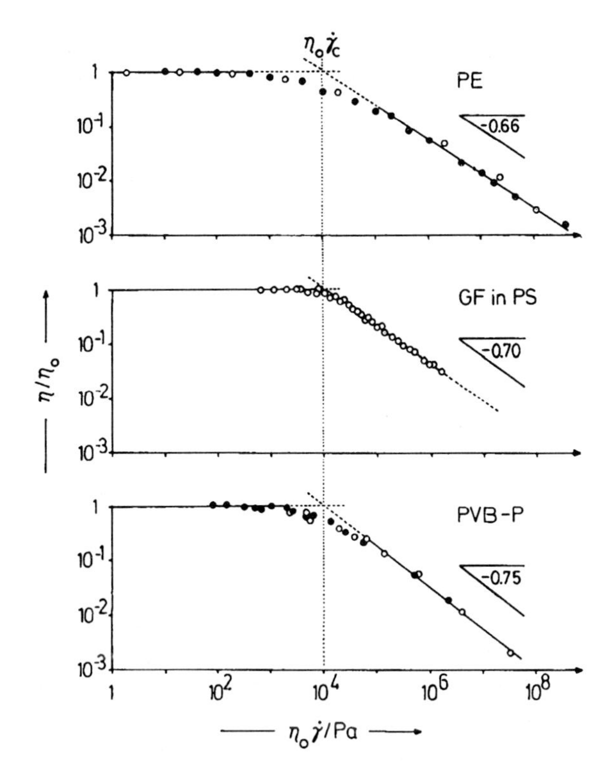

Fig. 6-7 Logarithms of normalized shear viscosities as function of normalized shear stress (Vinogradov-Malkin plot) for PE = melts of poly(ethylene)s with different molar masses (O *MVI-1* = 1.7; ⊙ *MVI-1* = 7; ● *MVI-1* = 20) at 150°C; GF in PS = glass fiber-filled poly(styrene); PVB-P = plasticized poly(vinyl butyral) at 125°C.

therefore processed at the highest possible shear rates. The upper range is given by the processing method (calendering, extrusion, etc.) and the polymer properties (thermal degradation, melt fracture, etc.).

Surface roughnesses of barrel walls, diameter changes, etc., create additional rate components in extrusion that are dampened by the viscosity of the melt. These disturbances become stronger with increasing flow rates and can no longer be dampened at high shear rates. Finally, turbulence sets in. In entangled polymer melts, additional elastic vibrations occur due to the presence of physical cross-links. The resulting elastic turbulences lead to rough surfaces of the extrudate, which are subsequently frozen-in upon exit from the orifice. The polymer surface appears "fractured"; "melt fracture" thus does not refer to a breakage of the extrudate strand.

6.3.4. Temperature Dependence

Newtonian and non-Newtonian viscosities of common polymers decrease strongly with increasing temperature. The temperature dependence of first Newtonian and near-Newtonian viscosities can be fairly well described by a WLF-type of equation (Eq.(5-11)) if the shift factor lg a_T is replaced by the corresponding viscosity change (lg η - lg η_o):

$$\text{(6-8)} \qquad \lg \eta = \lg \eta_o - \frac{K[T - T_o]}{K' + [T - T_o]}$$

For viscosities, Vicat temperatures T_{VT} can be used as "iso-viscous" base temperatures; reference temperatures $T_o = T_{VT} + 50$ K are chosen for the midway of the usual validity range T + 100 K (Eq.(5-11)). This procedure generates a master curve for the change of viscosity with temperature.

CAMPUS® uses an Arrhenius-type equation ("power approximation") to decribe the temperature and shear rate dependence of non-Newtonian viscosities:

$$\text{(6-9)} \qquad \eta = A \dot{\gamma}^B \exp(CT)$$

where A, B, and C are adjustable power approximation constants. Examples are A = 838 151, B = - 0.65057, and C = - 0.01666 if η is measured in Pa s, $\dot{\gamma}$ in s^{-1}, and T in K (Lustran SAN 32 of Monsanto).

CAMPUS® also expresses the temperature and shear rate dependences of viscosities by a combined Carreau-WLF equation with five adjustable Carreau-WLF approximation constants K_1, K_2, K_3, K_4, and K_5:

$$\text{(6-10)} \qquad \eta = \frac{K_1 a_T}{(1 + K_2\, \dot{\gamma}\, a_T)^{K_3}}; \quad \lg a_T = \frac{8.86(K_4 - K_5)}{101.6 + K_4 - K_5} - \frac{8.86(T - K_5)}{101.6 + T - K_5)}$$

6.4. Extensional Viscosities

6.4.1. Melts

Polymer melts and solutions can be considerably elongated without being broken. This extensibility allows fiber spinning from melts and solutions, blow molding of hollow bodies, vacuum forming of parts, uniaxial and biaxial stretching of films, etc. Extensional viscosities are also important, yet not fully understood, components of flow in other processing methods if the geometry of the flow changes suddenly; examples are the flow through sudden contractions or out of an orifice.

Extensional and shear viscosities depend very differently on deformation rates. At low rates, extensional viscosities are independent of extension rates (Fig. 6-8), similar to the first Newtonian region of shear viscosities. The uniaxial extensional viscosity at rest (historically: "coefficient of viscous traction", later: "Trouton viscosity") is three

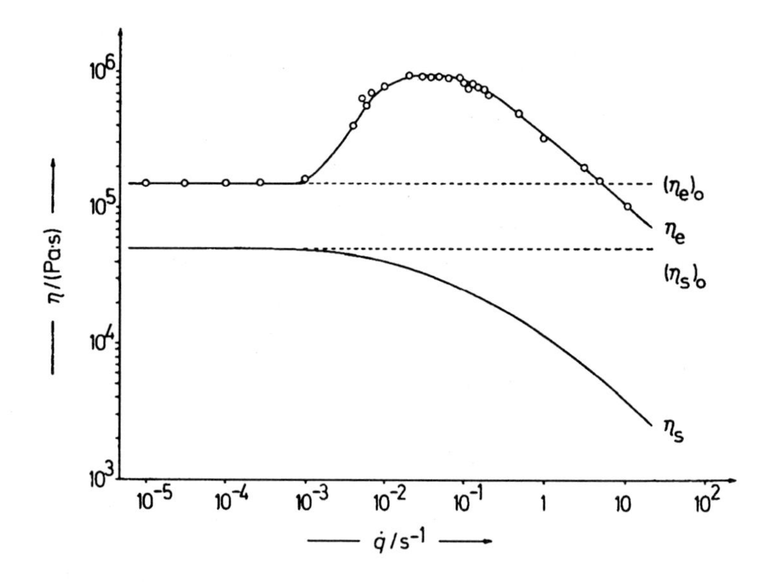

Fig. 6-8 Shear viscosity η_s as function of the shear rate $\dot{q} = \dot{\gamma}$ and uniaxial extensional viscosity η_e as function of the uniaxial extensional rate $\dot{q} = \dot{\varepsilon}$ of a branched poly(ethylene) at 150°C. Reprinted with permission by Steinkopff-Verlag, Darmstadt [6].

times the Newton viscosity ($\eta_{e,o} = 3\ \eta_{s,o} \equiv 3\ \eta_o$); the planar extensional viscosity at rest is 4 times and the biaxial extensional viscosity at rest is six times the shear viscosity (see below for definitions of the various extensional viscosities).

It is not possible to predict behavior in elongation from behavior in shear for non-Newtonian liquids because shear and extensional viscosities may respond very differently to increasing deformation rates. For example, a branched poly(ethylene) exhibited shear thinning with increasing shear rates but *tension-thickening* with increasing elongation rates (Fig. 6-8). The extensional viscosity then passes through a maximum and decreases with further increase of extension rates (*tension-thinning).* Melts and concentrated solutions of linear polymers, on the other hand, showed only tension-thinning. The maximum of the extensional viscosity increases with broader molar mass distributions and increased long chain branching, which indicates a strong influence of entanglements on extensional viscosities.

6.4.2. Types of Extensional Viscosities

Extensional viscosities are very diffult to measure. Elastomers and melts of entangled polymer coils can be stretched between rotating rollers. Extensional viscosities of solutions can be determined if two liquid jets streaming towards each other are redirected by rollers or syphoned off.

The type of deformation must always be indicated for extensional viscosities, contrary to shear viscosities. Three types of extensional viscosities can be distinguished:

– *uniaxial extensional viscosities* govern the extensions of rods, fibers, etc., in draw direction (x) with a simultaneous decrease of dimensions in the two directions (y, z) perpendicular to the direction of elongation;
– *biaxial extensional viscosities* control the simultaneous stretching of a thin sheet in two orthogonal directions (x, y), with a corresponding decrease in the thickness of the sheet (z);
– *planar extensional viscosities* apply if a thin flat sheet is stretched in one direction (x) with a corresponding contraction in its thickness (z), but with no change of its width (y).

The three principal deformation rates can be defined in such a way that $\dot{\varepsilon}_{11} \geq \dot{\varepsilon}_{22} \geq \dot{\varepsilon}_{33}$. The ratio $m = \dot{\varepsilon}_{22} / \dot{\varepsilon}_{11}$ characterizes the special type of elongational flow: theory gives $m = - 1/2$ for uniaxial elongation, $m = 1$ for equal-biaxial, and $m = 0$ for planar (pure shear). Uniaxial elongational viscosities are important for fiber spinning; they are the only ones which are used to characterize fluids. Biaxial elongational viscosities play an important role in blow and vacuum forming; very little is known about them.

The uniaxial extensional viscosity η_e (elongational viscosity) is given by the ratio of tensile stress σ_{11} in draw direction 11 to the elongational rate $\dot{\varepsilon}$:

$$(6\text{-}11) \quad \eta_e = \sigma_{11}/\dot{\varepsilon}$$

Rates of elongation must be calculated from the true strain $\varepsilon' = \ln(L/L_o)$ (Hencky strain) and not from the nominal strain $\varepsilon = (L - L_o)/L_o$ (Cauchy strain, engineering strain). The rate of elongation is thus $\dot{\varepsilon} = \mathrm{d}\varepsilon/\mathrm{d}t = \mathrm{d}\ln L/\mathrm{d}t = L^{-1}(\mathrm{d}L/\mathrm{d}t)$.

6.4.3. Melt Strength

Extensional viscosities may change with time (Fig. 6-9). For example, the uniaxial extensional viscosity of the melt of a standard poly(propylene) increases with time at constant extensional rate until it suddenly decreases whereas the extensional viscosity of a special high melt strength grade increases first slowly and then dramatically with time (Fig. 6-9). This behavior reflects the time dependence of melt strengths F per cross-sectional area A_o; the melt strength can be calculated from extensional viscosities η_e, strain rates $\dot{\varepsilon}$, and times t:

$$(6\text{-}12) \quad F = A_o \frac{\eta_e \dot{\varepsilon}}{\exp(\dot{\varepsilon} t)}$$

Several processing methods such as melt spinning and thermoforming require sufficiently high melt strengths over long times and for the required extensions. The melt should not break or convert into droplets. The melt strengths of the standard poly(propylene) of Fig. 6-9 increase slightly with time at constant strain rate before

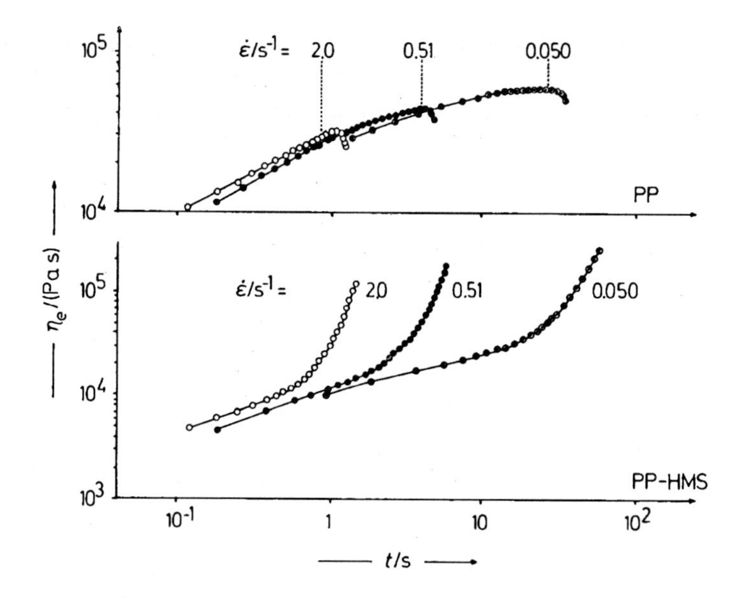

Fig. 6-9 Time dependence of uniaxial extensional viscosities η_e of a standard poly(propylene) PP and a high melt strength poly(propylene) PP-HMS at 180°C and constant elongational rates $\dot{\varepsilon}$. Reprinted with permission by Hanser-Verlag, Munich [7].

they suddenly drop (Fig. 6-10), whereas melt strengths of a high melt strength poly-(propylene) are either constant with time at low strain rates or increase with time at higher strain rates.

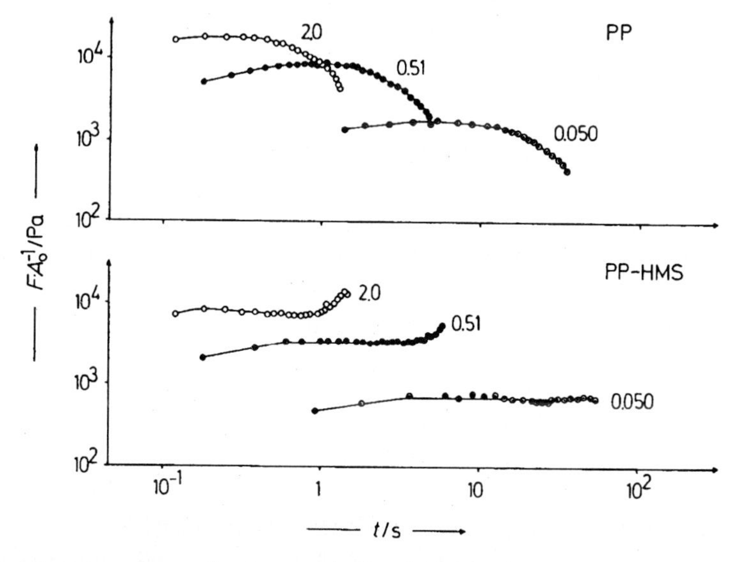

Fig. 6-10 Time dependence of melt strength F per initial cross-sectional area A_0 at constant strain rates of 2.0, 0.51, and 0.050 s^{-1}, respectively, for a standard poly(propylene) PP and a high melt strength poly(propylene) PP-HMS at 150°C. Data points calculated from Fig. 6-9.

Fillers act as physical cross-links and increase extensional viscosities. An example is the addition of mica to a melt of high density poly(ethylene) (Fig. 6-11). The uniaxial extensional viscosities of the unfilled poly(ethylene) go through maxima with increasing extension at constant deformation rates whereas the filled polymer exhibts extended plateaus at much higher extensional viscosities.

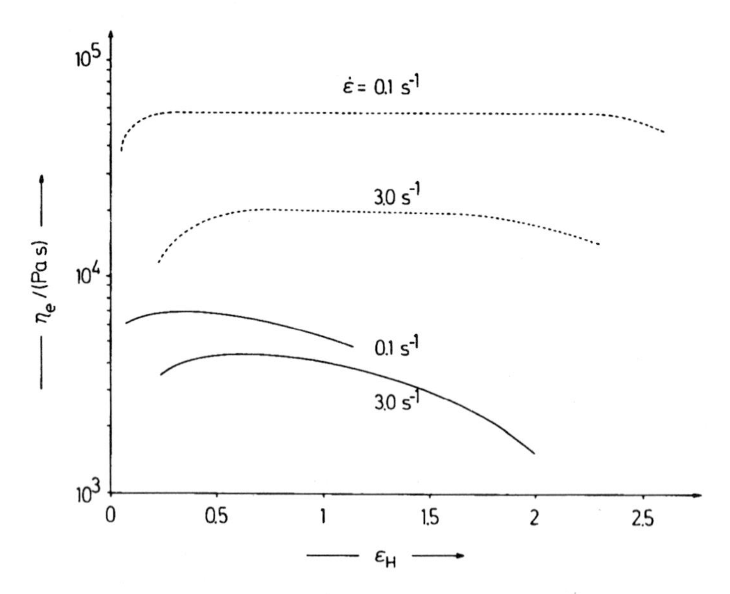

Fig. 6-11 Variation of uniaxial viscosities η_e with Hencky strain ε_H at 150°C and constant deformation rates for a high density poly(ethylene) (------) and the same polymer filled with 25 wt-% (ϕ = 0.098) mica (- - - -). Selected data from graphs by L.A.Utracki and J.Lara [8].

6.4.4. Solutions

Rod-like molecules and segments are increasingly oriented in the extensional direction with increasing elongational rates. The molecular axes are no longer distributed at random, and the solution becomes anisotropic and thus birefringent. A limiting value is asymptotically reached if all molecule axes are completely aligned in the flow direction.

Flexible molecules in solution are only slightly deformed and oriented at comparable extension rates because elastic (entropic) forces cause the chains to return to the thermodynamically favored coil shape. At high critical extension rates, these retraction forces can be overcome and the chain axes become oriented in the flow direction. Only an incremental increase of extension rates is needed to orient the chains completely. Molecules are more and more stressed above the critical extension rates until they finally break. This fracture occurs primarily in the middle of the molecule so that degradation products possess 1/2, 1/4, 1/8, etc., of the initial molar mass. Such degradation occurs under very low deformation rates for rigid macromolecules, for example, during pipetting of dilute solutions of high molar mass nucleic acids.

These degradations by extensional flow are not caused by turbulence because they happen at lower Reynolds numbers than those of pure solvents. Chain degradations by turbulence do occur on shearing of very dilute solutions of flexible coil molecules, (e.g., 10^{-4} g/mL aqueous solutions of poly(oxyethylene)s). The degradation reduces the frictional resistance of liquids up to 75 %. This "Toms effect" by added small amounts of such polymers eases the flow of crude oil through pipe lines and increases the distance and height to water can be directed at fires.

Literature

6.1. Introduction and Overviews

P.G.de Gennes, Scaling Concepts in Polymer Physics, Cornell Univesity Press, Ithaca (NY) 1979
R.T.Bailey, A.M.North, R.A.Pethrick, Molecular Motion in High Polymers, Oxford Univ.Press, Oxford 1981
M.Doi, S.F.Edwards, The Theory of Polymer Dynamics, Oxford University Press, Oxford 1986
H.A.Barnes, J.F.Hutton, K.Walters, An Introduction to Rheology, Elsevier, Amsterdam 1989

6.2. and 6.3. Shear Viscosities

M.Reiner, Deformation and Flow, Lewis and Co., London 1949
F.R.Eirich, ed., Rheology, Theory and Applications, Academic Press, New York 1956-1969 (5 volumes)
A.S.Lodge, Elastic Liquids, Academic Press, London 1964
S.Middleman, The Flow of High Polymers, Interscience, New York 1968
J.A.Brydson, Flow Properties of Polymer Melts, Iliffe Books, London 1970
W.W.Graessley, The Entanglement Concept in Polymer Rheology, Adv.Polym.Sci. **16** (1974) 1
G.Astarita, G.Marruci, Principles of Non-Newtonian Fluid Mechanics, McGraw-Hill, London 1974
R.Darby, Viscoelastic Fluids, Dekker, New York 1976
C.D.Han, Rheology in Polymer Processing, Academic Press, New York 1976
L.E.Nielsen, Polymer Rheology, Dekker, New York 1977
R.B.Bird, R.C.Armstrong, O.Hassager, Dynamics of Polymeric Liquids, vol. 1, Fluid Mechanics; Wiley, New York 1977
R.B.Bird, O.Hassager, R.C.Armstrong, C.F.Curtiss, Dynamics of Polymeric Liquids, vol. 2, Kinetic Theory, Wiley, New York 1977
R.S.Lenk, Polymer Rheology, Appl.Sci.Publ., Barking, Essex 1978
R.S.Porter, A.Casale, Polymer Stress Reactions, Academic Press, New York 1978/79 (2 vols.)
K.Murakami, K.Ono, Chemorheology of Polymers, Elsevier, Amsterdam 1979
K.Walters, Rheometry: Industrial Applications, Res.Studies Press (Wiley), Chichester 1980
G.Astarita, ed., Rheology, Plenum, New York, 3 vols. 1980
G.V.Vinogradov, A.Ya.Malkin, Rheology of Polymers, Mir Publ., Moscow; Springer, Berlin 1980
R.W.Whorlow, Rheological Techniques, E.Horwood, Chichester 1980
F.N.Cogswell, Polymer Melt Rheology, Wiley, New York 1981

M.Bohdanecky, J.Kovar, Viscosity of Polymer Solutions, Elsevier, Amsterdam 1982
H.Janeschitz-Kriegl, Polymer Melt Rheology and Flow Birefringence, Springer, Berlin 1983
M.Ballauf, B.A.Wolff, Thermodynamically Induced Shear Degradation, Adv.Polym.Sci. **85** (1988) 1
R.Larsen, Constitutive Equations for Polymer Melts and Solutions, Butterworths, Stoneham, MA 1988
S.W.Churchill, Viscous Flows: The Practical Use of Theory, Butterworths, Stoneham (MA) 1988
N.W.Tschoegl, The Phenomenological Theory of Linear Viscoelastic Behavior, Springer 1989
J.M.Dealy, K.F.Wissbrun, Melt Rheology and Its Role in Plastics Processing, Van Nostrand-Reinhold, New York 1990

6.4. Extensional Viscosities

J.W.Hill, J.A.Cuculo, Elongational Flow Behavior of Polymeric Fluids, J.Macromol.Sci.Revs. C **14** (1976) 107
J.C.S.Petrie, Elongational Flows (Research Notes in Mathematics, vol. 29), Pitman, London 1979
J.Ferguson, N.E.Hudson, Extensional Flow of Polymers, in R.A.Pethrick, ed., Polymer Yearbook **2**, Harwood Academic Publ., Chur 1985, p. 155
A.Keller, J.A.Odell, The Extensibility of Macromolecules in Solution; A New Focus for Macromolecular Science, Colloid Polym.Sci. **263** (1985) 181

References

[1] Data of D.S.Pearson, G.Ver Strate, E.von Meerwall, F.C.Schilling, Macromolecules **20** (1987) 1133, Table II.
[2] H.-G. Elias, Makromoleküle, vol. I: Grundlagen, Hüthig and Wepf, Basle, 5th ed. 1990, Fig. 24-5.
[3] H.-G. Elias, Makromoleküle, vol. I: Grundlagen, Hüthig and Wepf, Basle, 5th ed. 1990, Fig. 24-6.
[4] BASF, Kunststoff-Physik im Gespräch, BASF, Ludwigshafen/Rhein, 7th ed. 1988, figure on p. 112.
[5] G.Menges, in H.Batzer, ed., Polymere Werkstoffe, vol. 2, G.Thieme, Stuttgart 1984, Fig. 3.22.
[6] H.M.Laun, H.Münstedt, Rheol.Acta **17** (1978) 415, Fig. 5.
[7] E.M.Phillips, K.G.McHugh, K.Ogale, M.B.Bradley, Kunststoffe **82** (1992) 671, Figs. 2A and 2B.
[8] L.A.Utracki, J.Lara, Polym.Comp. **5** (1984) 44, selected data of Figs. 1 and 2.

7. Mechanical Properties

7.1. Introduction

7.1.1. Deformation of Polymers

Mechanical properties of a polymer include the deformation of polymers or their surfaces, the resistance to deformation, and the ultimate failure, all under static or dynamic loads. Deformations can be caused by drawing, shearing, compression, bending, or/and torsion; they may be reversible or irreversible.

Properties are not only effected by testing conditions but also by the size, shape, and preparation of test specimens. In injection moldings, melt temperature, mold temperature, and flow front velocities are important (ISO 294; DIN 16 770 T2); in compression moldings, press temperature, and cooling rate (ISO 293; DIN 16 770 T1).

Reversible deformations are due to the presence of elasticity. Irreversible deformations are also called "inelastic"; they are further subdivided into deformation by viscous flow, plasticity, phase transformations, craze formation, cracking, viscoelasticity, creep, etc. An inelastic deformation of metals by viscoelasticity is known as anelasticity; in polymer physics, anelasticity denotes a reversible elasticity with retardation, that does not lead to energy dissipation. Deformation of the upper layers of a polymer body is characterized by its "hardness," which influences friction and abrasion.

The term elasticity refers to either energy or entropy elasticity. These elasticities differ in their molecular mechanisms and the resulting phenomenological behavior. In *energy-elastic bodies* (steel, plastics, and elastomers at low strains), deformation changes torsion angles and bond angles and enlarge bond distances, whereas macroconformations of chains remain basically the same. A deformation of *entropy-elastic bodies* (elastomers at high strains), on the other hand, leads to entropically unfavorable positions of chain segments which, however, cannot slip irreversibly from each other because of the cross-linking ("rubber elasticity"). The deformation thus changes the macroconformation (molecular picture), decreases entropy (thermodynamics), and creates normal stresses (mechanics). The molecular changes on deformation are reflected in the properties of energy- and entropy-elastic materials (Table 7-1).

Table 7-1 Energy- and entropy-elastic bodies.

Property or behavior	Energy-elastic	Entropy-elastic
Elastic moduli	large	small
Reversible deformation	small	large
Temperature change on deformation	cooling	warming
Length change on heating	expansion	contraction

The deformation behavior of thermoplastics depends on the molar mass M of their constituting polymers and on the testing temperature: That is, whether the molar mass M is greater than the one needed for the establishment of entanglements (temporary physical cross-links) between chains ($M > M_{ent}$) and whether the testing temperature T is lower than the glass temperature T_G of the polymers ($T < T_G$). All mechanical deformations recover for $M > M_{ent}$ and $T < T_G$ as long as no chains are broken because chain entanglements cannot reorganize below the glass temperature (memory effect). Yielded plastics and crazed plastics recover on (not too long) heating above the glass temperature. These plastics thus do not show true plastic flow.

The deformation behavior of thermosets is essentially controlled by the density of their chemical cross-links and thus by the molar mass of segments between cross-links. Thermosets are rigid if these segments are small and the testing temperature is below the glass temperature.

7.1.2. Tensile Tests

Mechanical properties of polymers are most commonly evaluated by tensile tests in which a rectangular or dumbell-shaped test specimen is drawn with constant speed. The type of test specimen depends on the properties of polymers: different test specimens are used for plastics with (a) moderate elongations at break, (b) high elongations at break, and (c) thermosetting moldings (see ISO 527 and DIN 53455 and 53457); fibers and elastomers require other types of test specimens.

The tensile stress σ_{11} (usually as σ) is recorded as a function of time t, draw ratio (strain ratio) $\lambda = L/L_o$, or tensile strain (elongation) $\varepsilon = (L - L_o)/L_o = \lambda - 1$. If a specimen is extended to a length of $L = 2.5\ L_o$ of the original length L_o, then it is said to have been drawn by 150 % ($\varepsilon = 1.5$).

Thermoplastics (and most fibers) exhibit often stress-strain curves such as shown in Figure 7-1. For small strains (up to the point ε_H, σ_H), they follow Hooke's law

$$(7\text{-}1) \qquad \sigma_{11} = (F/A_o)\varepsilon = E\varepsilon$$

where A_o is the minimum original cross-section of the specimen, F is the force, and E is the tensile modulus (Young's modulus). The point ε_H is the *proportionality limit* (elastic limit); it is defined for a remaining strain of 0.1 % after removal of the stress. *Young's modulus* is the ratio of tensile stress to lengthwise strain at diminishing strains; in CAMPUS® data, it is a secant modulus between elongations of 0.05 % and 0.25 %.

The *yield point* Y is the first point on the load/extension curve at which an increase in extension occurs without an increase in load; in Figure 7-1, it is the first maximum of the stress-strain curve. The yield point is sometimes called the upper yield point, maximum yield, or intrinsic yield point; it gives the stress at yield (yield stress; yield strength) and the strain at yield (yield strain). Difficult to measure upper yield points are expressed as *offset yield points*, i.e., the yield stress at a specified percentage elongation (a secant value). Some authors call the proportionality limit a yield point.

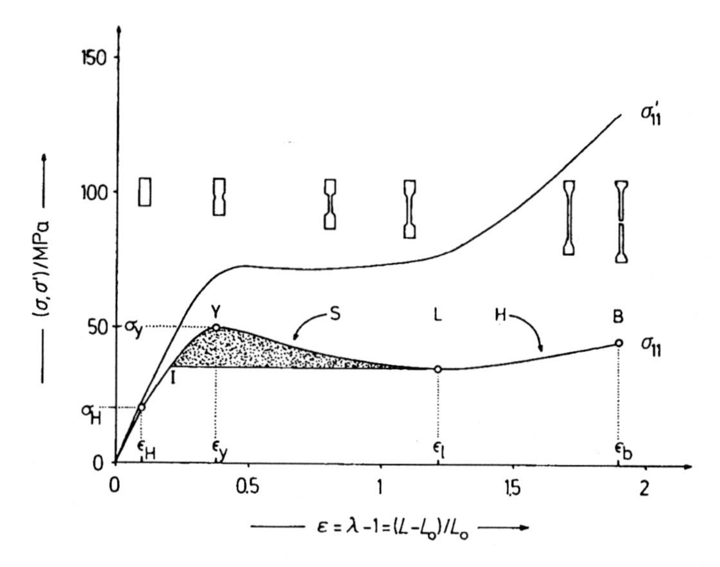

Fig. 7-1 Nominal stress σ_{11} and true stress σ_{11}' as function of the engineering (Cauchy) strain ε. Nominal stress-strain curve $\sigma_{11} = f(\varepsilon)$: the curve was drawn for a thermoplastic with E = 200 MPa, σ_y = 50 MPa at ε_y = 0.38 (38 %), and σ_b = 45 MPa at ε_b = 1.9 (190 %).
True stress $\sigma_{11}' = f(\varepsilon)$: the true stress was calculated from the nominal stress assuming the indicated necking behavior of a rectangular specimen.

Y = upper yield point S = stress softening L = lower yield point
H = stress hardening B = break

Brittle polymers break before they reach a yield point. Tough polymers continue to extend and the stress either remains constant (see below) or decreases with increasing strain (as in Fig. 7-1), sometimes passing through a minimum at the *lower yield point* L. The decrease of stress with strain is called *stress softening*. It is typical for specimen with neck formation (telescope effect) and it is only nominal since it disappears if the nominal stress (engineering stress) $\sigma_{11} = F/A_o$ as force F per initial minimum area A_o is replaced by the true stress $\sigma_{11}' = F/A = \sigma_{11}(L/L_o)$, which considers the true area A and true length L after an elongation. Similarly, the engineering strain (Cauchy strain) ε has to be replaced by the true strain (Hencky strain) $\varepsilon' = \ln (L/L_o) = \ln (A_o/A)$, which is obtained by integration over the differential dL/L.

The region between points I-Y-L is known as the *ductile region*. The total area under the stress-strain curve measures the absorbed energy; it describes the *toughness* of the specimen. A subsequent increase of stress with strain is called *stress hardening* (work hardening).

At point B, the polymer breaks; it exhibits an *ultimate strength* (fracture strength; tensile strength at break) σ_b and an *ultimate elongation* ε_b (elongation at break; elongation at rupture; strain at break). The tensile strength at break is sometimes simply called tensile strength but the latter term is also used for the highest tensile stress of the stress-strain curve (which may be the yield strength!).

Tensile tests are short term experiments. Standard tensile tests (ISO, DIN) measure Young's moduli at draw rates of 1 mm/min, yield stresses and strains at 50 mm/min, and tensile strengths and strains at break at 5 mm/min. The greater the drawing speed,

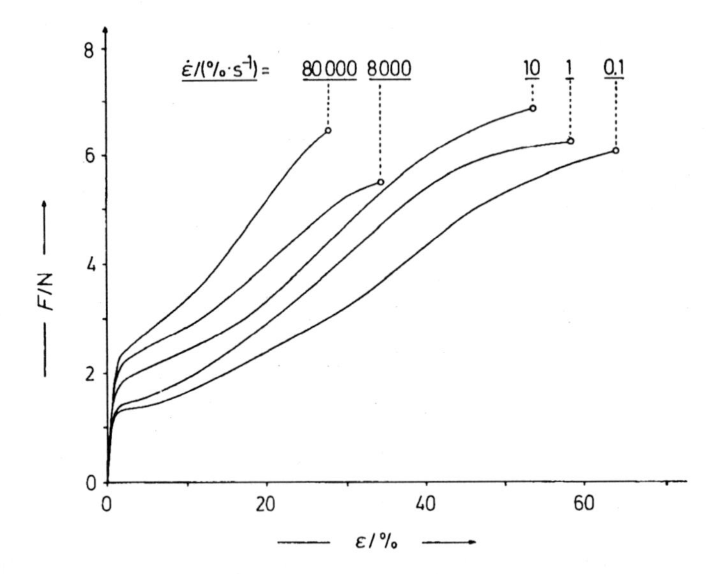

Fig. 7-2 Effect of drawing speed (in % per second) on force-strain curves of multifilament yarns from a poly(ethylene terephthalate) at 220°C. o Fracture. Reprinted with permission by Hüthig and Wepf Publ., Basle, Switzerland [2].

the steeper are the stress-strain curves and the smaller are the ultimate elongations (Fig. 7-2): the polymer has less time to flow and behaves as a more rigid material. Higher testing temperatures promote flow (Fig. 7-3). Poly(methyl methacrylate) is a very stiff material at 4°C (high modulus). At 60°C, PMMA has a far lower Young's modulus and shows considerable stress softening. At temperatures approaching the static glass temperature of 110°C, PMMA behaves rubbery.

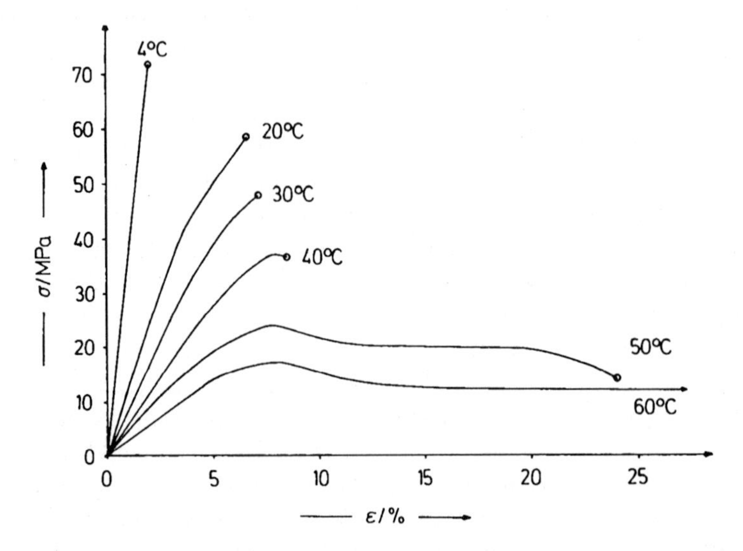

Fig. 7-3 Temperature dependence of stress-strain curves of a poly(methyl methacrylate) at constant strain rates. o Fracture. Reprinted with permission by Oliver and Boyd, London [3].

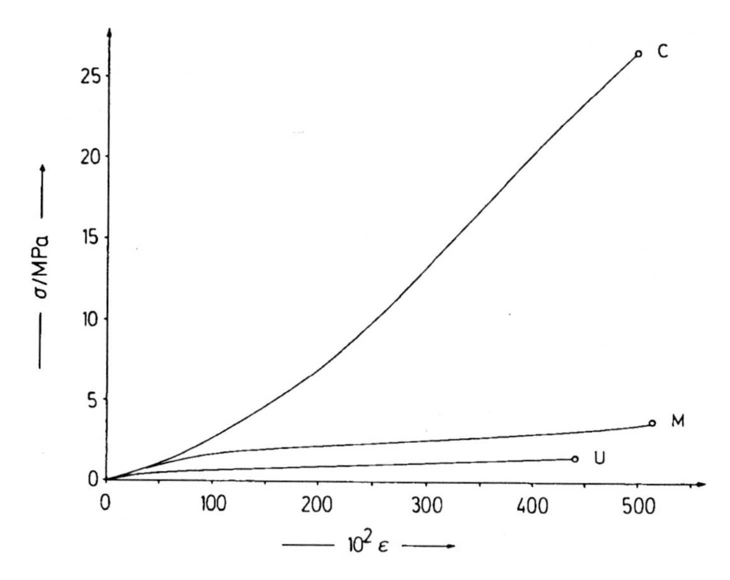

Fig. 7-4 Stress-strain curves of SBR elastomers: U = unfilled vulcanizate; M = mineral-filled vulcanizate; C = carbon black filled vulcanizate. Elastomers rupture at points o. Reprinted with permission by Hüthig and Wepf Publ., Basle, Switzerland [1].

Figure 7-4 shows such stress-strain curves for an elastomer from styrene-butadiene rubber (SBR). The tensile stress σ increases continuously with increasing tensile strain ε until the rubber ruptures at an elongation ε_r and a stress σ_r. Stresses are higher for filled elastomers because the filler particles act as additional cross-linking sites.

Comparable mechanical data can thus be only obtained by standardized test conditions (see Table 7-2 for CAMPUS® data).

Table 7-2 Mechanical properties in CAMPUS® data bases. All data at 23°C (unless noted otherwise) and 50 % relative humidity from test specimen with dimensions ≥10 mm × ≥10 mm × 4 mm (tensile and creep tests) and 80 mm × 10 mm × 4 mm (impact tests). Specimen preparation according to ISO 294 (DIN 16770 T2) for injection molded specimen and ISO 293 (DIN 16770 T1) for compression molded specimen. (*) Also at -30°C.

Property	Condition	Unit	ISO	DIN	Remarks
Young's modulus	1 mm/min	MPa	527	53457	0.05 % < ε < 0.25 %
Stress at yield (σ_y)	50 mm/min	MPa	527	53455	
Stress at 50 % elongation (σ_{50})	50 mm/min	MPa	527	53455	if σ_y not possible
Tensile strength	5 mm/min	MPa	527	53455	if σ_y and σ_{50} not possible
Strain at yield	50 mm/min	%	527	53455	
Strain at break	50 mm/min	%	527	53455	
Strain at break	5 mm/min	%	527	53455	
Creep modulus	1 h	MPa	899	53444	at $\varepsilon \leq 0.5$ %
Creep modulus	1000 h	MPa	899	53444	at $\varepsilon \leq 0.5$ %
Impact strength (Izod)	23°C (*)	kJ/m²	180/1C	-	
Notched impact strength (Izod)	23°C (*)	kJ/m²	180/1A	-	
Notched tensile impact strength	23°C	kJ/m²	8256	53448/1B	

7.1.3. Classification of Plastics

The stress-strain behavior discussed sofar is typical for tensile experiments on plastics. For example, atactic poly(styrene) PS is a hard-brittle polymer under tension T (Fig. 7-5, PS(T)); it fractures because of microvoid formation upon drawing. No voids can be formed, however, under an all-sided compression C and PS appears now as a hard-tough material (PS(C)). Rubber modification of PS leads to high-impact poly(styrene) (HIPS) that behaves under tension quite differently from PS (see Section 7.6.4).

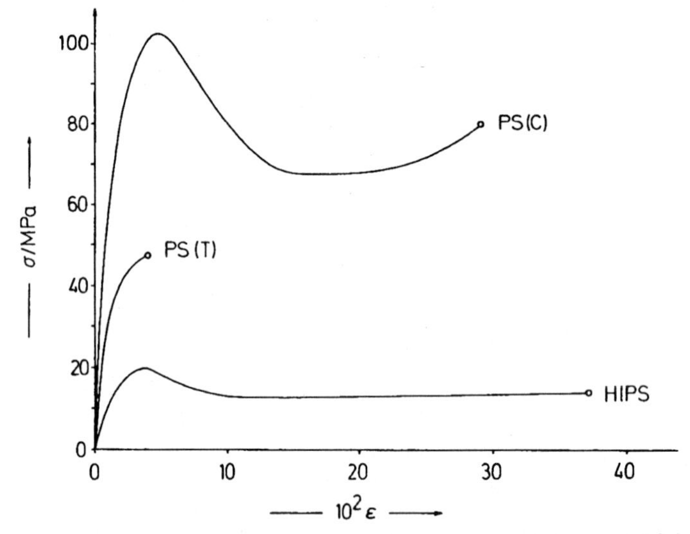

Fig. 7-5 Stress-strain curves at room temperature of a standard poly(styrene) PS in tension (T) and in compression (C) and of a high impact-modified poly(styrene) HIPS in tension.

The stress-strain behavior in tension may differ considerably for various types of plastics. Plastics with high Young's moduli (steep initial slopes) are usually called *hard plastics*; those with low moduli, soft. Typical hard plastics are phenolics (phenol-formaldehyde plastics, PF), polyacetal resins (poly(oxymethylene)s, POM), polycarbonates (PC), and unstretched poly(ethylene terephthalate)s (PET-u) (Fig. 7-6).

This "hardness" should not be confused with surface hardness. Hard plastics are therefore better called *rigid plastics* (ASTM); they are defined as plastics with elastic moduli $E > 700$ MPa as measured by specific ASTM methods. Semirigid plastics have moduli of 70-700 MPa; nonrigid ("soft") plastics are those with $E < 70$ MPa.

Polymers are further characterized by their stress-strain behavior between upper yield and failure. Nonyielding polymers cannot absorb energy and thus break easily; they are *hard-brittle* polymers such as phenolic resins PF (Fig. 7-6) and poly(styrene) PS (Fig. 7-5)). Polycarbonates PC, on the other hand, show an extended ductile region and a fairly high fracture strain; they are called *hard-tough*. Poly(ethylene) PE is similar to polycarbonate with respect to ductile behavior and strain hardening; the modulus is however much lower, and PE is said to be *soft-tough*.

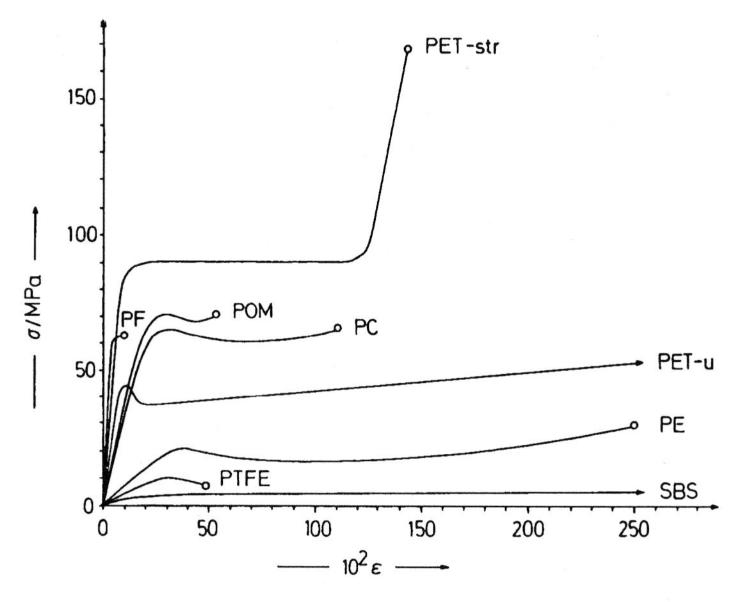

Fig. 7-6 Stress-strain curves of polymers at 23°C. PC = Bisphenol A polycarbonate; PE = poly(ethylene); PET = poly(ethylene terephthalate), u = unstretched or str = biaxially oriented; PF = phenol-formaldehyde thermoset; POM = poly(oxymethylene); PTFE = poly(tetrafluoroethylene); SBS = poly(styrene)-*block*-poly(butadiene)-*block*-poly(styrene) (thermoplastic elastomer).

The term "toughness" is also used to characterize the behavior of plastics against impact (see below); a brittle plastic cannot be tough under impact and vice versa. The toughness under tensile stress is therefore better called *ductility.*

Ductility, toughness, and brittleness obviously depend on the speed at which chains can adjust their macroconformations under deformation (strain or impact). This speed is determined by the mobility of chains and chain segments under testing conditions, and thus also on the testing temperature T relative to the glass temperature T_G as well as by the modulus of elasticity E. Polymers can thus be approximately classified as

brittle	$T \ll T_G$	E/GPa > 4.5
semiductile	$T < T_G$	2 < E/GPa < 4
ductile (tough)	$T < T_G$	E/GPa < 1.5
rubber-elastic	$T > T_G$	

No yield value is found if drawn polymers are further subjected to tensile tests. In biaxially stretched poly(ethylene terephthalate) (PET-str) some chain segments are already oriented, whereas other segments remain in their original positions. Biaxially stretched films are thus under stress, which is utilized in shrink films. On heating above the glass temperature T_G and below the melting temperature T_M of such crystalline polymers, segments become more mobile. Molecules attempt to attain their random coil dimensions and the films shrink, covering the goods tightly. Such films are used for the packaging of goods.

7.2. Energy Elasticity

7.2.1. Moduli and Poisson Ratios

Elasticities can be described by three types of elastic moduli: tensile modulus or Young's modulus E, shear modulus G, and bulk modulus or compressive modulus K. Their values are inversely proportional to the corresponding compliances for static deformations (but not to dynamic ones):

Moduli	Compliances
$E = \sigma_{11}/\varepsilon$	$D = 1/E$
$G = \sigma_{21}/\gamma$	$J = 1/G$
$K = p/(-\Delta V/V_o)$	$B = 1/K$

The three simple moduli are related to each other for small deformations of simple isotropic bodies:

(7-2) $E = 2\,G(1 + \mu) = 3\,K(1 - 2\mu)$

where $\mu = (\Delta d/d_o)/(\Delta L/L_o)$ is the Poisson ratio (Poisson number; lateral strain contraction ratio), d the diameter, and L is the length of the specimen. Poisson ratios can only adopt values $0 < \mu < 1/2$ for isotropic bodies but may assume $\mu > 1/2$ for anisotropic ones. Equation (7-2) is invalid for anisotropic bodies and/or viscoelastic materials; $E/3 < G < E/2$ and $0 < K < E/3$ are, however, still valid for these materials.

Common polymers behave more like liquids than metals with respect to μ, E, G, and K (Table 7-3). Ultradrawn and self-reinforcing polymers may however exhibit tensile moduli that exceed those of steel (Section 7.2.2).

Table 7-3 Experimental Poisson ratios and elastic constants of various materials at 25°C.

Material	μ	E/GPa	G/GPa	K/GPa
Water	0.50	≈ 0	≈ 0	≈ 2
Gelatin (80 % water)	0.50	0.002		
Natural rubber	0.495	0.0009	0.0003	≈ 2
Poly(ethylene), low density	0.49	0.20	0.070	3.3
Polyamide 6.6	0.44	1.9	0.70	5.1
Poly(styrene)	0.38	3.4	1.2	5.0
Granite	0.30	30	12	25
Steel	0.28	211	80	160
Glass	0.23	60	25	37
Quartz	0.07	101	47	39
Graphite	0	1000	500	333
Aluminum oxide fibers	0	2000	1000	667

7.2.2. Theoretical Moduli

The tensile moduli of common polymers are far lower than the theoretical moduli deduced from their chemical and physical structures (Table 7-4). Such theoretical moduli can be calculated from bond lengths, valence angles, conformational angles, and the force constants for the deformation of these quantities, assuming infinitely high molar masses of the polymer molecules.

Table 7-4 Modulus of elasticity E and cross-sectional areas A_m of polymers.

Polymer	Confor-mation	A_m/nm^2	$E_{\parallel}$/GPa Theory	Lattice	Tensile	$E_{\perp}$/GPa Lattice
Poly(*p*-phenylene benzbisthiazole)	trans		640			
Poly(ethylene)	trans	0.183	340	325	< 1	3.4
Poly(*p*-phenylene terephthalamide)	trans	0.203	182	200	132	10
Poly(oxymethylene), orthorhombic	9_5	0.182	220	189	< 2	7.8
it-Poly(propylene)	3_1	0.343	50	42	< 3	2.9
Poly(4-methylpentene-1)	7_2	0.864	6.7		1	2.9

The theoretical moduli agree well with the microscopic lattice moduli, which are determined experimentally by X-ray diffraction (change of Bragg reflexes), Raman spectroscopy, or inelastic neutron scattering under load. The longitudinal theoretical moduli $E_{\parallel}$ in chain direction are far greater than the transverse moduli $E_{\perp}$ because the former are controlled by covalent bonds and the latter by van der Waals forces or hydrogen bonds between chains. Longitudinal moduli from tensile testing of conventionally processed polymers are far lower than the theoretical or lattice moduli because chain segments are oriented randomly in space.

Moduli approaching theoretical values have been realized by ultradrawing of mats of poly(ethylene) single crystals, giving polymers with $E_{\parallel}$ = 240 GPa. Industrially, high modulus poly(ethylene) fibers with $E_{\parallel}$ = 90 GPa $\hat{=}$ 95 N/tex are manufactured by gel-spinning of ultrahigh molecular poly(ethylene). Fiber moduli are reported as forces per linear mass instead of forces per area (modulus of elasticity in GPa divided by density in g/cm^3 = textile modulus in N/tex; 1 tex = 1 g/km).

Theoretical moduli depend on cross-sectional areas and macroconformations of chains. The applied force F is distributed to fewer chains per unit area A_o, the fewer chains are present per area A_o, i.e., the higher the cross-sectional area A_m per chain. The modulus thus decreases with increasing cross-sectional area A_m for a series of polymers with the same macroconformation (Fig. 7-8).

Poly(ethylene) with a cross-sectional area of A_m = 0.182 nm^2 has the highest theoretical modulus of all chains with σ-bonds between chain atoms ($E_{\parallel}$ = 340 GPa); a similar value (312 GPa) can be estimated from the modulus of diamond ($E_{\parallel}$ = 1160 GPa) with its "naked" carbon "chains" (A_m = 0.049 nm^2). Even higher moduli than that of poly(ethylene) are achieved if σ-bonded chain atoms are replaced by less deformable units such as the rigid benzbisthiazole units and 1,4-phenylene rings.

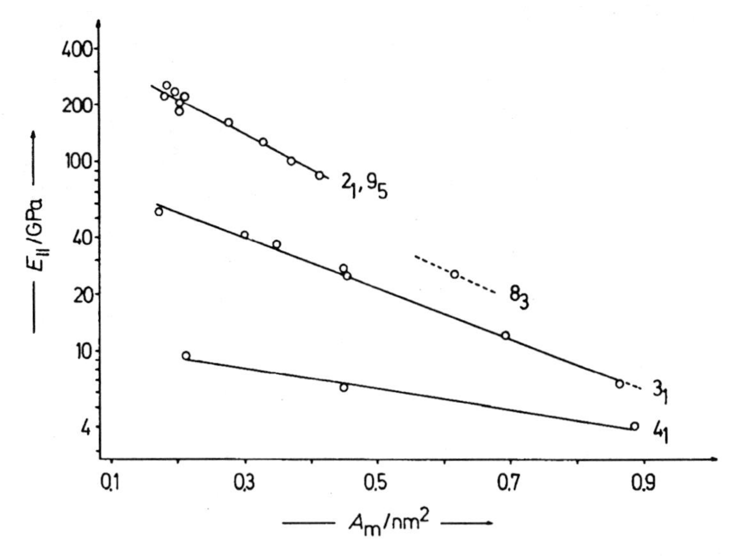

Fig. 7-7 Longitudinal lattice moduli $E_{\parallel}$ as function of cross-sectional area A_m of polymer chains in all-*trans* (2_1) or helical conformations (9_5, 8_3, 3_1, 4_1) of their main chains. Reprinted with permission by Hanser-Verlag, Munich [4].

Chains in all-*trans* conformations always exhibit higher theoretical moduli than helical chains with the same cross-sectional area since the elongation of the former deforms bond angles, whereas the extension of the latter needs less energy to change torsion angles. The orthorhombic crystal modification of poly(oxymethylene) has the highest theoretical modulus of helical chains ($E_{\parallel}$ = 220 GPa) because its chain atoms are in all-*gauche* conformations which are tightly packed in helices with small cross-sectional areas (see Fig. 12-3).

The theoretical longitudinal moduli of poly(ethylene) and some other polymers exceed those of conventional engineering materials such as steel, concrete, and wood (Tab. 7-4). The moduli of conventionally processed polymers are much lower. However, materials are not used on a per weight but on a per volume basis. Property per density is thus a more meaningful parameter. On this basis, engineering plastics such as polysulfone compete successfully with steel, glass, concrete, and wood.

Tab. 7-5 Densities ρ, moduli of elasticity $E_{\parallel}$, and reduced moduli $E_{\parallel}/\rho$ of various materials.

Material	$\frac{\rho}{\text{g cm}^{-3}}$	$\frac{E_{\parallel}}{\text{GPa}}$	$\frac{E_{\parallel}\rho^{-1}}{\text{GPa cm}^3\text{ g}^{-1}}$
Poly(ethylene), theory	1.00	340	340
conventionally processed PE-HD	0.96	1.1	1.15
Polysulfone	1.24	26	21.0
Steel	7.86	200	25.4
E-Glass	2.54	72	28.3
Concrete	2.3	30	13.0
Wood	≈ 0.5	≈ 11	≈ 22

7.2.3. Real Moduli

The low tensile moduli of conventionally processed polymers result from their disordered physical structures. In amorphous polymers, chain segments are oriented at random. Even in partially crystallized polymers, amorphous layers exist and chain axes (stems of lamellae) are distributed at random. The moduli of flexible polymers can be increased somewhat by processing under external force fields, for example, by partial orientation of chain segments during fiber spinning (Table 7-6, column L) or by stretching of films. Extrusion of solid polymers is a particularly effective method: the longitudinal modulus of poly(oxymethylene) increased to 24 GPa from 2 GPa on hydrostatic extrusion.

Rod-like mesogens of liquid semiflexible polymers align in mesophase domains. The domains orient themselves in shear fields; the orientations can be frozen-in to LCP glasses. Such self-orienting polymers possess much higher moduli in longitudinal direction than conventionally processed flexible polymers (cf. Table 7-4). The transverse moduli of self-orienting polymers are also higher because of intermolecular dipole-dipole interactions.

Table 7-6 Moduli and fracture strengths of conventional polymers as isotropic molding masses (I) or in draw direction of fibers (L) and of thermotropic (TT) and lyotropic LCP glasses longitudinal (L) and transverse (T) to draw direction DD compared to isotropic polymers (I).

Pe-LD = Low density poly(ethylene);
PA 6.6 = Poly(hexamethylene adipamide);
PET = Poly(ethylene terepthalate);
X7G™ = Poly(p-hydroxybenzoate-*co*-ethylene terephthalate);
Vectra™ = Poly(*p*-hydroxybenzoate-*co*-2-hydroxy-6-naphthalate);
Kevlar 49™ = Poly(*p*-phenylene terephthalamide);
PPBT = 30 % Poly(*p*-phenylene benzbisthiazole) in poly(2.5-benzimidazole).

Polymer		E/GPa			σ_B/MPa		
DD	Type	L	T	I	L	T	I
--	PE-LD	?	?	0.15	?	?	23
--	PA 6.6	13	?	2.5	1000	?	74
--	PET	19	?	0.13	1400	?	54
TT	X7G	54	1.4	2.2	151	10	63
TT	Vectra	11	2.6	5.0	144	54	97
LT	Kevlar	138	7	?	2800	?	?
LT	PPBT	120	17	62	1500	680	700

7.2.4. Temperature Dependence

Young's moduli change characteristically with temperature for the various classes of conventionally processed polymers (Fig. 7-8). Five characteristic ranges can be distinguished: the behavior of uncross-linked, noncrystalline polymers changes with increasing temperature from glassy (GL) to leather-like (LE), rubbery (RE), viscoelastic (RF), and finally to viscous (VF).

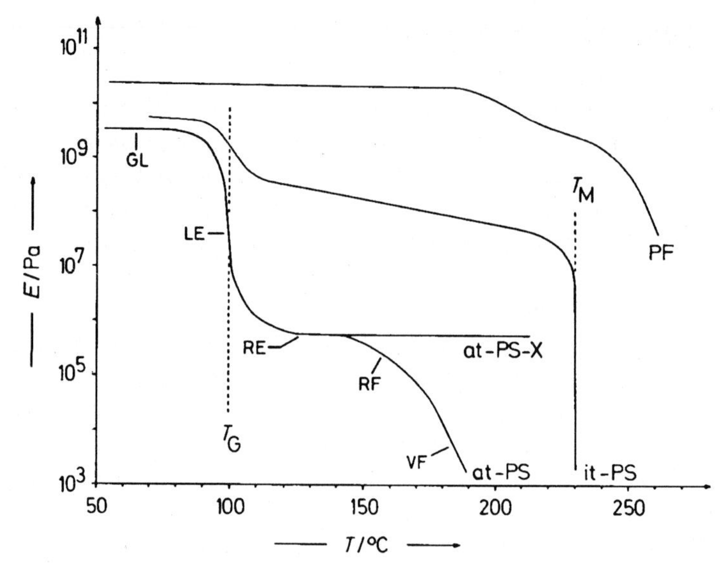

Fig. 7-8 Temperature dependence of Young's moduli of some polymers. GL = glass-like, LE = leather-like, RE = rubber-like (entangled), RF = viscoelastic (rubber-like flow), VF = viscous flow.
at-PS = Amorphous (atactic) poly(styrene) ($M > M_c$);
at-PS-X = its slightly cross-linked product;
it-PS = partially crystalline isotactic poly(styrene);
PF = hardened phenol-formaldehyde resin.
Reprinted with permission by Hüthig and Wepf Publ., Basle, Switzerland [5].

Amorphous Polymers. Moduli (0.1-1 GPa) are practically independent of temperatures below the glass temperature T_G but drop to 10^5-10^6 Pa at $T \approx T_G$; the polymers appear leathery around T_G and rubbery above the glass temperature. The rubbery region is maintained for entangled (high molar mass) polymers but is nonexistent for low molar masses ($M < M_c$). Moduli decrease on further temperature increase until the polymers behave like viscous liquids ($E \approx 10^3$ Pa).

Elastomers. At use temperatures, these polymers are above their T_G's and thus behave like lightly cross-linked thermoplastics above the T_G's of the latter. The decrease in modulus at T_G is limited by (and a measure of) the degree of cross-linking. At higher temperatures, elastomers decompose and the moduli decrease drastically.

Crystalline polymers have only a weak leathery region around T_G since their amorphicity is small and their crystallites act as physical cross-linkers. The modulus at the beginning of the rubbery "plateau" is a measure of the degree of crystallinity. The moduli decrease slowly during the rubbery "plateau" because more and more crystallites are molten. The remaining crystallites practically disappear at the melting temperature: the moduli drop drastically and the polymers behave like viscous liquids.

Thermosets. Thermosets are highly cross-linked polymers. They have high Young's moduli below T_G, usually ten times higher than thermoplastics. A very weak glass transition (due to short segment lengths between cross-links) is accompanied by a somewhat leathery behavior. The subsequent rubbery region is not very marked because the cross-linking density is high.

7.3. Entropy Elasticity

Elastomers show pronounced entropy elasticities (rubber elasticities). They exhibit simultaneously some characteristics of solids, liquids, and gases. Like solids, they display Hookean behavior at not too high deformations (i.e., they are show no permanent deformation after removal of the load). Moduli and expansion coefficients, on the other hand, resemble those of liquids. Like compressed gases, stresses increase with increasing temperature for elastomers at $T > T_G$.

The elastic behavior can be modeled with various theories. In the simplest case, chains are assumed to be infinitely thin (*phantom chains*) and dislocations of network junctions are supposed to be affine to the macroscopic deformation of the network. The tensile stress σ_{11} varies with the elongation $\lambda = L/L_o$ according to

(7-3) $\sigma_{11} = RT[M_c](V_o/V)^{-1/3}(\lambda - \lambda^{-2})$

The tensile stress thus depends on the volumes before (V_o) and after (V) deformation. True Young's moduli can be obtained from the limiting value of $\sigma_{11}/(\lambda - \lambda^{-2})$ at $\lambda \rightarrow 1$, for example, for volume-constant deformations ($V_o/V = 1$):

(7-4) $RT[M_c] = \lim_{\lambda \rightarrow 1} [\sigma_{11}/(\lambda - \lambda^{-2})] = \sigma_{11}/[3(\lambda - 1)] = \sigma_{11}/(3\ \varepsilon) = E_o$

The Young's modulus is thus directly proportional to the molar concentration of network junctions (cross-linking sites) and independent of the chemical nature of the latter. Chemically and physically cross-linked polymers have thus higher moduli than their uncross-linked counterparts (see Fig. 7-8).

On shearing, elastomers behave like Hookean bodies since

(7-5) $RT[M_c] = \sigma_{21}/\gamma = G$

They are, however, non-Hookean for elongations because $E_o \neq \sigma_{11}/\varepsilon$ (Eq. 7-4).

7.4. Viscoelasticity

7.4.1. Fundamentals

Most polymers do not revert "instantaneously" to their initial states after the removal of loads; they are neither ideal energy-elastic (Section 7.2) nor ideal entropy-elastic (Section 7.3). These processes take certain times, i.e., time-independent elastic and time-dependent viscous properties work together to produce a viscoelastic

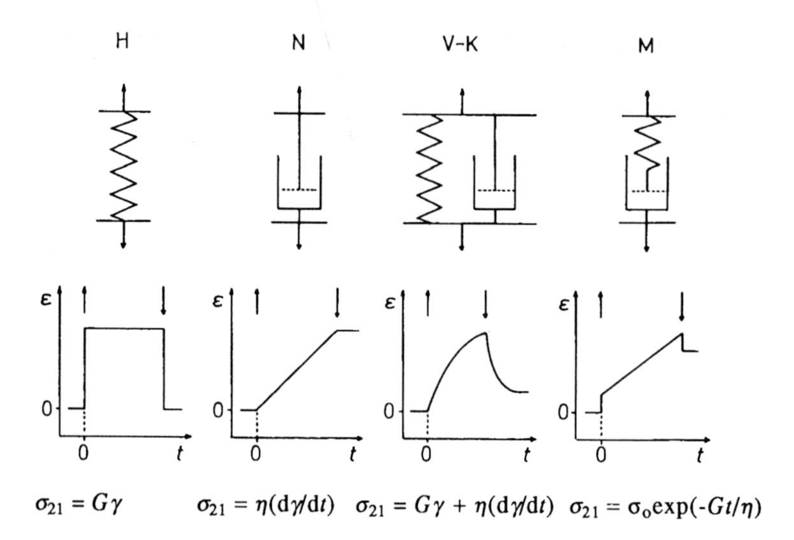

Fig. 7-9 Time dependence of deformations according to various models.
First row: H = Hookean spring; N = Newtonian dashpot; V-K = Voigt-Kelvin element; M = Maxwell element.
Second row: Elongation ε upon application of tension at ↑ and removal at ↓.
Third row: Equations for shear stress σ_{21} as function of shear modulus G, deformation γ, kinematic viscosity η, and time t.

behavior. If stress, strain, and strain rate can be combined linearly, then the process is said to be linear viscoelastic. In addition, polymers may be irreversibly deformed.

The two ideal cases of response to deformations can be well described by mechanical models: a spring for a Hookean body (instantaneous response) and a dashpot for a Newtonian liquid (linear time-dependence of response) (Fig. 7-9). The Maxwell element combines spring and dashpot in a series; the Voigt-Kelvin element, in a parallel manner.

The Maxwell element describes a *relaxation* (i.e., the decrease of stress at constant deformation (Table 7-7). Linear combination of elastic deformation rates $d\gamma_e/dt = (1/G)(d\sigma/dt)$ and viscous deformation rates $d\gamma_v/dt = \sigma/\eta$ yields, after integration of the resulting expression and indexing the time for this particular behavior for $d\gamma/dt = 0$:

$$(7\text{-}6) \qquad \sigma = \sigma_o \cdot \exp(-\,G \cdot t_e/\eta) = \sigma_o \cdot \exp(-\,t_e/\tau)$$

The relaxation time $\tau = \eta/G$ indicates the time after which the stress has fallen to the e^{-1}th fraction of its initial value. The ratio τ/t_e of the relaxation time τ to the time scale t_e of the experiment is called the Deborah number ($De = \tau/t_e$). By definition, it is 0 for Newtonian fluids and ∞ for Hookean solids. The Deborah numbers of other materials are 10^{-12} for water, 10^{-6} for lubricant oils, 1-10 for polymer melts, and 10^9-10^{11} for glasses (see also Section 5.3.2.).

Retardation is defined as the increase of deformation with time at constant stress. It is characterized by a "creep" of the material. Since this phenomenon was first observed on seemingly solid polymers at room temperature, it is also called "cold flow".

Table 7-7 Simple mechanical models for the deformation of polymers. S = Solid-like behavior, L = liquid-like behavior.

Model	Function		Behavior initial	final
Newtonian liquid	σ	$= \eta \dot{\gamma}$	L	L
Voigt-Kelvin element	σ	$= \eta \dot{\gamma} + G \dot{\gamma}$	L	S
Maxwell element	$\sigma + (\eta/G)\dot{\sigma}$	$= \eta \dot{\gamma}$	S	L
Hookean body	σ	$= G \dot{\gamma}$	S	S
Jeffrey's body	$\sigma + (\eta/G)\dot{\sigma}$	$= G \dot{\gamma} + \eta \dot{\gamma}$	S	S

In principle, retardation phenomena can be described by a Maxwell element. Because of mathematical difficulties in the solution of the equations, a special model is prefered (Voigt-Kelvin element) (Fig. 7-9), which delivers for the deformation γ_r at constant stress σ_o after indexation for retardations r:

$$(7\text{-}7) \qquad \gamma_r = (\sigma_o/G_r)\cdot[1 - \exp(-G_r t_r/\eta)]$$

G_r is the retardation modulus, often also called relaxation modulus. The retardation time t_r indicates the time at which the deformation has reached (1 - 1/e) = 0.632 of the final deformation σ_o/G_r. Retardation times and relaxation times are of the same magnitude but not equal because they rely on different models.

The term "viscoelasticity" is sometimes used to describe the reversible deformation according to Equation (7-7). However, it is often applied to the total deformation, which is composed of contributions by Equation (7-7), a Hookean body with $\gamma_e = \sigma_o/G_o$, and a Newtonian liquid with $\gamma_v = (\sigma_o/\eta_o)t$:

$$(7\text{-}8) \qquad \gamma_{tot} = \{(1/G_o) + (t/\eta_o) + (1/G_r)[1 - \exp(-t/\tau)]\}\cdot\sigma_o = \gamma_e + \gamma_v + \gamma_r$$

The three deformation terms γ_e (elastic), γ_v (viscous), and γ_r (viscoelastic) are sometimes not explicitly evaluated. The time-dependent viscous and viscoelastic parts are rather combined to a new parameter $\gamma_c = \gamma t^n$ (Findlay law). The resulting function for the creep curve allows an extrapolation to long-time behavior from short-time experiments:

$$(7\text{-}9) \qquad \gamma_{tot} = \gamma_e + \gamma_c = \gamma_e + \gamma t^n$$

7.4.2. Time-Temperature Superposition

Deformations ε, shear moduli G, and shear compliances J are time and temperature dependent (Figs. 7-9 and 7-10). The moduli vary about one decade for six decades in time at constant temperature, and up to one decade for each ten kelvins at constant frequency. Since no single experimental method can cover the 15-20 decades of frequencies that are required for a good characterization of a polymer, time-temperature data from various techniques are usually combined with the help of the Boltzmann superposition principle.

This principle states that the deformation (or recovery) caused by an additional load (or removal thereof) is independent of previous loads or their removal. The $G = f(t,T)$ curves can be combined if (1) the relaxation time spectrum is temperature independent and (2) the thermal activation is the same over the entire time and temperature range (no transition or relaxation temperatures). A reference temperature is chosen close to the static glass transition temperature (115°C vs. 105°C in Fig. 7-10) and the G values are horizontally shifted with the help of a shift factor from the WLF equation (Eq. 5-9).

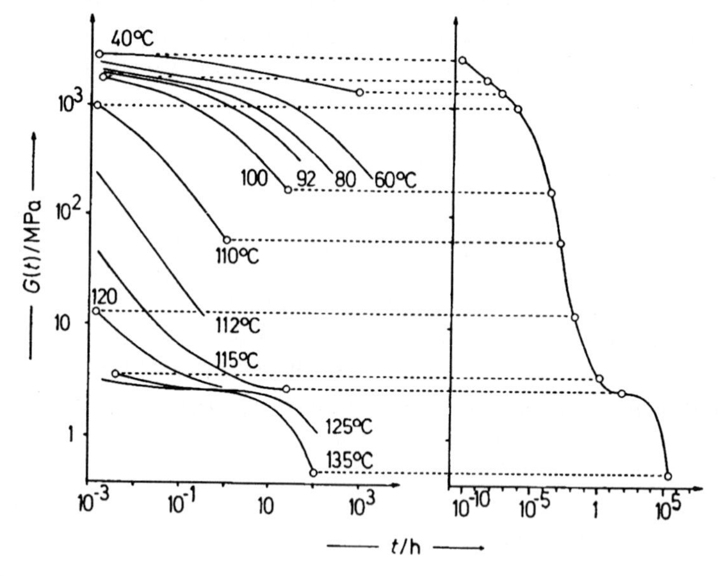

Fig. 7-10 Time dependence of the shear modulus $G(t)$ from measurements of the stress relaxation of a poly(methyl methacrylate) (viscosimetric molar mass = 3 600 000 g/mol) at various temperatures (left) and the resulting time-temperature superpositions for a reference temperature of 115°C (right) [6]. Circles represent equivalent positions.

7.5. Dynamic Behavior

7.5.1. Fundamentals

Dynamic-mechanical methods expose the specimen to periodic stresses. The polymer either is put under torsion once and then oscillates freely (torsion pendulum) or is subjected continuously to forced oscillations (e.g., Rheovibron™). In addition, ultrasound, dielectric, and NMR methods can be used to study dynamic properties.

In the simplest case, stress σ is applied sinusoidal with a frequency ω (i.e., $\sigma_t = \sigma_o \cdot \sin \omega t$). The deformation γ of ideal-elastic bodies follows the stress instantaneously ($\gamma_t = \gamma_o \cdot \sin \omega t$) but the deformation of viscoelastic polymers experiences a delay ($\gamma_t = \gamma_o \cdot \sin (\omega t - \vartheta)$). The stress vector is assumed to be a sum of two components: one component is in phase with the deformation ($\sigma' = \sigma_o \cdot \cos \vartheta$) and the other is not ($\sigma'' = \sigma_o \cdot \sin \vartheta$).

Each of these two components possesses a modulus. The *real modulus* G' (*shear storage modulus, elastic modulus, in-phase modulus*) measures the stiffness and shape stability of the specimen:

(7-10) $G' = \sigma'/\gamma_o = (\sigma_o/\gamma_o) \cdot \cos \vartheta = G^* \cdot \cos \vartheta$

whereas the *imaginary modulus* G'' (*shear loss modulus, viscous modulus, 90° out-of-phase modulus*) describes the loss of usable mechanical energy by dissipation into heat:

(7-11) $G'' = \sigma''/\gamma_o = G^* \cdot \sin \vartheta$

The same quantities can also be derived if complex variables are introduced

(7-12) $G^* = G' + iG'' = [(G')^2 + (G'')^2]^{1/2}$

The loss factor Δ is the ratio of imaginary (loss) to real (storage) modulus. It is the same for shear moduli G and Young's moduli E but not for compression moduli K:

(7-13) $\Delta = \tan \vartheta = G''/G' = E''/E' < K''/K'$

7.5.2. Molecular Interpretations

The *shear storage moduli* G' of melts of polymers with low molar mass and narrow molar mass distribution increase continually with increasing frequency ω (Fig. 7-11). At high normalized frequencies $a_T\omega$, all storage moduli asymptotically approach a limiting line, regardless of the molar mass of the polymers. This part of the relaxation spectrum thus originates from the mobility of chain segments; it is called the transition range T.

High molar mass polymers show a corresponding frequency dependence of *loss moduli* at very low frequencies (called end range or termination relaxation zone), followed by a plateau at higher frequencies (plateau modulus G_N^o), and finally the transition range. The end ranges of these spectra are molar mass dependent; this behavior must come from long-range conformational changes. The storage modulus in the termination relaxation zone for the polymer melt is a mirror of the cumulative molar mass distribution: the modulus axis can be transformed into the mass fraction $w_i = 1 - [G'(\omega_i)/G_N^o]^{1/2}$ and the inverse frequency into the molar mass ($1/\omega \sim \overline{M}_w^{3.4}$).

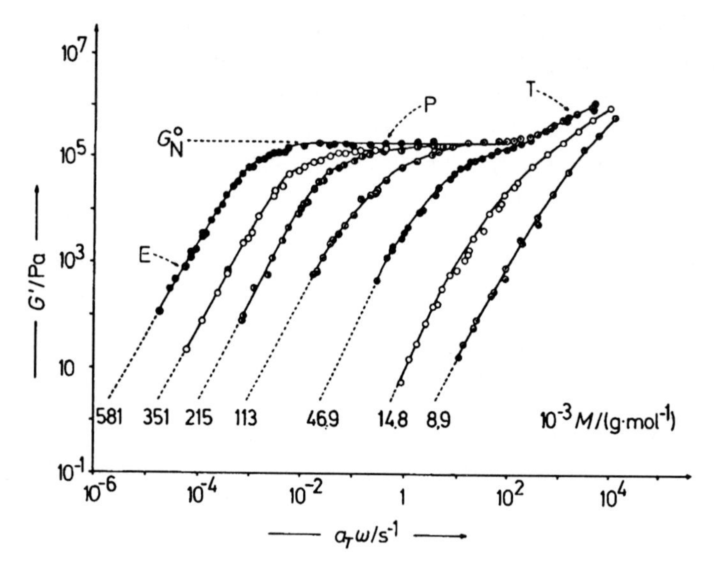

Fig. 7-11 Shear storage modulus G' of melts as function of normalized frequency ω. A shift factor α_T was used to convert to 160°C data measured at various temperatures. E = End range, P = plateau, T = transition range.

The specimens are narrow-distribution poly(styrene)s of molar masses $M/(g\ mol^{-1})$ = 581 000, 351 000, 215 000, 113 000, 46 900, 14 800, and 8900.

Reprinted with permission by American Chemical Society, Washington, D.C. [7].

Since the transition range characterizes the viscous behavior and the end range the viscoelastic one, the plateau range must reflect the rubbery behavior (see Fig. 7-8 for the temperature dependence).

The *rubbery behavior* of polymer melts can be described by the theories of entropy elasticity according to which the shear modulus of chemically cross-linked polymers depends on the molar concentration of network junctions. These theories are equilibrium theories for permanent cross-links that here are applied to the dynamic phenomenon of entanglements due to temporary junctions.

The plateau modulus G_N^o of melts is asumed to indicate the concentration of temporary junctions (entanglements). The molar mass M_e of segments between such junctions of polymers with volume fractions ϕ_p in solution ($\phi_p = 1$ for melts) and polymer melt densities ρ_p is given by rubber theory as

$$(7\text{-}14) \quad M_e = RT\rho_p\phi_p/G_N^o$$

(reptation theory gives the right-hand side of Equation (7-14) as (4/5) $RT\rho_p\phi_p/G_N^o$). These dynamic critical entanglement molar masses M_e are a factor 2.0 ± 0.2 lower than the corresponding molar masses M_c from rest viscosities (Table 7-8); reasons are unclear (frequency dependence of entanglement concentrations, different definitions of segments, different averaging?).

The plateau is not well developed or may even be absent for polymers with broad molar mass distributions; a complicated dependence of shear compliances on higher molar mass averages has been predicted by reptation theory.

Table 7-8 Critical molar masses for entanglement from shear modulus (M_e) and rest viscosity (M_c)

Polymer	$\frac{T}{°C}$	$\frac{M_c}{\text{g mol}^{-1}}$	$\frac{M_e}{\text{g mol}^{-1}}$	$\frac{M_c}{M_e}$
Poly(ethylene)	190	3 800	1 790	2.1
Poly(isobutylene)	25	15 200	8 800	1.7
Poly(dimethylsiloxane)	25	24 500	10 500	2.4
Poly(vinyl acetate), at	57	24 500	12 000	2.0
Poly(α-methylstyrene), at	100	28 000	13 500	2.1
Poly(styrene), at	190	35 000	18 100	1.9

7.6. Strength

7.6.1. Overview

Polymers are subjected to very different stress conditions in typical applications; they thus experience different failure modes. *Failure* is defined as the cessation of proper performance caused by mechanical or nonmechanical causes. The type of failure depends on the chemical and physical structure of the polymers; the environment such as humidity, solvents, chemical agents, temperature, and irradiation; and the type, magnitude, duration and/or frequency of deformation. Some polymers yield, others break. Some do it immediately; others are unchanged even after months. The fracture surface can be smooth or splintery; the elongation at fracture, less than 1 % or greater than 1000 %.

Test methods try to simulate complex real-life situations by standardized procedures. They include long-term experiments such as static deformations under constant load by tension, compression, or bending; short-term methods such as tensile tests under various speeds; impact tests with unnotched or notched specimens; dynamic testing with various numbers of loadings-unloadings, impacts, vibrations, etc.

Two failure modes can be distinguished: brittle and tough (ductile). *Tough failures* are caused by viscous flow ("plastic flow"). This flow process may involve the slipping of chain segments past each other (amorphous polymers) or the movement of crystalline domains (partly crystalline polymers). Polymer chains may also de-entangle if small stresses act during long times. A yielding polymer fails for load-bearing applications but yielding may be desirable for packaging films.

Brittle fractures are rare for ideal solids since many bonds must be severed simultaneously. Real solids contain however many small imperfect regions that act as "nuclei" for the formation of microcracks. Brittle polymers usually possess "natural" microvoids, which may also appear in drawn amorphous polymers or through separation of crystal lamellae in hard-tough polymers. Brittle polymers fracture perpendicularly to the stress direction, tough polymers longitudinally. A polymer is defined as brittle if its elongation at break is less than 20 %.

7.6.2. Theoretical Fracture Strength

The fracture of brittle polymers generates free radicals. Since the probabilities of such homolyses depend on bond strengths, which also determine tensile moduli, relationships must exist between the theoretical moduli and the theoretical fracture strengths of polymers.

Bonds are severed if atoms are separated from each other by certain distances L_b that are greater than their equilibrium distances L_o. The necessary theoretical strength $\sigma_{\parallel}{}^{o}$ can be calculated by assuming a sine function with wavelength λ for the stress-distance curve; differentiation of the function and insertion of Hooke's law delivers the Frenkel equation:

$$(7\text{-}15) \quad \sigma_{\parallel}^{o} = \frac{E_{\parallel}^{o}\lambda}{2\,\pi L_o} = \frac{E_{\parallel}^{o}(L_b - L_o)}{\pi L_o} = K\,E_{\parallel}^{o}$$

Fracture occurs if atoms have moved away from the equilibrium distance L_o by $\lambda/2 = L_b - L_o$. Polymer main-chain bonds break at approximately the same relative distance $L_b \approx 1.3\,L_o$ because lengths and strengths of bonds are not too different for bonds such as C-C, C-O, C-N, etc. Thus, $K \approx 0.095_5$ and the theoretical fracture strength $\sigma_{\parallel}{}^{o}$ should be ca. 9.6 % of the theoretical tensile modulus $E_{\parallel}^{o}$ in chain direction, regardless of the chemical nature of the polymer.

Poly(ethylene) with a longitudinal theoretical modulus of 340 GPa (Table 7-5) should thus have a theoretical fracture strength of ca. 32 GPa in chain direction (i.e., much higher than the theoretical strength of steel (ca. 20 GPa)). Industrially manufactured ultradrawn poly(ethylene) fibers have already higher experimental fracture strengths than steel (2.9 GPa $\hat{=}$ 3.1 N/tex (PE) vs. 2.5 GPa (steel)).

The theoretical fracture strength-tensile modulus relationship of Equation (7-15) has been realized for certain ultradrawn poly(ethylene)s where the predictions of Equation (7-15) were confirmed experimentally: a proportionality constant K = 0.096 and a dependence of $\sigma_{\parallel}$ on the first power of E (Fig. 7-12). The same first power dependences (but smaller constants K) are also observed for ultradrawn it-poly(propylene)s and for films and fibers from molecular composites of poly(*p*-phenylene-2,6-bisbenzthiazole) (PBTZ) with rod-like mesogens in coil-like polybenzimide (PBI), both for heterogeneous and for homogeneous composites. However, a power dependence $\sigma = K' \cdot E^n$ was found in other ultradrawing experiments.

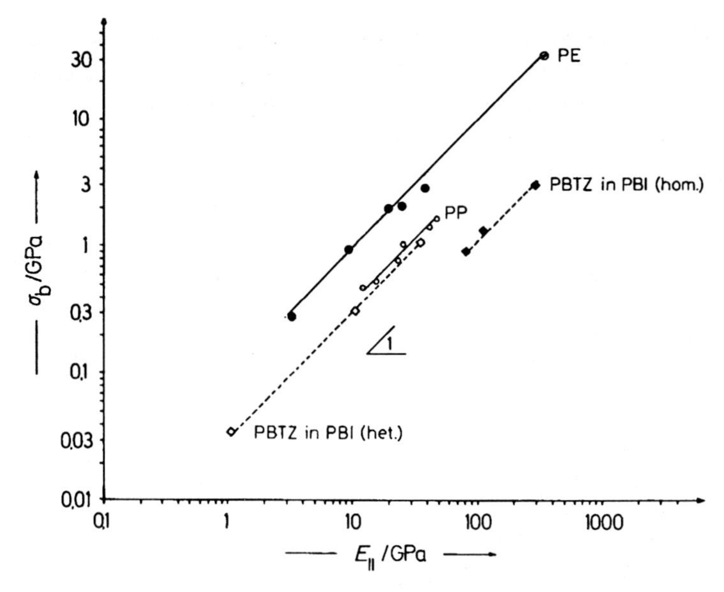

Fig. 7-12 Tensile strength at break σ_b as function of longitudinal tensile modulus $E_{\parallel}$. The solid line corresponds to $\sigma_b = 0.095E_{\parallel}$.

- ● Experimentally ultradrawn ultrahigh-modulus poly(ethylene)s;
- **O** theory for perfectly aligned poly(ethylene) chains;
- o ultradrawn poly(propylene)s;
- ◇ heterogeneous molecular composites of PBTZ fibers or films in PBI matrix;
- ◆ homogeneous molecular composites of PBTZ fibers or films in PBI matrix.

PBTZ PBI

7.6.3. Fracture Strengths of Unaligned Chains

Plastics have considerably lower moduli and strengths than those predicted for theoretical longitudinal and transverse chain directions (see Tables 7-4 and 7-6). Theoretical fracture strengths relate to infinitely long, completely aligned, immobile polymer chains, i.e., to perfect extended chain crystals. Chain segments of amorphous polymers and of amorphous regions in semicrystalline polymers are distributed at random, however. Fracture occurs mainly between segments of different chains; the fracture strength is very low because only intermolecular bonds are severed.

Fracture strengths increase with the inverse molar mass M of polymers with medium degrees of polymerization (Fig. 7-13). The effect was assumed to be caused by a diminishing role of endgroups. This explanation cannot account for the lack of fracture strength below a critical molar mass and the constancy of strength at high M.

The molar mass dependence of fracture strengths seems rather predominantly due to entanglements. The higher the molar mass, the more often a polymer chain can become entangled. Fracture strengths thus increase linearly with the inverse number-average molar mass of the polymers above the critical molar mass M_c for entanglements (Fig. 7-13) until additional entanglements no longer contribute to the strength of the physical network created by the entanglements. The fracture strength then becomes practically independent of the molar mass (see PC in Fig. 7-13).

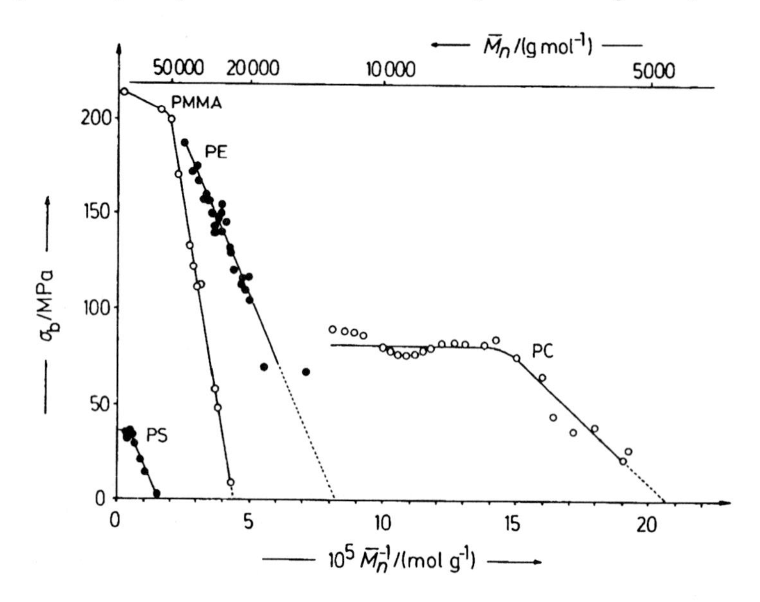

Fig. 7-13 Fracture strengths of various polymers as function of their inverse number-average molar masses. PS: atactic poly(styrene)s with narrow molar mass distributions at room temperature; PMMA: atactic poly(methyl methacrylate)s at -196°C; PE = poly(ethylene)s at -196°C; PC = radiated bisphenol A-polycarbonates at room temperature [8].

Both the onset of fracture strength and the onset of the plateau region are affected by the molar mass distribution of the polymers and the preparation of the test specimen (Fig. 7-14). Injection-molded specimens have higher fracture strengths than compression-molded ones of the same molar mass and molar mass distribution because injection molding leads to segment orientation in machine direction which in turn adds to fracture strength. Polymers with broad molar mass distributions have higher fracture strengths than those with narrow ones of the same number-average molar mass because they contain more longer chains with a greater probability of entanglements. Young's moduli do not show the transition region between onset and plateau because they reflect the resistance to rather small deformations.

All data can be fitted by a master curve $\sigma_b/E = f(\varepsilon_b)$, regardless of the broadness of the molar mass distribution and the type of specimen preparation (Fig. 7-15). The normalized fracture strength σ_b increases linearly with the ultimate elongation ε_b up to $\sigma_b/E = \varepsilon_b \approx 0.01$. The ratio σ_b/E then increases further with elongation and finally becomes independent of ε_b.

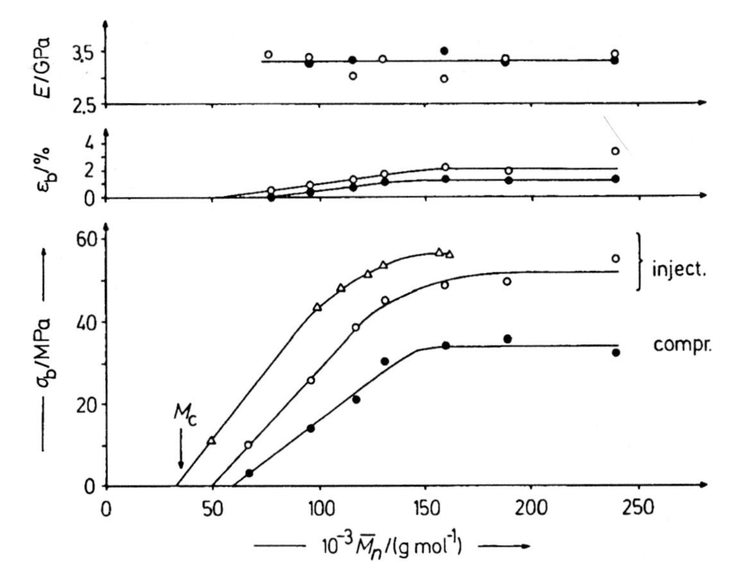

Fig. 7-14 Young's moduli E, elongations at break ε_b, and tensile strengths at break σ_b as function of the number-average molar masses of compression-molded (●) or injection-molded (○,△) poly-(styrene)s with narrow (○,●) and broad (△) molar mass distributions [9].

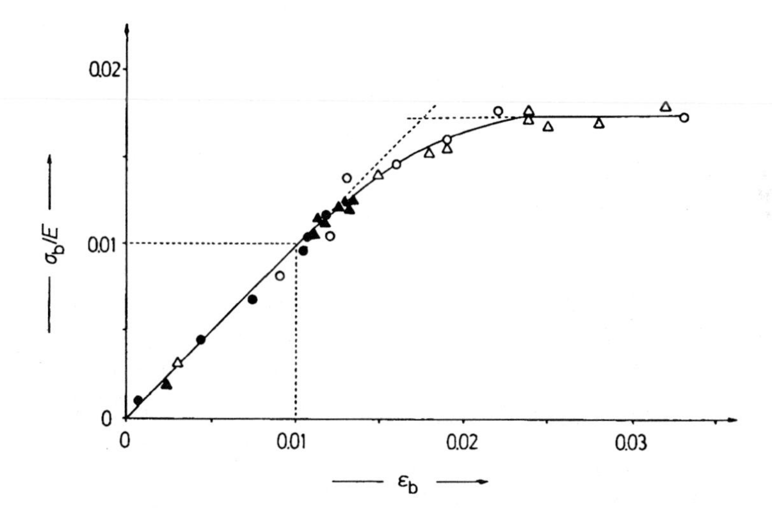

Fig. 7-15 Normalized fracture strength σ_b/E as function of the elongation ε_b at break for compression-molded (●,▲) and injection-molded (○,△) poly(styrene)s with narrow ((○,●) or broad (△,▲) molar mass distributions. Data of Fig. 7-14 and [9].

The plateau value of $\sigma_b/E = 0.0172$ for atactic poly(styrene)s is lower than the theoretical value of $K = 0.0955$ predicted by Equation (7-15) and the experimental value for fully aligned polymer chains. Because lower values than 0.096 are found for all plastics, various theories have been advanced for explanations:

The *Ingles theory* was originally introduced in order to explain the fracture properties of silicate glasses. A brittle fracture across such polymers will create two new surfaces; the necessary work per area is obtained by integration of the model stress function (see Section 7.6.2) from 0 to $\lambda/2$ (i.e., $\int \sigma \mathrm{d}L = \lambda \sigma_b^o/\pi$). Each of the two new surfaces has a surface energy of γ_{lv}; thus, $\lambda \sigma_b^o/\pi = 2\gamma_{lv}$. Insertion of Equation (7-15) leads to the theoretical fracture strength for brittle, energy-elastic bodies

$$(7\text{-}16) \quad \sigma_b^o = (E\gamma_{lv}/L_o)^{1/2}$$

The fracture strengths predicted by Equation (7-16) are however much higher than the experimentally found fracture strengths σ_b of plastics and fibers (Fig. 7-16). The ratios σ_b^o/σ_b of molded (unoriented) plastics decrease with increasing elongation ε_b at break, whereas those of drawn fibers (partially oriented chains) increase.

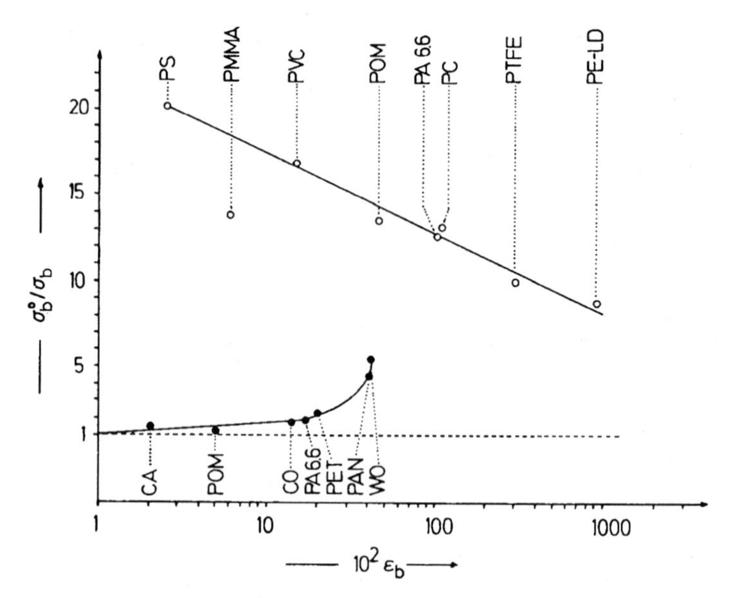

Fig. 7-16 Ratio of theoretical (σ_b^o) to experimental (σ_b) fracture strengths as function of elongation at break ε_b.
O Unoriented polymers (molded plastics);
● oriented polymers (fibers).
CA = Hemp (canabis), CO = cotton, WO = wool (for other abbreviations see Appendix).
Reprinted with permission by Hüthig and Wepf, Basle [10].

The *Griffith* theory explains the deviation of σ_b from σ_b^o as effect of the formation of new surfaces; the resulting microvoids act on drawing as nuclei for cracks. A crack can only grow if the energy required for the fracture of chemical bonds is just surpassed by the stored elastic energy. This theory predicts a dependence of the fracture strength σ_o on crack length L and surface energy γ_{lv}:

$$(7\text{-}17) \quad \sigma_o = [(2\,E\gamma_{lv})/(\pi L)]^{1/2}$$

This functionality is indeed observed for artificially introduced long cracks. The predicted fracture strengths are however much smaller than those found for polymers because the ultimate strength of plastics is not dominated by the cleavage of chemical bonds but by other types of energy absorption (crazing; shear flow).

7.6.4. Yield Strength

Upon drawing, an amorphous or a semicrystalline polymer reaches the upper yield point (see Fig. 7-1), where it either breaks (brittle polymers) or yields (ductile polymers). Yielding is always followed by stress-softening (Figs. 7-1, 7-3, and 7-6). This behavior resembles superficially the plasticity of metals. Plastics, despite their name, do not show plasticity, however:

True *plasticity* is characterized by an energy elastic behavior up to the yield point, followed by a liquid-like flow under application of high distortional stress; i.e., a continuous deformation without relaxation beyond the yield point. The specimen does not increase its volume and the work is totally converted as heat. Plastics, on the other hand, already show deviations from the energy elastic behavior before reaching a yield point. The stress-softening occurs with volume increase and only a part of the work is liberated as heat, for example, less than 50 % in poly(ethylene), which undergoes considerable stress-softening (see Fig. 7-6).

Plastics show this behavior because their stress-softening can be caused by shear flow or crazing or a combination of both. Shear flow is a viscous process; the volume of the specimen does not change. Crazing involves the formation of microfibril-filled "microvoids" (see below), i.e., an increase of volume. It is expedient to distinguish the phenomena at the yield point from the happenings after the yield point.

At the yield point, glass temperatures of polymers may be below or above the testing temperature. Two groups of polymers can thus be distinguished:

Mobilities of chain segments are low below glass temperatures. Such polymers include polyamides PA 6 and 6.6 ($T_G = 50°C$), poly(vinyl chloride) PVC ($T_G = 85°C$), certain styrene-butadiene copolymers SB ($T_G > 73°C$), a copolyamide 6I/XT ($T_G \approx 106°C$), and a polyethersulfone PESU ($T_G > 200°C$). The yield strength σ_y of these polymers is directly proportional to their secant moduli E (Eq.(7-15); Fig. 7-17). The proportionality constant K allows to calculate a separation ratio $L_b/L_o = 1 + \pi K$ from $\lambda/2 = L_b - L_o$ and $K = \lambda/(2\ \pi L_o)$ (Eq.(7-15)). These separation ratios follow the known bond strengths of physical bonds: high for hydrogen bonds (1.093 for PA 6I/XT; 1.085 for PA 6), high to medium for dipole-dipole interactions (1.087 for PESU; 1.068 for PBT; 1.060 for PVC), and small for π-π interactions (1.050 for PS; 1.039 for SB). They are considerably smaller than those of covalent bonds (ca. 1.3).

A second group comprises those polymers whose glass temperatures are lower than the testing temperature: conditioned PA 12 (T_G of conditioned polymer lower than $T_G = 42°C$ of the dry polymer); poly(oxymethylene) POM (- 30°C); high density poly(ethylene) PE-HD (- 80°C); and poly(propylene) (- 10°C). The fracture strength

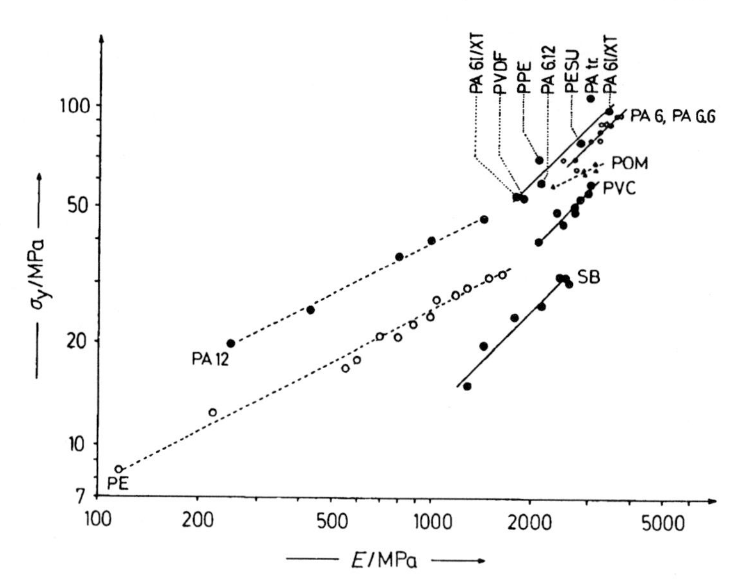

Fig. 7-17 Logarithm of yield strengths σ_y of plastics as function of the logarithm of Young's modulus E at 25°C. Slopes of lines are 1/2 (- - -) (Eq.(7-17) or 1 (-----) (Eq.(7-15)) [11]

of these polymers varies with the square root of Young's moduli. Application of Equation (7-17) gives "crack" lengths L between 14 nm (PA 12) and 39 nm (PE-HD).

These two types of $\sigma_y = f(E)$ functions probably represent the initiation of yielding since the yield point marks the conditions near the onset of yielding. The yielding itself occurs beyond the yield point on further deformation and the phenomena observed far beyond the yield point are not necessarily identical with those directly at the yield point.

Immediately beyond the yield point, specimens undergo stress-softening. This easy deformability without additional heating was found first at room temperature and has therefore been called *cold flow*. Stress-softening is accompanied by necking (telescope effects, see Fig. 7-2). The temperature increases in neck zones, sometimes up to several tens of degree Celsius. This local warming decreases the viscosity, which promotes flow. Necking is also observed in isothermal experiments, however. The effect must thus have another primary reason:

The width of test specimens varies slightly. Greater tensile stresses thus reign at the smaller cross sections, causing a relative decrease of extensional viscosities. The flow leads to friction, which generates heat. The liberated heat is stowed locally, which in turn stabilizes the necking zone.

On normal stress yielding, the whole specimen yields by *shear flow*, either homogeneously or heterogeneously (localized). In the latter case, shear bands are formed at angles of 38-45° to the stress direction. Chain segments are arranged at angles between shear bands and stress directions.

Crazing is the yield mechanism for polymers with $T_G < T$, regardless of whether they are amorphous, semicrystalline, linear, or weakly cross-linked. Crazes can be up to 100 μm long and 10 μm wide; their long axes are parallel to the stress direction (Fig. 7-18). They are not voids since their interior is filled with amorphous microfibrils of 0.6-30 nm diameter. These microfibrils are oriented in the stress direction, i.e., perpendicular to the long axes of the crazes.

On further deformation, microvoids are formed and the material stress-softens. The volume of the polymer thus increases upon crazing whereas it stays constant for shear flow.

The difference in refractive indices between microfibrils and spherulithic structures in crazed semicrystalline polymers and/or polymer and air leads generally to extensive light scattering because the diameters of these phases exceed on half of the wavelength of incident light. The phenomenon is also known as *stress-whitening*.

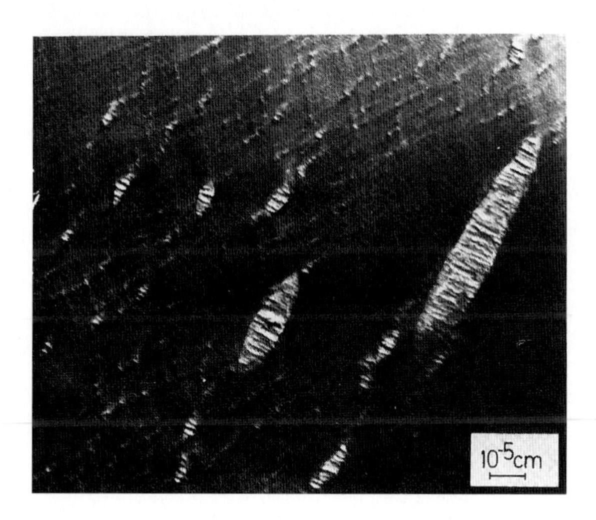

Fig. 7-18 Crazes in a poly(styrene) drawn to 25 %. Reprinted with permission by S.Wellington and E.Baer [12].

Continued drawing generates more microfibrils, which increase the resistance against strain and thus lead to strain-hardening. Low molar mass organic materials, metals, and thermosets lack the ability to form microfibrils; crazing is thus a unique process in thermoplastics.

Shear flow and crazing are energy absorbing processes for thermoplastics. Shear flow can spread in principle through the whole specimen whereas crazing is localized. Shear flow thus absorbs more energy in total; it is also less sensitive to environmental effects.

Although less efficient, crazing is still capable to raise the crack resistance of thermoplastics relative to thermosets. It is utilized in the rubber-reinforcement of amorphous polymers with high glass temperatures such as poly(styrene) (T_G = 100°C) (see Chapter 13). Rubber-modified poly(propylene), on the other hand, deforms mainly by shear flow because of the high mobility of of the PP chain segments in amorphous regions (T_G = -10°C).

7.6.5. Impact Resistance

Impact strength is the resistance of a material against impact. It is one of the many quantities used to characterize the strength of a material under (the usually complex) use conditions; all test methods are thus standardized. Most test methods measure the energy required to break a notched or unnotched specimen (Izod, Charpy, high-speed tensile). Impact speeds range from 10^{-5}-10^{-1} m/s in conventional tensile tests to 20-250 m/s for high speed tensiles; elongation speeds are usually from 10^{-3}-10^{4} s^{-1}.

Impact strengths depend on experimental conditions (Table 7-9). In *Charpy* tests, a specimen lies horizontally on two supports and is struck from above by a pendulum. Notches are at the bottom side because it is the side with the largest tensile stress. Notched impact strengths are usually given in energy per cross-sectional area of the specimen, e.g., in J/m^2 (Europe) or in ft lbf/in^2 (USA).

Table 7-9 Notched impact strengths I of rectangular test specimens with lengths L_x, widths L_y, and thicknesses L_z of a polycarbonate. h = Residual height; cr. = cracked. Reprinted by permission of Hanser Publ., Munich [12].

Method	Standard	Specimen Type	Size in mm	Notch shape	h/mm	I/(kJ m^{-2})
Charpy	DIN 53 453	NKS	50 × 6 × 4	U	2.7	30
Charpy	ISO 179	ISO	80 × 10 × 4	U	2.7	40 cr.
Charpy	ISO 179	ISO	80 × 10 × 4	V	3.2	25
Izod	ISO 180	ISO	80 × 10 × 4	V	8.0	70 cr.
Izod	ASTM D 256	ASTM	63 × 12.7 × 3.2	V	10.1	85 cr.

In *Izod* tests, vertically arranged specimens are clamped at the bottom and struck at the upper part by a pendulum. Notches are in the center of the specimen, below the hitpoint of the pendulum. The smaller the radius of the notch, the higher is the stress concentration at the tip and the lower is the impact strength at constant temperature (Fig. 7-19). Impact strengths are also affected by the thickness of the specimens. The impact strength of infinitely thick specimens is governed by the stress field; the deformation of the whole specimen is unimportant. Infinitely thin specimens, on the other hand, are strongly deformed but there is no stress field.

Notched impact strengths are reported in different units: in the United States as fracture energy per width of notch (ft lbf/in → kJ/m); in Europe, as fracture energy per width of notch and thickness of specimen (kJ/m^2). For finite thicknesses, the former is more indicative of crack initiation, the latter more of crack propagation.

Impact strengths from Charpy and Izod tests are not intrinsic material properties because they depend on external conditions such as specimen thickness and notch shape, depth, and radius. They are helpful for the comparison of various materials but are useless for design calculations. For engineering purposes, impact strengths are characterized by a critical stress intensity factor K_{1c} or a fracture toughness G_{1c} (Table 7-10). These parameters are interrelated by:

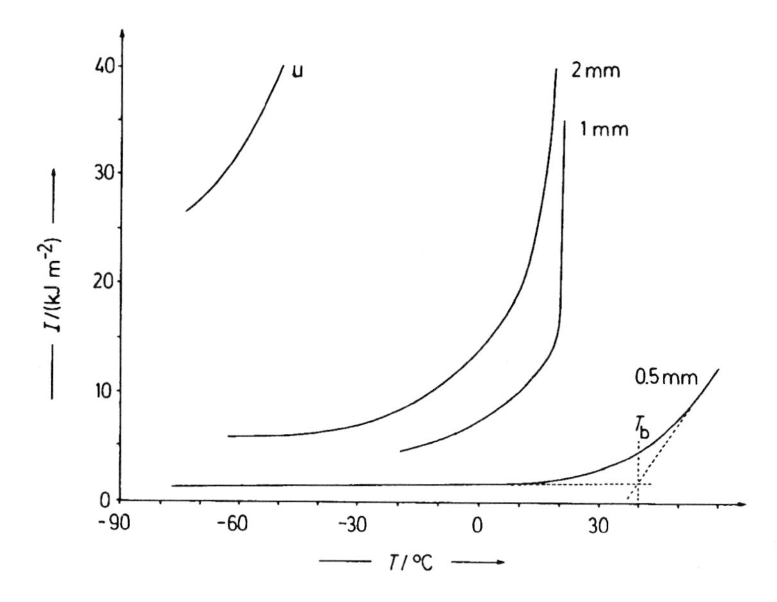

Fig. 7-19 Temperature dependence of impact strengths of unnotched (u) and notched test specimens of a poly(vinyl chloride) (numbers indicate notch radii). T_b = brittle temperature for the indicated notch radius. The glass temperature is 80°C. Reproduced with permission by Plastics Institute, London [14].

$$(7\text{-}18) \quad K_{1c}^2 = EG_{1c} = Y^2\sigma^2(L/2)$$

where E is Young's modulus, σ is the stress, L is the crack length, and Y is a calibration factor that depends on the geometry of the specimen. G_{1c} equals two times the surface work for a perfectly brittle solid (no plasticity).

Tab. 7-10 Glass temperatures T_G and Young's moduli E, critical stress intensity factors K_{1c}, and fracture toughnesses G_{1c} of various materials at room temperature.

Material	$\frac{T_G}{°C}$	$\frac{E}{GPa}$	$\frac{K_{1c}}{MPa\ m^{1/2}}$	$\frac{G_{1c}}{kJ\ m^{-2}}$
Natural rubber, vulcanized	- 73	0.003	0.2	13
Poly(ethylene), medium density	- 80	0.89	5.0	28
Poly(propylene), isotactic	- 15	1.37	4.6	16
Polyamide 6	50	3.0	2.8	2.6
Poly(vinyl chloride)	82	3.6	2.5	1.7
Poly(methyl methacrylate), atactic	105	3.4	1.7	0.5
Polycarbonate A	150	3.2	2.2	1.5
Poly(styrene), atactic	100	3.3	1.0	0.3
Poly(styrene), atactic, rubber-modified	?	2.0	2.0	2.0
Epoxide, hardened	?	2.5	0.5	0.1
Epoxide, rubber-modified, hardened	?	2.4	2.2	2.0
Glass	?	70	0.7	0.007
Steel	-	210	150	107

All polymers are brittle at very low temperatures. The mobility of chain segments increases with increasing temperature, allowing stresses to be relieved by shear-band or craze formation: impact strengths increase with temperature, especially near the glass temperature (Fig. 7-19). Polymers with additional transition temperatures below the glass temperatures are for the same reason almost always more impact resistant than polymers without such transitions. Nonentangled polymers exhibit very low impact strengths because no crazes can be formed. The impact behavior of polymers can be improved considerably by modification with rubber (see Section 13.4.3).

7.6.6. Stress Cracking

Stress cracking (stress corrosion, stress crazing) is the formation of crazes under the physical action of chemicals, especially surfactants or solvents. Stress crazing starts at polymer surfaces and proceeds into the interior until the polymer finally cracks. The appearance and the extent of stress cracking depends on the interaction polymer-reagent and the magnitude of the stress.

Effects are small in nonwetting liquids but strong in polymer-liquid systems with solubility parameters of polymers and liquids matching each other (Fig. 2-14) and even more dramatic under tension in the presence of surfactants. Stress cracking decreases with increasing molar mass of the polymer since entanglements allow stresses to relax elastically. Cross-linked polymers are less prone to stress cracking for the same reason. Stress cracking is also reduced if polymer plasticizers are present in plastics because these additives increase the mobility of chain segments and thus the ability to relieve stresses. The same action is responsible for the fact that no stress corrosions are observed above glass temperatures.

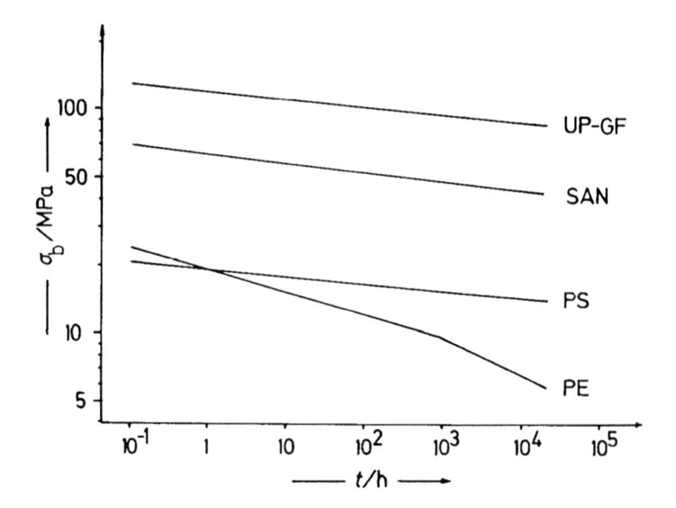

Fig. 7-20 Time dependence of tensile strengths of a poly(ethylene) PE, a poly(styrene) PS, a styrene-acrylonitrile copolymer SAN, and a glass fiber-filled cross-linked unsaturated polyester UP-GF under constant load. Reprinted with permission by BASF, Ludwigshafen/Rhein [15].

7.6.7. Fatigue

Materials may be damaged not only "instantaneously" (i.e., on impact) but also by static or periodic loads after certain times or number of loadings. This fatigue is characterized by the *fatigue limit* (endurance) at which the plastics are not damaged even after infinite time and the *fatigue strength*, which indicates the load at which damage sets in after a certain time. Fatigue is caused by flow of polymers, growth of crazes and cracks, recrystallization phenomena, time-dependent environmental interactions, and the like.

Plastics may be subjected to static loads for certain times t after which their fracture strengths σ_B are measured by tensile tests. The logarithms of strengths of amorphous polymers usually decrease linearily with logarithms of time due to viscous flow (Fig. 7-20). Partially crystalline polymers show a bend of these lines after certain times, which indicates a change from ductile fracture (short times) to brittle fracture (large times), probably caused by recrystallization phenomena. These *creep strengths* vary widely: at a load of 40 MPa, standard poly(styrene)s fracture after only 0.1-10 hours whereas impact grades survive 10 000 hours.

Fatigue can also be measured under cyclic loading such as the number N of loadings (bending vibration tests) or the number of torsions (torsion vibration tests). The results of bending vibration tests are usually plotted as logarithms of flexural strengths σ_b vs. the number N of loadings at constant flexural stress. The resulting "Wöhler curves" show dramatic differences for the various plastics (Fig. 7-21). For example, at flexural strengths of 40 MPa, standard poly(styrene)s fracture after N = 300 loadings whereas impact-modified poly(styrene)s can be subjected to more than 1 million loadings. Cyclic loading leads to smaller flexural strengths than static loading. Similar results are obtained by torsion vibration tests.

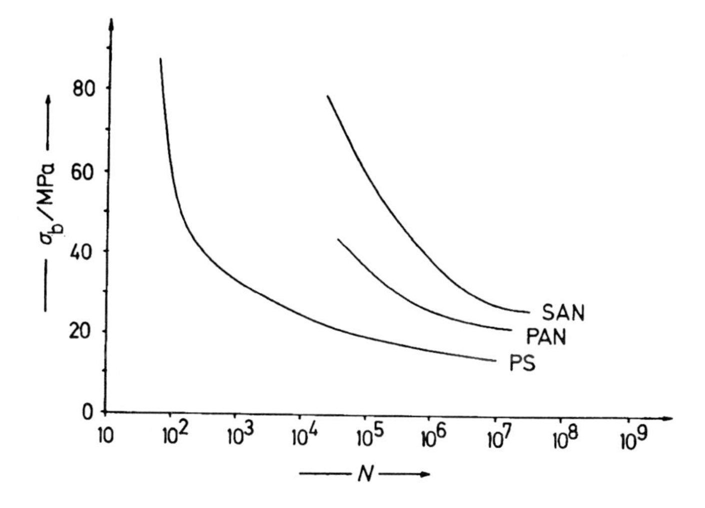

Fig. 7-21 Dependence of flexural strengths of a poly(styrene) PS, a poly(acrylonitrile) PAN, and a styrene-acrylonitrile copolymer SAN on the number N of loadings. Reprinted with permission by BASF, Ludwigshafen/Rhein [15].

7.7. Surface Mechanics

7.7.1. Hardness

The hardness of a material is its resistance against the penetration by another body. Hardness is a very complex quantity; it depends on Young's modulus, yield stress, and stress hardening. A general definition of hardness, applicable to all materials, does not exist; neither does a universally applicable testing method. The various technical test methods thus emphasize the one or the other factor that contributes to the hardness of a specific class of materials.

The hardness of *hard plastics* is normally characterized by various Rockwell (ISO, ASTM) or ball indentation hardnesses (DIN, ISO). These methods measure the indentation of a polymer by a steel sphere under load and thus the compression set *and* the recoverable deformation. The plastic deformations of polymers increase with time (creep), whereas those of metals are time independent. Because of the short duration of the hardness test, polymers exhibit relatively high Rockwell hardnesses.

The hardness of *soft plastics* is characterized by their durometer (ASTM) or various Shore hardnesses (ISO, DIN). These methods measure the resistance to penetration by a truncated cone (static methods). Hardness properties of metals and hard plastics are evaluated by another Shore hardness, which uses the rebound of a small steel sphere (dynamic methods).

All methods measure the hardness of surfaces, not of the interior of the specimen. The surface may, for example, be plasticized by the humidity of the air. Crystallizable polymers may have lower surface hardnesses than interior hardnesses (if the plastic had been injected into a cold mold) or the reverse may be true (if transcrystallization occured). Because hardness tests are affected by the elasticity of the specimen, degrees of hardness depend also on the type and the thickness of the support.

7.7.2. Friction

Friction is the resistance against the relative movement of two bodies contacting each other. It is measured by the friction coefficient $\mu = R/L$ (i.e., the ratio of friction R to total load L). Friction depends in a complex and not understood way on both the surface roughness and the mechanical properties of the specimen.

The rolling of *hard bodies on soft materials* is almost exclusively determined by the deformation of the soft base (i.e., its viscoelastic properties). Elastomers thus have fairly high friction coefficients of $0.5 < \mu < 3.0$, depending on the contacting body and its type of movement (rolling, sliding).

The sliding of *hard bodies on other hard bodies* occurs on the tops of the microscopic surfaces: the true contact area is much smaller than the geometric one. The applied load thus acts on very small effective areas. Local stresses are high and the tops are leveled. Large adhesion forces exist between chemical groups of the resulting effective contact areas of both bodies, which have to be overcome by breaking the bonds or by shearing one of the materials. The adhesive friction $R = A_w\sigma_b$ is given by the effective contact surface A_w and the shear strength σ_b. Soft materials possess high effective surfaces (large A_w) and are easily sheared (small σ_b), whereas the opposite is true for hard materials. Plastics, metals, and ceramics therefore often exhibit very similar friction coefficients (Table 7-11).

Table 7-11 Friction coefficients of various sliding bodies.

Plastic	Friction coefficient of Plastic on plastic	 Plastic on steel	 Steel on plastic
Poly(methyl methacrylate)	0.8	0.5	0.45
Poly(styrene)	0.5	0.3	0.35
Poly(ethylene), low density	0.3	-	0.80
Poly(ethylene), high density	0.1	0.15	0.20
Poly(tetrafluoroethylene)	0.04	0.04	0.10

7.7.3. Abrasion and Wear

Abrasion is the loss of material from surfaces by friction. It is thus affected by both the friction properties and the hardness of specimens. The abrasion coefficient K is given by the applied force F, the linear speed v of the contacting body, the total time t, and the volume loss ΔV of the abraded material

$$K = \Delta V/(Fvt) \qquad (7\text{-}19)$$

Abrasion coefficients vary widely with polymer type and state (resting, mobile) (Table 7-12). The best resistance against abrasion is shown by polyureas, followed by polyamides and polyacetals. It can be enhanced greatly by addition of certain fillers (e.g., short fibers).

Table 7-12 Abrasion coefficients K for moving plastics against resting materials.

Resting material	$\frac{10^{10}\,K}{\text{MPa}^{-1}}$	Moving plastics	$\frac{10^{10}\,K}{\text{MPa}^{-1}}$
Polycarbonate A	200 000	Polyamide 6.6	11 000
Polyamide 6.6	250	Polycarbonate A	9 800
Polyamide 6.6	220	Polyamide 6.6	510
Polyamide 6.6	10	Polyacetal	12
Polyacetal	11	Polyamide 6.6	15
Steel	-	Polyamide 6.6	8 600
Steel	-	Polyamide 6.6 with 30 % glass fibers	1

Literature

7.1. Introduction: General Texts

L.Nielsen, Mechanical Properties of Polymers and Composites, Dekker, New York, 2 vols. (1974 and 1976)
R.G.C.Arridge, Mechanics of Polymers, Clarendon Press, Oxford 1975
D.W. van Krevelen, Properties of Polymers - Correlation with Chemical Structures, Elsevier, Amsterdam, 3rd ed. 1989
J.S.Hearle, Polymers and Their Properties. Vol. 1: Fundamentals of Structure and Mechanics, Wiley, New York 1982
I.M.Ward, Mechanical Properties of Solid Polymers, Wily, New York, 2nd ed. 1983
J.I.Kroschwitz, ed., Polymers. An Encyclopedic Sourcebook of Engineering Properties (= reprints of articles in H.F.Mark et al., Encyclopedia of Polymer Science and Engineering), Wiley, New York, 2nd ed. 1987
A.W.Birley, B.Haworth, J.Batchelor, eds., Physics of Plastics, Hanser, Munich 1991

7.1.1. and 7.1.2. Deformation of Polymers, Testing

G.C.Ives, J.A.Mead, M.M.Riley, Handbook of Plastics Test Methods, Iliffe, London 1971
S.Turner, Mechanical Testing of Plastics, Butterworths, London 1973
A.Y.Malkin, A.A.Askadsky, V.V.Kovriga, A.E.Chalykah, Experimental Methods in Polymer Physics, Mir Publ., Moscow; Prentice Hall, Englewood Cliffs (NJ), 1978 and 1979 (2 vols.)
J.G.Williams, Stress Analysis of Polymers, Wiley, Chichester, 2nd ed. 1980
J.F.Rabek, Experimental Methods in Polymer Chemistry. Physical Properties, Wiley-Interscience, New York 1980
I.I.Perepechko, Low-Temperature Properties of Polymers, Pergamon, London 1980
R.A.Pethrick, R.W.Richards, ed., Static and Dynamic Properties of the Polymeric Solid State, Reidel, New York 1982
V.Shah, Handbook of Plastics Testing Technology, Wiley, New York 1984
G.Kämpf, Characterization of Plastics by Physical Methods, Hanser, Munich 1986
R.P.Brown, ed., Handbook of Plastics Test Methods, Wiley, New York, 3rd ed. 1989

7.2. Energy Elasticity

I.M.Ward, Mechanical Properties of Solid Polymers, Wiley, New York, 2nd ed. 1983
A.E.Zachariades, R.S.Porter, eds., The Strength und Stiffness of Polymers, Dekker, New York 1983
A.E.Zachariades, R.S.Porter, eds., High Modulus Polymers – Approaches to Design and Development, Dekker, New York 1987

7.3. Entropy Elasticity

L.R.G.Treloar, The Physics of Rubber Elasticity, Clarendon Press, Oxford, 3rd ed. 1975
G.Heinrich, E.Straube, G.Helmis, Rubber Elasticity of Polymer Networks: Theories, Adv.Polym.Sci. **85** (1988) 33
J.E.Mark und B.Erman, Rubberlike Elasticity: A Molecular Primer, Wiley, New York 1988

7.4. Viscoelasticity and 7.5. Dynamic Behavior

N.G.McCrum, B.R.Read, G.Williams, Anelastic and Dielectric Effects in Polymeric Solids, Wiley, London 1967
R.M.Christensen, Theory of Viscoelasticity: An Introduction, Academic Press, New York 1970
T.Murayama, Dynamic Mechanical Analysis of Polymeric Materials, Elsevier, Amsterdam 1978
J.D.Ferry, Viscoelastic Properties of Polymers, Wiley, New York, 3rd ed. 1980
R.A.Pethrick, R.W.Richards, eds., Static and Dynamic Properties of the Polymeric Solid State, Reidel, New York 1982
R.T.Bailey, A.M.North, R.A.Pethrick, Molecular Motion in High Polymers, Clarendon Press, Oxford 1981
J.J.Aklonis, W.J.MacKnight, Introduction to Polymer Viscoelasticity, Wiley-Interscience, New York, 2nd ed. 1983
M.Doi, S.F.Edwards, The Theory of Polymer Dynamics, Oxford University Press, Oxford 1986
M.Nagasawa, ed., Molecular Conformation and Dynamics of Macromolecules in Condensed Systems (First Toyota Conference), Elsevier, Amsterdam 1988
S.Matsuoka, ed., Relaxation Phenomena in Polymers, Hanser, Munich 1992
R.A.Dickiem M.Shelef, eds., Prediction of Deformation Properties of Polymers, ACS, Washington, D.C., 1993

7.6. Fracture

E.H.Andrews, Fracture in Polymers, Oliver and Boyd, Edinburgh 1968
S.Rabinowitz, P.Beardmore, Craze Formation and Fracture in Glassy Polymers, Crit.Revs.Macromol.Sci. **1** (1972) 1
R.P.Kambour, A Review of Crazing and Fracture in Thermoplastics, J.Polymer Sci. [Revs.] **D 7** (1973) 1
J.A.Manson, R.W.Hertzberg, Fatigue Failure in Polymers, Crit.Revs.Macromol.Sci. **1** (1973) 433
E.H.Andrews, P.E.Reed, Molecular Fracture in Polymers, Adv.Polym.Sci. **27** (1978) 1
L.G.E.Struik, Physical Aging in Amorphous Polymers and Other Materials, Elsevier, Amsterdam 1978
R.W.Hertzberg, J.A.Manson, Fatigue of Engineering Plastics, Academic Press, New York 1980
H.H.Kausch, ed., Crazing in Polymers, Adv.Polym.Sci. **52/53** (1983)
A.J.Kinlock, R.J.Young, Fracture Behaviour of Polymers, Appl.Sci.Publ., London 1983
A.S.Argon, R.E.Cohen, O.S.Gebizlioglu, C.E.Schwier, Crazing of Block Copolymers and Blends, Adv.Polym.Sci **52/53** (1983) 275
J.G.Williams, Fracture Mechanics of Polymers, Wiley, New York 1984
W.Brostow, R.D.Corneliussen, ed., Failure of Plastics, Hanser, München 1986
A.G.Atkins, Y.W.Mai, Elastic and Plastic Fracture, Halsted Press, New York 1986
H.H.Kausch, Polymer Fracture, Springer, Heidelberg, 2nd ed. 1987

7.7. Surface Mechanics

J.J.Bikerman, Sliding Friction of Polymers, J.Macromol.Sci. [Revs.] **C 11** (1974) 1
L.H.Lee, ed., Advances in Polymer Friction and Wear, Plenum, London 1975 (2 vols.)
P.A.Engel, Impact Wear of Materials, Elsevier, Amsterdam 1978
B.J.Briscoe und D.Tabor, Friction and Wear of Polymers (= Chapter 1 of "Polymer Surfaces" D.T.Clark und W.J.Feast,ed.), Wiley, New York 1978

K.Sato, The Hardness of Coating Films, Progr.Org.Coatings **8** (1980) 1
G.M.Bartenev, V.V.Lavrentev, Friction and Wear of Polymers, Elsevier, Amsterdam 1981
G.Heinicke, Tribochemistry, Hanser, München 1984
F.J.Balta-Calleja, Microhardness Relating to Crystalline Polymers, Adv.Polym.Sci. **66** (1985) 117
I.M.Huchings, Tribology. Friction and Wear of Engineering Materials, CRC Press, Boca Raton (FL) 1993

References

[1] Modification of a graph by G.Kraus, Angew.Makromol.Chem. **60/61** (1977) 215, fig. 12.
[2] M.Beyer, E.Schollmeyer, Angew.Makromol.Chem. **60/61** (1989) 53, Fig. 3a.
[3] E.H.Andrews, Fracture in Polymers, Oliver and Boyd, London 1968, Fig. 1.14.
[4] Updated graph of H.-G.Elias, F.Vohwinkel, Neue polymere Werkstoffe für die industrielle Anwendung, Hanser, Munich 1983, fig. 17-1; New Commercial Polymers II, Gordon and Breach, New York, 1986, fig. 17-1.
[5] H.-G.Elias, Makromoleküle, Vol. I: Grundlagen, Hüthig and Wepf, Basle, 5th ed. 1990, Fig. 25-6.
[6] J.R..McLoughlin, A.V.Tobolsky, J.Colloid Sci. **7** (1952) 555, Figs. 3-5.
[7] S.Onogi, T.Masuda, K.Kitagawa, Macromolecules **3** (1972) 109, Fig. 2.
[8] D.T.Turner, ACS Polymer Preprints **19** (1978)
[9] H.W.McCormick, F.M.Brower, L.Kin, J.Polym.Sci. **39** (1959) 87, Tables 1-3.
[10] H.-G.Elias, Makromoleküle, vol. II, Technologie, Hüthig and Wepf, Basle, 5th ed. 1992, Fig. 14-10
[11] H.-G.Elias, submitted to Polymer
[12] S.Wellington, E.Baer, private communication.
[13] K.Oberbach, L.Rupprecht, Kunststoffe **77** (1987) 783, Table 2.
[14] P.I.Vincent, Impact Tests and Service Performance of Thermoplastics, Plastics Institute, London 1971, Fig.
[15] -, Kunststoff-Physik im Gespräch, BASF, Ludwigshafen/Rh., 2nd ed. 1968, Fig. on p. 38

8. Electrical Properties

8.1. Fundamentals

Matter behaves differently in an electric field: on application of an electric potential difference U (measured in volt V), some materials conduct electricity and some resist the flow of an electric current I (measured in ampere A). The electric resistance $R = U/I$ is measured in ohms $\Omega = \text{V A}^{-1}$ and the electric conductance $G = 1/R$ in siemens $\text{S} = 1/\Omega = \text{A/V}$. Because the latter is the inverse of the former, it is often given as mho = ohm^{-1}.

The *resistivity* ρ is defined as the electric resistance R between the opposite sides of a cube of the material with side lengths L; it is therefore conventionally called a *volume resistivity*. It is proportional to the cross-sectional area $A = L^2$ of the material and inverse proportional to its length L. The volume resistivity is thus $\rho = RL^2/L$; its physical units are Ω cm. Because the volume resistivity is "specific" to the material, it is conventionally called "specific volume resistivity" although it is not a specific quantity as defined by modern nomenclature (it is not normalized by mass).

The inverse of volume resistivity ρ is the *conductivity* σ (e.g., in S cm^{-1}). Conductivity was formerly called specific conductance in science and is still named specific conductivity in technology. CAMPUS® test conditions are shown in Table 8-1.

The electric resistance R of a material depends on the distance between electrodes. It is a resistance of the measuring system and not that of the material per se and thus an *electric surface resistance*, which cannot be converted into the volume resistivity ρ. The electric surface resistance is conventionally called "specific surface resistivity".

Table 8-1 Test conditions for electrical properties (at 23°C and 50 % relative humidity) in the CAMPUS® system. TO = Transmission oil. IEC = International Electrotechnical Commission; VDE = Verband deutscher Elektriker (Association of German Electricians).

Property	Conditions	Unit	IEC	DIN VDE	Remarks
Relative permittivity	50 Hz	1	250	0303 T4	Plate (1 ± 0.1) mm
Relative permittivity	1 MHz	1	250	0303 T4	Plate (1 ± 0.1) mm
Dissipation factor	50 Hz	1	250	0303 T4	Plate (1 ± 0.1) mm
Dissipation factor	1 MHz	1	250	0303 T4	Plate (1 ± 0.1) mm
Specific volume resistivity	contact electrodes	Ω cm	93	0303 T30	Plate (1 ± 0.1) mm
Specific surface resistivity	contact electrodes	Ω	93	0303 T30	Plate (1 ± 0.1) mm
Dielectric strength	P25/P75	kV/mm	243-1	0303 T21	in TO, IEC 296
Comparative tracking index	test soln. A	-	112	0303 T1	($\geq$15·15·4) mm^3
CTI 100 drops value	test soln. A	-	112	0303 T1	($\geq$15·15·4) mm^3
Comparative tracking index	M, test soln. B	-	112	0303 T1	($\geq$15·15·4) mm^3
CTI M 100 drops values	M, test soln. B	-	112	0303 T1	($\geq$15·15·4) mm^3
Electrolytic corrosion	-	-	426	0303 T6	(30·10·4) mm^3

Matter is subdivided according to its electric conductivity σ into electric insulators (σ = 10^{-22}-10^{-14} S/cm), semiconductors (10^{-9}-10^{-2} S/cm), conductors (> 10^{3} S/cm), and superconductors ($\approx 10^{20}$ S/cm). Most plastics are electric insulators (Section 8.2), but certain polymers are intrinsic semiconductors and some are conductors after doping (Section 8.3).

8.2. Dielectric Properties

8.2.1. Relative Permittivity

Metals and plastics behave differently in electric fields. Metals possess many electrons in their outer atomic shells that move on the application of an electric field. These electrons are shared by all nuclei. The electric conductivities of metals thus do not change very much with the chemical nature of the metals: Silver 590 000 S/cm, aluminum 330 000 S/cm, iron 100 000 S/cm, and mercury 10 200 S/cm.

In nonionic polymer molecules, however, electrons are tightly bound to atomic nuclei. The binding electron pairs of covalent bonds between two atoms are not shared equally by the atoms. They are rather nearer to one of the atoms than to the other. The resulting permanent dipole becomes oriented in field direction if an electric field is applied. In certain atomic groups, negatively charged electrons are shifted in the opposite direction of positively charged atomic nuclei to give induced dipoles.

Polarizations are difficult to determine directly. They are usually measured by the ratio of capacitances of a condensator in vacuo and in the specimen, i.e., the relative permittivity ε_r of the specimen; formerly called the dielectric constant. Relative permittivities are low for apolar polymers (PTFE, PE), higher for polymers with polarizable groups (PS, PC), and still higher for polar materials (dry PA) (Table 8-2).

Relative permittivities of apolar polymers decrease with increasing temperature. Higher temperatures let the volume expand; this widens the distance between atoms, which in turn decreases dispersion forces.

Relative permittivities of polar polymers increase however with increasing temperature because dipoles become more mobile, especially above glass temperatures. Plasticization also makes dipoles more mobile. The relative permittivity also increases with increasing water content of plastics ($\varepsilon_r = 81$ for water). It also increases with increased segmental mobility of the polymer molecules: $\varepsilon_r = 13.0$ for *cis*-1,4-poly(isoprene) vs. 3-8 for plastics (Table 8-2). Rubber-modified plastics thus have higher relative permittivities than conventional plastics. Fillers have high relative permittivities. They increase the relative permittivities according to mixing rules: up to 170 (metal-filled thermoplastics) and up to 18 000 (metal-filled elastomers). Expanded plastics and elastomers are composites with air ($\varepsilon_r = 1.000\ 58$) and subsequently have low relative permittivities of ca. 1-2.

Table 8-2 Electrical properties of plastics. * Depends on impurities from polymerization (e.g., emulsifier residues). For polymer abbreviations, see Appendix.
Δw = Water absorption at 50 % relative humidity;
ε_r = relative permittivity ("dielectric constant");
ρ = resistivity = inverse conductivity = "specific electrical volume resistivity";
R = surface resistance = inverse conductance = inverse "specific electrical surface resistivity";
tg δ = dissipation factor (dielectric loss tangent) (at 1 MHz), S = dielectric strength;
U = tracking resistance (method KC).

Polymer	Δw/%	ε_r	ρ/Ω cm	R/Ω	tg δ	S/kV mm^{-1}	U/V
PTFE	0	2.15	10^{18}		0.0001	40	> 600
PE	0.05	2.3	10^{17}	10^{13}	0.0007	70	600
PS	0.1	2.5	10^{18}	10^{15}	0.0002	140	500
ABS		3.2	10^{15}	10^{13}	0.02	15	600
PVC	< 1.8*	< 3.7*	10^{15}	10^{13}	0.015	< 50	< 600
PA 6, dry	0	3.7	10^{15}		0.03	< 150	600
PA 6, conditioned	9.5	7	10^{12}		0.3	80	600
CA	4.7	5.8	10^{13}		0.03	35	
UP		3.4	10^{13}	10^{12}	0.01	50	500
PF, unfilled		8	10^{14}		0.05	12	
PF, mineral-filled			10^{9}		< 0.5	10	125

8.2.2. Resistivity

Small electric fields cause an orientation of the dipoles of a dielectric material. Higher fields remove electrons from some atoms and ions result. It is these ions that conduct electricity in dielectric polymers and not the electrons as in metals.

Resistivities thus decrease with increasing content of polar groups and increasing mobilities of chain segments. The resistivity (specific volume resistivity) of dielectric polymers decreases with increasing relative permittivity and finally becomes constant at $\varepsilon_r > 8$. It also decreases with temperature, whereas that of metals increases.

Resistances R (specific surface resistivities) depend on far more parameters than resistivities ρ (specific volume resistivities). Especially important is the effect of humidity since it leads to considerable ionic conductivies if small amounts of surface impurities are present. Resistances are thus often 2-3 decades lower than resistivities.

8.2.3. Dielectric Loss

Dipoles try to follow the direction of electric fields if alternating currents are applied. The required adjustment times correspond to the orientation times of groups and molecules. The faster the alternation, the longer the orientation lags behind the field and the greater is the electrical energy consumed. Available power output is decreased because electric power is lost by conversion into thermal energy.

The ratio of power loss, N_v to total power output N_b is called the dielectric dissipation factor (power factor; loss tangent tan δ), which can also be expressed as the ratio of imaginary relative permittivity ε'' to real relative permittivity ε':

$$(8\text{-}1) \qquad N_v/N_b \equiv \tan\delta = \sin\delta/\cos\delta = \varepsilon''/\varepsilon'$$

Polymers with high loss factors $\varepsilon \cdot \tan\delta$ can be heated and thus welded by high frequency fields; PVC is an example. Polymers with low loss factors (PE, PS, PIB), on the other hand, are excellent insulators for high frequency conductors.

Real and imaginary relative permittivities depend on the frequencies ν of the alternating current (Fig. 8-1). The function $\varepsilon' = f(\nu)$ corresponds to a dispersion and the function $\varepsilon'' = f(\nu)$ to an absorption of energy. Transitions and relaxations consume energy and therefore inflection points (ε'-curves) and maxima (ε''-curves) are found at appropriate temperatures and frequencies.

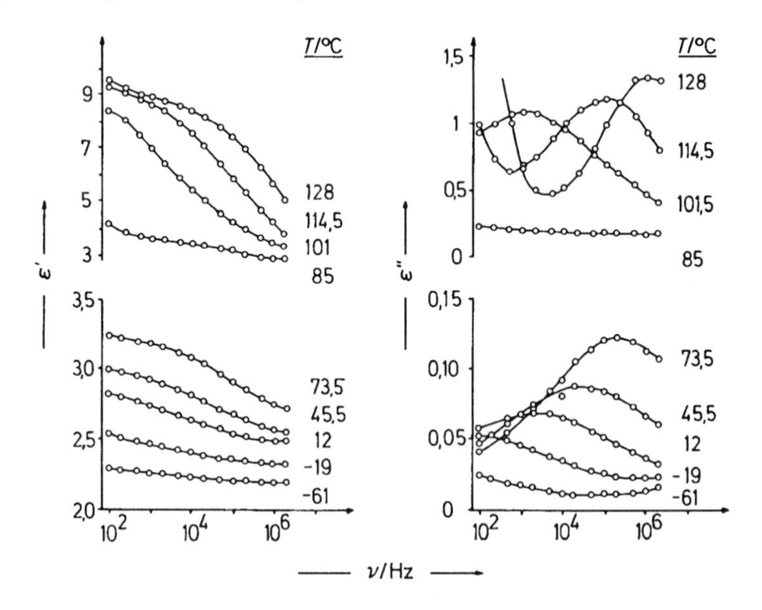

Fig. 8-1 Frequency dependence of real (ε') and imaginary relative permittivities (ε'') of a poly(vinyl chloride) at various temperatures. Upper row: β-dispersion; lower row: γ-dispersion. Reproduced with permission by Steinkopff -Verlag, Darmstadt, Germany [1].

8.2.4. Dielectric Strength and Tracking Resistance

The imaginary part of the relative permittivity is caused by the dissociation of polar groups that may be either inherent to the polymer and/or introduced by extraneous impurities. These polar groups must be of ionic nature since the electric conductivities of conventional plastics increase strongly with temperature whereas electronic conductivities vary far less with temperature.

Heat is caused to develop by the imaginary part of the relative permittivity. The low thermal conductivities of plastics do not allow this heat to dissipate, and the temperature raises. Ionic electric conductivities are thus strongly increased until finally a breakdown (arcthrough) occurs. The resistance against such a breakdown is measured by the dielectric strength S of a plastic (Table 8-1).

A breakdown can also occur through tracking on the surface of a plastic. The tracking resistance is difficult to measure because surface resistivities are 2-3 decades lower than volume resistivities (Table 8-1). The tracking resistance is thus determined by standardized methods, such as the the maximal voltage that does not cause tracking if 50 drops of an aqueous 0.1 wt-% NH_4Cl solution are applied between two platinum electrodes that are under an alternating current and 4 mm apart on the specimen surface.

A polymer has a good tracking resistance if it forms volatile decomposition products and no carbon upon degradation, for example, volatile monomer by depolymerization of PMMA, and volatile oligomers by degradation of PE or PA. Poly(*N*-vinyl carbazole) does not form volatile products and thus has a bad tracking resistance although it is a good insulator.

8.2.5. Electrostatic Charging

Static electricity originates from an excess or a deficiency of electrons on isolated or ungrounded surfaces. It can be created by rubbing two surfaces against each other (triboelectric charging) or by contact of a surface with ionized air. Matter is charged electrostatically if conductivities are smaller than ca. 10^{-8} S/cm and relative humidities lower than ca. 70 %. All conventional plastics can be charged electrostatically (Table 8-1). Charge densities may vary between, e.g., 8.2 C/g for poly(chlorotrifluoroethylene) and -13.9 C/g for phenolic resins; charges may range between 3000 V/cm (POM vs. PA 6) and -1700 V/cm (ABS vs. PA 6).

Static charging can be reduced by incorporation of conducting fillers into the plastics, such as carbon black or metal powders (internal antistatics). External antistatics reduce surface resistivities by increasing the polarity of the surface by application of humidity-absorbing additives or by reducing friction through lubricants or coating with PTFE. External antistatics wear out and have to be renewed from time to time.

8.3. Electrical Conductivity

Electrical conductivities in metals are caused by $N/V = 10^{21}$-10^{22} (quasi)free electrons per cubic centimeter with electric charges $e \equiv 1.6 \cdot 10^{-19}$ C and charge mobilities $\mu = (10\text{-}10^6)$ cm^2 V^{-1} s^{-1}. Electrical conductivities are given by:

(8-2) $\sigma = (N/V)\,\mu\,e$

and specific electrical conductivities of metals are therefore between ca. 630 000 (Ag) and 10 400 S/cm (Hg). Graphite exhibits σ = 10 000 S/cm in plane direction and 1 S/cm perpendicular to it. Semiconductors possess carrier mobilities similar to metals but at far lower carrier concentrations.

Charge transfer is fairly easy in metals and semimetals because atoms are tightly packed. Chain atoms of polymer molecules are also close in chain direction because of covalent bonds. Only van der Waals and/or dipole forces act between chains, however, resulting in large intermolecular atomic distances and very difficult charge transfers. All electrons of covalent polymer chains are furthermore localized. Conventional polymers are thus insulators.

Polymers with conjugated chains (e.g., *trans*-poly(acetylene) ~$(CH{=}CH)_n$~) are, for these reasons, merely semiconductors and insulators, even if the chains are planar. These low conductivities can be increased substantially, however, if the polymers are doped with substances such as I_2, AsF_5, BF_3, etc.: doping with AsF_5 raises the specific electrical conductivity of *trans*-poly(acetylene) from 10^{-9} to 1200 S/cm and of poly-(*p*-phenylene) from 10^{-15} to 500 S/cm. Such doped polymers can be processed like thermoplastics to any desired shape, which together with their light weight, makes them attractive for many applications.

The action of these dopants is quite different from those of small amounts of dopants in inorganic semiconductors. The doping of inorganic semiconductors such as GaP, InSb, or Ge generates quasi-free electrons (*n*-carriers) or defect electrons (*p*-carriers), whereas the doping of suitable organic polymers leads to oxidation (*p*-doping) or reduction (*n*-doping) reactions. Sizable effects in organics are thus only achieved if large amounts of dopants are used, often up to 1:1 molar ratios of dopants to repeating units.

Doped polymers exhibit neither Curie paramagnetism (localized charge carriers) nor Pauli paramagnetism (electrons delocalized over the whole system). Thus the electrical conductivity of such systems is assumed to be due to solitons or polarons:

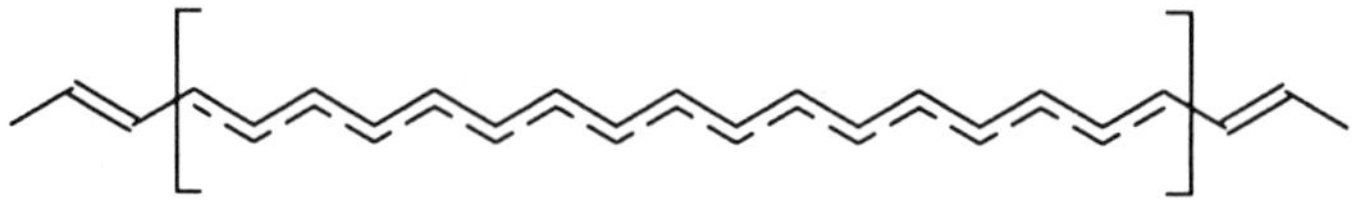

Soliton

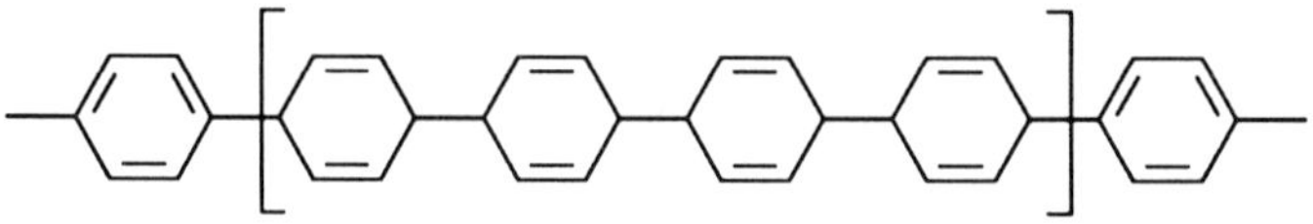

Polaron

Double and single bonds alternate in *trans*-poly(acetylene). Since these bonds are exchangeable, two low-energy states A and B with equal energy must exist. A *soliton* is a kind of topological kink that separates the A state from the B state with opposite bond alternation. Charges are delocalized: about 85 % of charges are distributed over 15 % of atoms. *Polarons* are similar kinks between aromatic and quinoid structures.

Kinks are small (about 14 chain atoms in *trans*-poly(acetylene)) and thus very mobile. Two types of solitons and polarons exist. Neutral solitons and polarons possess radicals that are produced as defects during isomerization of the polymers. Charged solitons and polarons are created by doping, either as carbonium ions or as carbanions.

Solitons can move only along the chain; there is no tunnel effect between chains. Because of this anisotropic behavior, doped polymers (and conducting low molar mass organic molecules) are called low-dimensional conductors, synthetic metals, or organic metals.

The most important condition for the existence of electrical conductivity in organic polymers seems to be the ability to form overlapping orbitals. The planar structure of *trans*-poly(acetylene) promotes the overlapping of its π and p orbitals. In poly(*p*-phenylene sulfide) PPS, $\sim[S\text{-}(p\text{-}C_6H_4)]_n\sim$, p and d orbitals of sulfur atoms probably overlap with the π systems of phenylene groups; PPS-AsF_5 shows an electrical conductivity of 10 S/cm, although the chain is not planar and the phenylene residues are arranged at angles of 45°C to the planar zigzag chain of the sulfur atoms.

Electrically conducting polymers are presently used in small batteries, for example, a complex of poly(2-vinylpyridine)/I_2 as cathode in Li/I_2 batteries for pacemakers ($\sigma = 10^{-3}$ S/cm). 3 V-Batteries have been developed on the basis of poly(aniline) and poly(pyrrole); these batteries find application in wrist watches and pocket calculators. Other possible applications of electrically conducting polymers are in solar cells, electrolysis membranes, microwave shielding, and integrated circuits.

8.4. Photoconductivity

Light generates radical ions and thus photoconductivities in certain systems. This effect is used in xerography™ to generate pictures of objects, e.g., copies of documents (Greek: *xeros* = dry; *graphein* = to write). In this dry-copy process, a photoconducting material is negatively charged in the dark by a corona discharge. On projection of the object unto the charged photoconductor, light causes the projected light-colored areas to discharge whereas the dark-calored areas remain negatively charged. The latent picture is sprayed with resin-coated, positively charged developer that attaches to the negatively charged areas, which now become positively charged. The resulting negative copy of the object is transferred to a negatively charged paper. Upon heating, the resins sinters and generates a positive copy.

The early photoconducting material As_2Se_3 has since been replaced by poly(*N*-vinylcarbazole) PVK. This polymer absorbs UV light and forms an exciton, which is ionized by an electric field. PVK is nonconducting in visible light; upon sensibilization by certain electron donors, charge transfer complexes are formed that are photoconductive. Another photoconducting system consists of polycarbonate A and triphenylamine.

Literature

8.1. Fundamentals

N.G.McCrum, B.E.Read, G.Williams, Anelastic and Dielectric Effects in Polymeric Solids, Wiley, London 1969
M.E.Baird, Electrical Properties of Polymeric Materials, Plastics Institute, London 1973
A.R.Blythe, Electrical Properties of Polymers, Cambridge University Press, Cambridge 1979
H.Block, The Nature and Appreciation of Electrical Phenomena in Polymers, Adv.Polym.Sci. **33** (1979) 93
D.A.Seanor, ed., Electrical Properties of Polymers, Academic Press, New York 1982
C.C.Ku, R.Liepins, Electrical Properties of Polymers. Chemical Principles, Hanser, Munich, 1987
J.I.Kroschwitz, ed., Electrical and Electronic Properties of Polymers (= reprints of articles in Encyclopedia of Polymer Science and Engineering, Wiley, New York, 2nd ed. 1988

8.2. Dielectric Properties

A.D.Moore, Electrostatics and Its Applications, Wiley, New York 1973
M.W.Williams, The Dependence of Triboelectric Charging of Polymers on Their Chemical Compositions, J.Macromol.Sci.-Revs.Macromol.Chem. **C 14** (1976) 251
P.Hedvig, Dielectric Spectroscopy of Polymers, Halsted, New York 1977
A.Bradwell, ed., Electrical Insulation, Peter Peregrinus Ltd., 1983
G.Heinicke, Tribochemistry, Hanser, Munich 1984
W.Tiller Shugg, Handbook of Electrical and Electronic Materials, Van Nostrand Reinhold, New York 1986
Y.Tabata, S.Nonogati, I.Mita, K.Horie, S.Tagawa, Polymers for Microelectronics, VCH, Weinheim 1990
E.Riande, E.Saiz, Dipole Moments and Birefringence of Polymers, Prentice Hall, Englewood Cliffs (NJ), 1992
L.A.Dissado, J.C.Fothergill, Electrical Degradation and Breakdown in Polymers, P.Peregrinus, Stevenage, Herts, 1992

8.3. Electrical Conductivity

W.E.Hatfield, ed., Molecular Metals, Plenum, New York 1980
M.Kryszewski, Semiconducting Polymers, PWN-Polish Sci.Publ., Warsaw 1980
J.S.Miller, ed., Extended Linear Chain Compounds, Plenum, New York 1982
E.K.Sichel, Carbon Black-Polymer Composites, Dekker, New York 1982
J.Mort, G.Pfister, ed., Electronic Properties of Polymers, Wiley, Chichester 1982
H.Kuzmany, M.Mehring, S.Roth, Electronic Properties of Polymers and Related Compounds, Springer, Berlin 1985
R.R.Chance, D.Bloor, eds., Polydiacetylenes, Nijhoff, Amsterdam 1985
T.A.Skotheim, ed., Handbook of Conducting Polymers, Dekker, New York 1986 (2 vols.)
N.C.Billingham, P.D.Calvert, Electrically Conducting Polymers: A Polymer Science Viewpoint, Adv.Polym. Sci. **90** (1989) 1
T.A.Skotheim, ed., Electroresponsive Molecular and Polymeric Systems, Dekker, New York, vol. 1 (1988), vol. 2 (1991)

8.4. Photoconductivity

J.H.Dessauer, H.E.Clark, Xerography and Related Processes, Focal Press, London 1965
R.M.Schaffert, Electrophotography, Focal Press, London, 2nd ed. 1975
M.Stolka, D.M.Pai, Polymers with Photoconductive Properties, Adv.Polym.Sci. **29** (1978) 1
M.E.Scharfe, Electrophotography, Principles and Optimization, Research Studies Press, Letchwort (UK) 1984
L.B.Schein, Electrophotography and Development Physics, Springer, Berlin 1987
W.Gerhartz, ed., Imaging and Information Storage Technology, VCH, Weinheim 1992

References

[1] Y.Ishida, Kolloid-Z. **168** (1960) 29, Figs. 2-6

9. Optical Properties

9.1. Refractive Index

Many optical properties of plastics depend on the *refractive index* n, for example, reflection, gloss, transparency, and hiding power. Light falling in vacuum on a transparent body under an angle of incidence α exits the body under an angle of refraction β. Refractive indices $n = \sin \alpha / \sin \beta = \sin \alpha' / \sin \beta'$ are given by the ratio of the sine of the angle of incidence ($\alpha = \alpha'$) and refraction ($\beta = \beta'$) of the light (definition of angles: see Fig.(9-1)). According to the Lorenz-Lorentz relationship

$$(9\text{-}1) \qquad (n^2 - 1)/(n^2 + 1) = (4/3)\ \pi\ q\ (N/V)$$

refractive indices are determined by the total polarizability $q(N/V)$ of the molecules, where N/V is the number concentration of molecules with the polarization q. The polarization q is a function of the dipole moments of all groups in a molecule (i.e., the mobility and number of electrons per molecule). Contributions to the refractive index are thus much higher for carbon atoms than for hydrogen atoms. Since the contributions of the latter can be neglected and since carbon atoms dominate polymer structures, all polymers possess approximately the same refractive index of 1.5 (e.g., at the D-Line (589.3 nm): PMMA 1.492, PP-it 1.53, PS 1.59). Deviations from this rule exist for strongly polarizable polymers (PTFE 1.37), polymers with bulky conjugated substituents (PVK 1.69), or polymers with a high content of noncarbon atoms (PDMS 1.40). According to the molecular structure of all known polymers, their refractive indices should always be between ca. 1.33 and 1.73.

Molecules are more tightly packed in crystalline polymers than in amorphous ones. Refractive indices thus increase with increasing crystallinity. Because crystalline polymers are always anisotropic, different polarizabilities and refractive indices are exhibited in chain direction and perpendicular to it.

Refractive indices are determined in the CAMPUS® system according to ISO 489 and DIN 53 491 at compression- or injection-molded specimens of (80·80·1) mm^3 .

9.2. Reflection

A part of the light falling in vacuum on a homogeneous, transparent body is reflected. *Reflectivity* is defined as the ratio $R = I_r/I_o$ of the intensities of reflected (I_r) and incident light (I_o), which depends on the angles of incidence α and refraction β according to Fresnel's law:

$$(9\text{-}2) \qquad R = \frac{I_r}{I_o} = \frac{1}{2}\left[\frac{\sin^2(\alpha-\beta)}{\sin^2(\alpha+\beta)} + \frac{\tan^2(\alpha-\beta)}{\tan^2(\alpha+\beta)}\right]$$

The reflectivity is small at small incident angles (R = 0.040 for n = 1.5 at 10°) and rises sharply at higher ones (R = 0.388 for n = 1.5 at 80°) (Fig. 9-1).

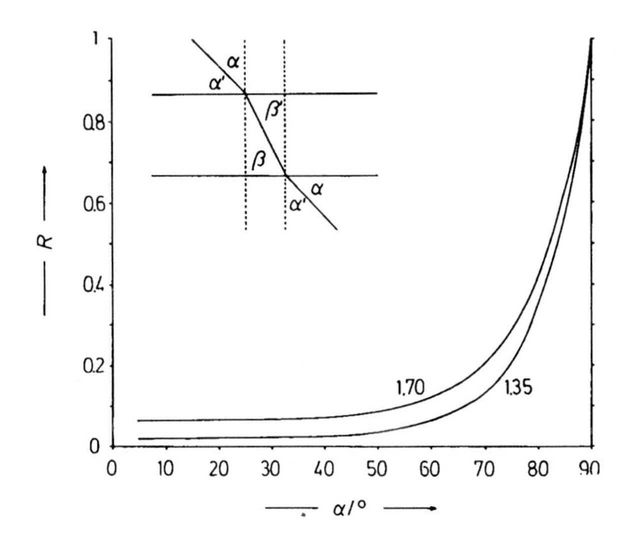

Fig. 9-1 Reflectivities R of bodies with refractive indices of n = 1.35 [for example, poly(tetrafluoroethylene)] and 1.70 [for example, phenol-formaldehyde resins] as function of the angles α of incident light. Insert: Definition of angles of incidence (α) and reflection (β).

Gloss is the ratio of the reflectivity R of a specimen to the reflectivity R_{st} of a standard, for example, a body with n_D = 1.567 in the paint industry. The maximum theoretical gloss $R_o = R/R_{st}$ is given by Equation (9-2). It increases with increasing refractive index of the specimen and with increasing angle of incidence (Fig. 9-2).

The maximum theoretical gloss is almost never achieved in practice because some light is always scattered by rough surfaces as well as optical inhomogeneities just below the surface. The reduction of the intensity of reflected light by light scattering depends on the angle of incidence of light. Ideal-matte surfaces reflect light equally well in all directions whereas ideal-glossy ones reflect preferentially in the direction of observation.

Gloss effects the appearance of subjects. Painted surfaces of various materials are, for example, very difficult to color-match because of different gloss. An example is the color-match of automobile fenders from plastics with the metal sheets of other automobile parts.

Glitter is caused by local glosses due to a high, directed light reflections or by contrasts in color or light intensity between the glittering particle and its surrounding. In plastics films and bodies, it may be caused by metal pigments, mica, etc. Fibers show glitter if they have triangular or trilobal cross sections.

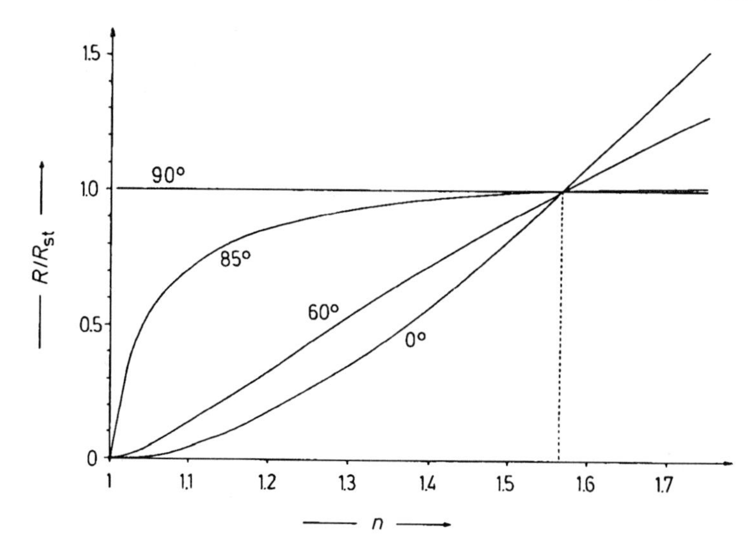

Fig. 9-2 Maximum theoretical gloss R/R_{st} as a function of the refractive index n of the specimen for angles of incidence of $\alpha = 0, 60, 85$ and 90° and a standard with $n = 1.567$.

9.3. Transparency

Fresnel's law also allows to calculate the maximum *transparency* of a plastic. A part of light falling perpendicularly on a optical homogeneous, planar plate is reflected, another part is transmitted. In this case, $\alpha \rightarrow 0$ and $\beta \rightarrow 0$ and Equation (9-2) reduces to

$$(9\text{-}3) \qquad R_o = (n - 1)^2/(n + 1)^2$$

The transparency is thus $\tau_i = 1 - R_o$. The maximum transparency of a polymer can adopt values between 98.0 % (n = 1.33) and 92.8 % (n = 1.73), i.e., between 2 and 7.2 % of the indicent light is maximally reflected at the polymer-air interface. These ideal transparencies are rarely achieved for polymers because a small portion of the light is always absorbed and/or scattered. Poly(methyl methacrylate), one of the most transparent plastics, never exceeds a transparency of ca. 93 % between wavelengths of 430 and 1100 nm (theory: 96.1 %). The transparency of PMMA decreases above wavelengths of ca. 1150 nm and (rapidly) below 380 nm. All polymers except halogenated poly(ethylene)s absorb infrared radiation. The degree of light transmission is measured at (80·80·1) mm^3 specimens according to DIN 5036 T3 (CAMPUS®).

A distinction is usually made between transparent and translucent polymers. *Transparent* polymers have transparencies $\tau_i > 90$ %; they look clear even at greater thicknesses. *Translucent* polymers ($\tau_i < 0.90$) look clear only as thin films. They are also called *contact clear* because films are clear in contact with packaged goods but hazy when viewed alone.

Clarity is not only lost by light absorption but also by light scattering. Electromagnetic waves lose part of their energy by scattering in inhomogeneous systems. The loss of contrast by forward scattering is called *haze*. The combined loss by forward and backward scattering causes *milkiness*.

A body appears opaque if local fluctuations of refractive indices and/or orientations of anisotropic volume elements are present. The different volume elements must also be larger than the wavelength of incident light. The clarity of a material can thus be increased considerably if the size of the different volume elements (e.g., microdomains) is decreased. Lamellar structures are optically less heterogeneous than spherulites of approximately the same diameter. Under certain conditions, clear poly(ethylene) films can be produced by quenching and orientation although these films contain crystalline lamellae with dimensions greater than the wavelength of light.

Adjustments of refractive indices of phases improve the clarity to a small extent. The difference in refractive indices of phases determines the appearance of the specimen. ABS-modified PVC, for example, looks milky yellowish if the ABS is dispersed in PVC and milky bluish if PVC is dispersed in ABS.

Transparent two-phase plastics can be generated by core-shell morphologies (i.e., if the dispersed particles consist of a core and a shell with a different refractive index). The materials are transparent if the mean refractive index of the core-shell particles matches the refractive index of the matrix.

Inhomogeneities of plastics lead to strong reflections if they are caused by fillers or differences in polymer composition. This phenomenon is related to the *hiding power* of surface coatings. The refractive index $n_2 = 1$ of a vacuum has to be replaced by the n_2 of, e.g., fillers in addition to n_1 of the polymer matrix and Equation (9-3) becomes

(9-4) $$R_o = (n_1 - n_2)^2/(n_1 + n_2)^2$$

The hiding power is also effected by light scattering. High filler (pigment) concentrations lead to multiple scattering of the light which reduces the relative scattering intensity and thus the hiding power. Larger filler particles cause higher scattering intensities but reduced backward scattering. The higher the backward scattering, the better is the hiding power, however. The two effects work against each other and the hiding power passes with increasing particle size through a maximum.

Literature

G.H.Meeten, ed., Optical Properties of Polymers, Elsevier, London 1986
E.Riande, E.Saiz, Dipole Moments and Birefringence of Polymers, Prentice Hall, Englewood Cliffs, NJ, 1992

10. Plastics Manufacturing

10.1. Compounds

10.1.1. Homogenization and Compounding

Polymerization reactors do not deliver polymers that can be used directly as raw materials for plastics. Polymer melts have to be filtered; melts, powders, beads, pellets, and granulates must be degassed; and all particulates must be dried and/or conditioned to the surrounding environment. Beads from suspension polymerizations are washed and polymer solutions concentrated.

Batchwise polymerizations often lead to slightly different products that are mixed (blended) to guaranty customers polymer grades according to specifications (*microhomogenization*). Blending of polymers with different molar masses and molar mass distributions leads to broadened molar mass distributions. The ratio of mass-average to number-average molar masses of the blends

$$(10\text{-}1) \qquad \frac{\overline{M}_w}{\overline{M}_n} = w_A Q_A \left[\frac{w_A R_n + w_B}{R_n} \right] + w_B Q_B \left[w_A R_n + w_B \right]$$

depends on four variables: mass fraction $w_A \equiv 1 - w_B$ of component A, mass-number molar mass ratios $Q_i = (\overline{M}_w / \overline{M}_n)_i$ of components (i = A or B), and ratio $R_n = (\overline{M}_n)_B / (\overline{M}_n)_A$ of number-average molar masses of components B and A. For example, the mixture of two polymers with unimodal Schulz-Flory distributions of molar masses (i.e., $[\overline{M}_w / \overline{M}_n]_i = 2$) at a ratio $(\overline{M}_n)_B / (\overline{M}_n)_A = 2$ leads to a polymer with a bimodal molar mass distribution and $(\overline{M}_w)/(\overline{M}_n) = 2.25$. Granulates and pellets are sometimes similarly blended by *macrohomogenization*.

Compounding is the mixing of polymers with additives and modifiers. It can be performed by adding single additives one at a time, by using additive systems, or by employing master batches. Additive systems are carefully adjusted mixtures of additives that are formulated to avoid mutually synergistic or antagonistic effects. Master batches are concentrates of additives in polymers; they facilitate the dosage of small amounts of additives. The resulting plastics compounds are (ideally!) physical mixtures and not chemical compounds from chemical reactions between the components.

The properties of compounds depend very much on the compounding process that is often carried out by specialized companies (compounders). Only heterogeneous compounds result when poly(vinyl chloride) is mixed with additives in regular mixers (premix); these compounds cannot be processed directly to end-use products. High-performance mixers deliver free-flowing powders; these "dry blends" can be extruded and injection molded.

10.1.2. Demixing

Polymers and added solids and liquids are often not in thermodynamic equilibrium. The components try to demix which is mostly unwanted because it leads to a detoriation of the appearance and/or other properties of the plastics. Demixing is desirable for some surface properties: lubricants and antistatics should be enriched at the surface because it is their locus of action.

Demixing processes have different names depending on the type of the material admixed with the polymer and the environment of the plastic. The deposition of admixed solids on the surface of plastics in air is called *efflorescence* or blooming. *Chalking* refers to the surface deposition of white solids (e.g., pigments, fillers), *bleeding* to that of colorants, *flooding* to the horizontal migration of pigments, and *migration* to that of plasticizers. The deposition of solids from the plastics on molds is a *plate-out* and the removal of plasticizers from plastics by liquids is an *extraction*.

10.2. Additives

10.2.1. Overview

Polymers are rarely used directly as plastics, elastomers, fibers, coatings, etc. They do not fulfill *per se* all technological requirements and become commercially useful only after they are admixed with certain polymer additives and/or modifiers. These added materials average ca. 23 % of the total weight of plastics (Table 10-1) but may range for individual plastics from 0 % (packaging films for foods) to 90 % (barium ferrite-filled ethylene-vinyl acetate copolymers for magnetizable sealing strips). The economic value of these added materials equals approximately that of plastics because some command prices that are a hundred times those of the plastics themselves.

Modern technical terminology distinguishes between additives and modifiers. *Additives* are added in small proportions, usually less than 5 %; they do not affect markedly the mechanical properties. *Modifiers* are used in greater proportions; they modify mechanical properties such as the impact strength.

Polymer additives and modifiers are to be distinguished from *polymer auxiliaries*. Additives and modifiers are added to polymers after the polymerization process, i.e., for the manufacture of plastics, whereas auxiliaries are used for the production of polymers themselves (e.g., as polymerization catalysts, emulsifiers, initiators, etc.). The free radical initiators used for the hardening of unsaturated polyesters are sometimes considered additives (added to the solution of unsaturated polyester molecules in monomers such as styrene) and sometimes auxiliaries (initiating and promoting the thermosetting reaction).

Additives and modifiers are also often subdivided according to their action into process additives and functional additives:

Process additives aid the processing of plastics by either stabilizing the chemical composition of polymers (processing stabilizers) or facilitating the processing itself (processing aids).

Functional additives stabilize the chemical composition against attacks by environmental agents (stabilizing additives) or improve end-use properties (modifiers).

Additives may also be classified according to their primary mode of action (chemical or physical). Some additives act in more than one way: Fillers or colorants may also be nucleating agents for crystallizations, a pigment may enhance discoloration, a parting agent may act as as slip agent, etc.

Table 10-1 Annual U.S. consumption of polymer additives and modifiers for plastics in 1980. Plastics production: 16 076 000 t/a.

Additive	Consumption in tons per year	Additive	Consumption in tons per year
Process additives		*Functional additives*	
Processing stabilizers		Stabilizing additives	
Heat stabilizers for PVC	43 000	Antioxidants	see left
Antioxidants	18 000	Metal deactivators	< 1 000
Processing aids		Light stabilizers	2 400
Lubricants, slip agents, etc.	50 000	Flame retardants	220 000
Polymeric processing aids	< 1 000	Biostabilizers	< 1 000
Mold release agents	< 1 000	Colorants	150 000
Nucleating agents	< 1 000	Modifiers	
		Fillers and reinforcing agents	
Radical initiators for thermosets	15 000	Mineral based	1 730 000
		Glass	462 000
		Natural organic	133 000
		Synthetic organic	2 000
		Plasticizers	1 000 000
		Blowing agents (except gases)	7 000
		Impact modifiers	< 5 000
		Antistatic agents	2 000
Total process additives	≈ 129 000	Total functional additives	3 715 400

10.2.2. Chemofunctional Additives

Polymers can be attacked by oxygen and ozone during processing and use. The attack generates radicals such as peroxy radicals $ROO^{\bullet}$, hydroperoxy radicals $HOO^{\bullet}$, oxy radicals $RO^{\bullet}$, hydroxy radicals $HO^{\bullet}$, and alkyl radicals $R^{\bullet}$. These radicals cause chain reactions that produce even more radicals. Ultimately, changes in chemical compositions, degradation and/or cross-linking result.

Oxidation is reduced if the accessibility of the oxidizable groups is limited and/or if radical formation is prevented. Oxygen diffusion into the polymer is slowed by certain coatings that act as mechanical barriers or by added antioxidants, which diffuse into surface layers and are preferentially oxidized there.

Antioxidants prevent radical formation, at least during processing and for the targeted life-time of the plastic. More than 90 % of all antioxidants are used for polymeric hydrocarbons such as poly(ethylene)s, poly(propylene)s, poly(styrene)s, and ABS polymers. Antioxidants are subdivided into deinitiators and chain terminators:

Chain terminators, such as hindered phenols, secondary amines, and zinc diethyldithiocarbamate, react with already formed radicals and terminate the kinetic chain; they are thus called chain-breaking or primary antioxidants. Examples are:

$$2,6\text{-}[(CH_3)_3C]_2\text{-}4\text{-}CH_3\text{-}C_6H_2\text{-}OH \qquad C_6H_5\text{-}NH\text{-}C_6H_4\text{-}NH\text{-}C_6H_5$$

$$[(CH_3)_3C]_2N\text{-}\underset{\underset{S}{\parallel}}{C}\text{-}S\text{-}Zn\text{-}S\text{-}\underset{\underset{S}{\parallel}}{C}\text{-}N[C(CH_3)_3]_2$$

Deinitiators prevent the formation of radicals (therefore also called preventive antioxidants) and are always used in combination with chain terminators (therefore also called secondary antioxidants). They are further subdivided into peroxide deactivators, metal deactivators, and UV absorbers.

Peroxide deactivators (phosphites, phosphonites, thioethers) convert peroxides and hydroperoxides into harmless compounds before they can form radicals. Examples:

$$[C_9H_{19}\text{-}C_6H_4\text{-}O]_3P \qquad H_{37}C_{18}\text{-}O\text{-}P(OCH_2)_2C(CH_2O)_2P\text{-}O\text{-}C_{18}H_{37}$$

$$H_{37}C_{18}\text{-}S\text{-}S\text{-}C_{18}H_{37} \qquad [H_{25}C_{12}\text{-}O\text{-}C\text{-}CH_2\text{-}CH_2]_2S$$

Metal deactivators are chelating agents that form inactive complexes with catalytically active metal species (mainly from Ziegler-Natta polymerizations). Example:

$$3,5\text{-}[(CH_3)_3C]_2\text{-}4\text{-}HO\text{-}C_6H_2\text{-}CH_2CH_2\text{-}\underset{\underset{O}{\parallel}}{C}\text{-}NH\text{-}NH\text{-}\underset{\underset{O}{\parallel}}{C}\text{-}CH_2CH_2\text{-}C_6H_2\text{-}4\text{-}OH\text{-}3,5\text{-}[C(CH_3)_3]_2$$

The center group $-CH_2CH_2-CO-NH-NH-CO-CH_2-CH_2-$ can also be replaced by $-CH=N-NH-CO-CO-NH-N=CH-$ or similar groups.

These hydrazine compounds contain metal deactivating and chain terminating groups. Such combinations of deinitiators and chain terminators often lead to synergistic effects, i.e., improvements of the antioxidative action over the simple additive action of the two additives (lengthening of the inhibition period and/or diminished rate of oxygen uptake). Antagonistic effects are also known, e.g., with carbon black as filler, which adsorbs antioxidants.

Light-induced degradation reactions can be prevented by the reduction of light absorption and/or addition of UV absorbers and quenchers. Less light is absorbed by the plastic if the surfaces are reflective and/or if certain pigments are added (e.g., carbon black). *Ultraviolet absorbers* must be used to stabilize transparent polymers. These UV absorbers either convert the incident light directly into harmless infrared radiation (*o*-hydroxybenzophenone) or are first transformed into other chemical compounds (phenyl salicylate into *o*-hydroxybenzophenone). Examples are 2,4-dihydroxybenzophenone (and derivatives) and 2-hydroxyphenyl benzotriazoles:

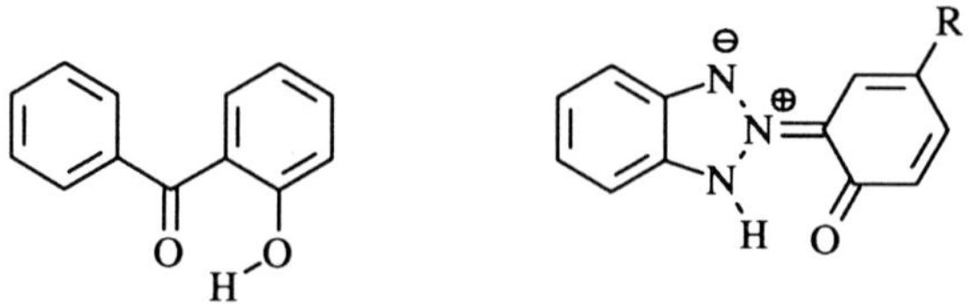

Quenchers deactivate excited states and become excited themselves; their effectiveness is probably due to light absorption. Examples are nickel dithiocarbamates and diamagnetic nickel compounds:

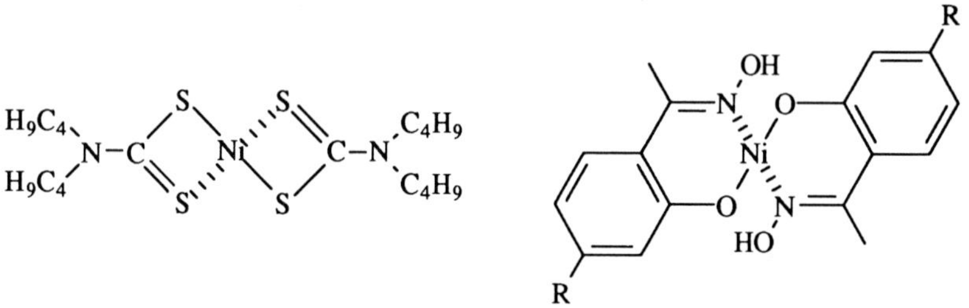

Heat stabilizers prevent chemical transformations of plastics at higher temperatures; their main application is for the stabilization of poly(vinyl chloride) against HCl elimination during processing. Examples ar organic compounds of barium, cadmium, zinc, and tin, and inorganic and organic lead compounds. These primary stabilizers are often combined with so-called secondary stabilizers such as organophosphites, dicyandiamides, or epoxidized vegetable oils.

Flame retardants either prevent oxygen access to burning plastics by formation of nonburning gases or by "poisoning" of radicals generated by the burning. Some plastics are "self-extinguishing" because either CO_2 (polycarbonates) or water vapor plus a protecting carbon layer are formed during burning (vulcanized fiber: a $ZnCl_2$-

treated cellulose). The same action can be obtained from added inorganic flame retardants. Aluminum hydroxide $Al(OH)_3$ decomposes into water and dialuminum trioxide Al_2O_3. Boric acid H_3BO_3 forms a glassy coating of B_2O_3. Phosphorus compounds are oxidized during the burning to nonvolatile phosphor oxides, which either form a protective layer or are converted by water into phosphorus acids that catalyze the elimination of water. Organic flame retardants are halogen-containing compounds that generate radicals during burning; these radicals combine with radicals from the degradation of plastics and stop the kinetic chain. They are either brominated aromatic compounds or chlorinated cycloaliphatics.

Flammabilities of plastics are often characterized by a *l*imiting *o*xygen *i*ndex LOI, the limiting value of the volume fraction of oxygen in an oxygen-nitrogen mixture that just allows the polymer to burn with a flame after ignition. Materials with LOI > 0.225 are called flame retardant; those with LOI > 0.27 are self-extinguishing. Low LOI values (high flammabilities) are exhibited by poly(oxymethylene)s (0.14), poly(olefin)s (0.17-0.18), and cellulose (0.20). High LOI values are shown by poly(vinyl chloride) (0.32), poly(benzimidazole) (0.48), and poly(tetrafluoroethylene) (0.95). LOI values are not absolute measures for the flammability and combustibility of plastics because these properties also depend on flame temperatures, heat capacities, heat conductivities, melting temperatures, and melt viscosities. Smoke formation and toxicity of evolving gases also add to the hazards of burning plastics.

10.2.3. Colorants

Colorants are subdivided into dyes (soluble in polymer matrix) and pigments (insoluble). Textile fibers are dyed mainly with dyes. Pigments are preferred for plastics because they have a higher lightfastness and are more stable against migration than dyes. Colorants for plastics are dominated by pigments, mainly titanium dioxide (60-65 %) and carbon black (20 %); only 2 % are dyes.

Pigment particles generally possess diameters of 0.3-0.8 µm; these sizes allow the pigmentation of films and fibers with thicknesses of more than ca. 20 µm. Very thin films and fibers are colored exclusively by organic pigments because these can be ground to much smaller diameters than inorganic pigments. The hiding power of pigments increases with increasing difference between refractive indices of pigment and polymer. Colored plastics are transparent if the diameters of pigment particles are smaller than ca. one half of the wave length λ of the incident light (for visible light: $0.19\ \mu m < \lambda/2 < 0.375\ \mu m$).

Pigments need not have a special affinity for polymers. They must be wettable by the polymer melt, however, which can be achieved by treating them with surfactants. The aggregation of pigments is mainly the result of air inclusions; it can be removed by the application of vacuum.

Pigments can be metered into plastics via master batches (in plastics), by color concentrates (in plasticizers), or electric charging of the surface of polymer particles in granulate mixers. Up to 1 wt-% of pigment can be mixed in by the last method.

10.2.4. Processing Aids

Processing aids facilitate the processing of plastics by either enhancing transport rates to and flow behavior in processing machines, achievement of ultimate properties during processing, or removal of shaped articles from machines or from each other.

Easy-flow grades of powdered poly(styrene)s, for example, often contain 3-4 % of mineral oil, which forms a low viscosity film on the surface of the particles and thus reduces friction; in polar polymers, amphiphilic compounds such as metal stearates or fatty acid amides are used for this purpose. Such *external lubricants* also reduce the friction between polymer particles and the walls of the processing machine and the friction of polymer melts at such walls; in addition, they prevent the cleavage of particles to smaller flow units. External lubricants are always incompatible with polymers and are thus found predominantly at polymer surfaces.

External lubricants are related to *release agents* (*parting agents*), which facilitate the separation of shaped articles from the molds, and *slip agents* (*abherents, antiblocking agents*), which prevent the sticking together of shaped articles caused by cold flow or static electricity. Abherents are thus sometimes also called lubricants. Abherents for poly(olefin) films are called *slip agents* and those for poly(vinyl chloride) films *antiblocking agents*.

Abherents are materials with low critical surface tensions that prevent wetting of molds by polymer melts. Silicones and fluoropolymers are used for apolar and slightly polar polymers such as poly(olefin)s, polyesters, epoxide resins, and polyurethanes. Fatty acids and waxes are slip agents for polar polymers.

Internal lubricants improve the flow behavior and homogeneity of polymer melts; they also reduce the Barus effect (see Chapter 11) and the melt fracture. Internal lubricants presumbly act by desegregating larger flow units (aggregates) that were probably formed during polymerization and are still present shortly after the melting of the polymer to a macroscopically homogeneous material. Typical internal lubricants are amphipolar compounds, e.g., modified esters of long-chain fatty acids.

Nucleating agents promote the crystallization of crystallizable polymers by generating many nuclei for crystallite formation. They prevent the formation of larger spherulites and thus improve the mechanical properties of plastics.

10.2.5. Blowing Agents

Foamed plastics (plastic foams, cellular plastics, expanded plastics) are blends of polymers with gases (Section 13.5.). They may be rigid (T_G of amorphous polymers, T_M of crystalline polymers higher than use temperature) or flexible; their cells can be open or closed. The gases may be air, N_2, CO_2, fluorinated hydrocarbons, etc.

Plastic foams can be produced by mechanical means (whipping, stirring), physical methods (shock volatilization of liquids, washing-out of solids), or chemical foaming, the latter either by internal foaming during polymerization or by external foaming with chemical blowing agents (CBAs) that were mixed into the polymer.

Chemical blowing agents (CBA) are chemical compounds that decompose at elevated temperatures with release of gases. The most widely used CBAs are:

AIBN	$(CH_3)_2C(CN)\text{-}N{=}N\text{-}C(CN)(CH_3)_2$
NTA	$CH_3\text{-}N(NO)\text{-}CO\text{-}(p\text{-}C_6H_4)\text{-}CO\text{-}N(NO)\text{-}CH_3$
ABFA	$H_2N\text{-}CO\text{-}N{=}N\text{-}CO\text{-}NH_2$

AIBN = N,N'-azobisisobutyronitrile is used mainly for natural rubber, NTA = N,N'-dinitroso dimethyl terephthalamide for plasticized PVC, and ABFA = 1,1'-azobisformamide for other plastics. Blowing agents are used in proportions of ca. 0.1 % to eliminate sinks in injection molding; 0.2-0.8 % for injection-molded structural foams; 0.3 % for extended profiles; 1-15 % for vinyl plastisol foaming; and 5-15 % for compression-molded foam products.

10.3. Modifiers

10.3.1. Extending and Reinforcing Fillers

Fillers are solid inorganic or organic materials that are added to modify the properties of plastics. Some fillers are added mainly in order to improve the economics of expensive polymers; they are *extenders*. Extenders are usually particulate materials of corpuscular nature such as chalk particles and glass spheres (aspect ratio ca. 1).

Other fillers are *reinforcing agents* ("active fillers"); they improve certain mechanical properties. Reinforcing agents possess aspect ratios higher than 1; they may be short fibers, platelets (e.g., kaolin, talc, mica), or long fibers (continuous filaments). Active fillers are sometimes subdivided into property enhancers (aspect ratio < 100) and true reinforcing fillers (aspect ratio > 100). No sharp dividing line exists between extenders and reinforcing agents, nor is the term "reinforcement" unambiguously defined (e.g., it may denote an increase in breaking or impact strength or a decrease of brittleness). The effect of fillers is discussed in greater detail in Chapter 13.

Many different fillers are used for plastics (Table 10-2). Glass fiber-reinforced plastics often carry the abbreviation GRP (glass fiber-reinforced plastic) or FRP, those reinforced by synthetic organic fibers, CRP (chemical fiber-reinforced plastic) (see also Chapter 10.3.2.). *Syntactic plastics* are polymers filled with hollow glass spheres. The proportion of fillers added varies widely: in general, industry standards are about 30 wt-% for thermoplastics and 60 wt-% for thermosets.

Fillers act very differently in polymers. Carbon black is a chemical cross-linker in elastomers. Other fillers can adsorb polymer chains on their surfaces; that is, physical bonds are introduced between fillers and polymers. On impact, adsorbed chain segments may take up energy and slip from the surface, which increases impact strength. Other fillers act also as nucleating agents in crystallizable polymers, leading to trans crystallizations (epitaxial overgrowth on the filler surface). Fillers furthermore constitute impenetrable walls to polymer coils. They restrict the number of conformational positions of chain segments near the filler surface; chains become less flexible, and tensile strengths and moduli of elasticity increase.

Table 10-2 Fillers for thermoplastics, thermosets, and elastomers. For definition of acronyms and abbreviations of names of individual polymers, see Appendix.

Filler	Application in	Conc. in %	Improved property
Inorganic fillers			
Chalk	PE, PVC, PPS, PB,UP	< 33	Price, gloss
Potassium titanate	PA	40	Dimensional stability
Heavy spar	PVC, PUR	< 25	Density
Talcum	PUR, UP, PVC, EP, PE PS, PP		White pigment, impact strength, plasticizer uptake
Mica	PUR, UP	< 25	Dimensional stability, stiffness, hardness
Kaolin	UP, vinyls	< 60	Demolding
Glass spheres	Thermoplastics, thermosets	< 40	Modulus, shrinkage, compressive strength, surface properties
Glass fibers	Thermoplastics, thermosets	< 40	Fracture strength, impact strength
Fumed silica	Thermoplastics, thermosets	< 3	Tear strength, viscosity (increase)
Quartz	PE, PMMA, EP	< 45	Heat stability, fracture
Sand	EP, UP, PF	< 60	Shrinkage
Al, Zn, Cu, Ni, etc.	PA, POM, PP	< 100	Conductivity (heat and electricity)
MgO	UP	< 70	Stiffness, hardness
ZnO	PP, PUR, UP, EP	< 70	UV stability, heat conductivity
Organic fillers			
Carbon black	PVC, HDPE, PUR, PI, PE, elastomers	< 60	UV stability, pigmentation, cross-linking
Graphite	EP, MF, PB, PI, PPS, UP, PMMA, PTFE	< 50	Stiffness, creep
Wood flour	PF, MF, UF, UP	< 50	Shrinkage, impact strength
Starch	PVAL, PE	< 7	Biological degradation

10.3.2. Fibers

Fibers for the reinforcement of plastics (Chapter 13) are always textile fibers, i.e., fibers that can be processed by the methods of textile industry. Only three types of reinforcing fibers are important for plastics: glass (ca. 1 400 000 t/a), carbon (ca. 4000 t/a), and aramide (ca. 3500 t/a).

Glass fibers are produced from molten glass. Often used is E glass (originally for *e*lectrical purposes), composed of ca. 54 wt-% SiO_2, 15 wt-% Al_2O_3, 22 wt-% MgO, 8 wt-% B_2O_3, and 1 wt-% of other metal oxides and F. Other compositions lead to high strength fibers (R and S glass), fibers for good dielectric properties (D glass), etc.

Table 10-3 Properties of short reinforcing fibers with density ρ, initial length L, diameter d, modulus of elasticity E, and tensile strength at break σ_b. Aramide and carbon fibers are offered with different lengths L; the initial length of chopped glass fibers is reduced by molding. N = normal, HM = high modulus, HT = high tenacity.

Type	$\frac{\rho}{\text{g cm}^{-3}}$	$\frac{L}{\text{mm}}$	$\frac{d}{\mu\text{m}}$	$\frac{L}{d}$	$\frac{E}{\text{GPa}}$	$\frac{\sigma_b}{\text{GPa}}$
E glass fiber, milled	2.59	0.2	10	20	73	3.4
E glass fiber, chopped	2.59	4.5	14	321	73	3.4
Aramide fiber N	1.44	6	12	500	80	2.8
Aramide fiber HM	1.45	6	12	500	115	2.8
Carbon fiber HT	1.78	6	6	1000	238	3.2
Carbon fiber HM	1.85	6	6	1000	400	2.3

Molten glass is either drawn to *filaments* (endless fibers) of constant fiber diameter (Table 10-3) or drawn or spun to *staple fibers* with varying lengths ((5-50) mm) and diameters ((5-15) μm). The surface of glass fibers is treated with sizes, lubricants, coupling agents, etc., for a good bonding to polymers *and* for easy processing.

Filaments can be assembled without intentional twist to *glass strands* and with twist to single glass *filament yarns*. Two or more yarns can be wound together without twist to *plied yarns* and folded or cabled with twist to *twisted yarns*.

Parallel strands are combined without twist to *glass rovings* that may either (1) woven into *roving fabric* (*roving cloth*) or assembled to *glass mats* (*nonwoven fabric*) or (2) cut by chopping into *chopped glass strands* (ca. 4.5 mm long) or by milling into *milled glass fibers* (ca. 0.2 mm long). Molding of reinforced plastics reduces the length of chopped strands to ca. (0.6-0.7) mm; the length of milled fibers remains the same. Staple fibers may be similarly converted to yarns, fabrics, cloths, and mats.

Carbon fibers contain (80-95) % carbon. Their graphite-like structures result from the pyrolysis of poly(acrylonitrile) PAN or pitch, for example, from PAN:

(10-2)

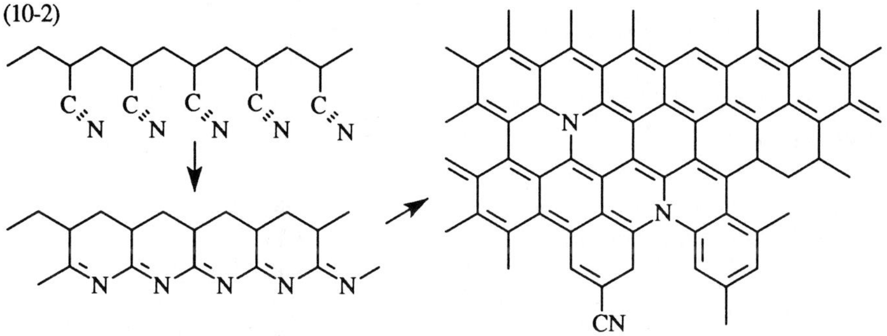

Filaments have round, oval, or kidney-shaped cross sections of (5-7.5) μm diameter. Short fibers are 3 mm or 6 mm long. Filament yarns with finenesses of 67 tex to 900 tex are usually preferred for the reinforcement of plastics.

Aramide fibers are spun from lyotropic solutions of poly(*p*-phenylene terephthalamide) ~[NH-(*p*-C_6H_4)-NH-OC-(*p*-C_6H_4)-CO]$_n$~ in sulfuric acid into a coagulation bath. Polymer chains are not entangled; the good orientation of "normal" aramides along the fiber axis can be increased by post-stretching to "high-modulus" grades. The filaments have circular cross sections. Short fibers are usually 6 mm long; shorter and longer grades are available. Spun fibers have lengths of 40 mm or 60 mm. Filament yarns are delivered in finenesses of 20 tex to 805 tex (1 tex = 1 g/km).

10.3.3. Rubbers and Impact Modifiers

Many plastics have low resistances against impact; their polymers are thus modified by rubbers or rubber-like impact modifiers (see Section 13.7.3.). Rubbers for toughening include:

ABS	Acrylonitrile + butadiene + styrene	for PVC, PC, PPE
ACR	Acrylic ester + cross-linkable comonomer	for PVC
ACS	acrylic ester + styrene + acrylonitrile	for PVC, PC, SAN
MBS	Methyl methacrylate + butadiene + styrene	for PVC, PBT, PET, PA 6
NBR	Acrylonitrile + butadiene	for PVC
PE-C	Chlorinated poly(ethylene)	for PVC
EPDM	Ethylene + propylene + nonconjugated diene	for PP
-	Ethylene copolymers	for PE films and sheets

These modifiers form domains that are either bound physically to the continuous matrix of the polymer by added diblock copolymers or chemically by grafting reactions during in-situ polymerizations of rubber solutions in the plastics forming monomers.

Impact-modified poly(styrene)s are almost exclusively prepared by in-situ polymerizations of styrenic rubber solutions: PS-I (IPS) and PS-HI (HIPS) from solutions of styrene-butadiene rubbers SBR or cis-butadiene rubbers BR in styrene; ABS from solutions of acrylonitrile-butadiene rubbers NBR in styrene-acrylonitrile; ACS from solutions of acryl ester rubbers ACR in styrene-acrylonitrile.

All other toughened thermoplastics are produced by melt blending of the matrix polymers with rubbers. The high shear forces during melt blending in mixers break chains and generate polymer radicals that in turn lead to grafting and cross-linking of rubber chains. Insufficient in-situ grafting can be complemented by adding compatibilizers (diblock or graft copolymers).

10.3.4. Plasticizers

Plasticizers are added to polymers to improve the flexibility, processibility, and/or foamability of plastics (see Section 13.3.). About 500 different types of plasticizers are marketed. They are generally low molar mass liquids; polymer(ic) plasticizers are used in much smaller amounts. Nonpolymeric plasticizers are sometimes called *monomer(ic) plasticizers,* although they are not monomers.

80 % to 85 % of all plasticizers for thermoplastics are used for poly(vinyl chloride); the most important plasticizer is di(2-ethylhexyl)phthalate ("dioctyl phthalate", DOP). The main plasticizer for elastomers is mineral oil; tires contain up to 40 %.

Polymer(ic) plasticizers are mainly aliphatic polyesters and polyethers. The former are prepared by polycondensation. They thus possess fairly broad molar mass distributions; they also contain monomeric and oligomeric molecules. Since their number-average molar mass is low (ca. 4000 g/mol), they are also called *oligomer plasticizers.* Plasticization can also be achieved by copolymerization of the parent monomer with certain other monomers (Fig. 13-7); this effect is called *internal plasticization* in analogy to *external plasticization* by added high and low molar mass plasticizers.

External plasticizers are subdivided into primary and secondary ones. *Primary plasticizers* interact directly with chains by way of solvation. *Secondary plasticizers* are merely extenders; they can be used only in combination with a primary plasticizer. A plasticizer may be primary or a secondary. Mineral oils are, for example, primary plasticizers for poly(diene)s but extenders for poly(vinyl chloride).

Literature

10.1.1. Homogenization and Compounding

G.A.R.Matthews, Polymer Mixing Technology, Applied Science Publ., London 1982
J.A.Biesenberger, ed., Devolatilization of Polymers, Hanser, Munich 1983
C.Rauwendaal, ed., Mixing in Polymer Processing, Dekker, New York 1991
J.M.Vergnaud, Drying of Polymeric and Solid Materials, Springer, London 1992

10.2. Additives and 10.3. Modifiers: General

J.Stépek, H.Daoust, Additives for Plastics, Springer, New York 1983
J.T.Lutz, ed., Thermoplastic Polymer Additives: Theory and Practice, Dekker, New York 1988
R.Gächter, H.Müller, Eds., Plastics Additives Handbook, Hanser, Munich, 3rd ed. 1990
M.Ash, I.Ash, Handbook of Industrial Chemical Additives (An International Guide by Product, Trade-Name, Function and Supplier, VCH, New York 1991
J.Edenbaum, ed., Plastics Additives and Modifiers Handbook, Van Nostrand Reinhold, New York 1992
E.W.Flick, Plastics Additives. An Industrial Guide, Noyes Data, Park Ridge (NJ), 2nd ed. 1993 (4000 additives from 164 manufacturers)

10.2.2. Chemofunctional Additives

G.Geuskens, ed., Degradation and Stabilization of Polymers, Appl.Sci.Publ., Barking, Essex 1975
B.Rånby, J.F.Rabek, Photodegradation, Photo-Oxidation and Photostabilization of Polymers, Wiley, New York 1975
M.Lewis, S.M.Atlas, E.M.Pearce, eds., Flame-Retardant Polymeric Materials, Plenum, New York 1976
W.C.Kuryla, A.J.Papa, ed., Flame Retardancy of Polymeric Materials, Dekker, New York 1978 (6 vols.)
W.Schnabel, Polymer Degradation, Hanser, Munich 1981
M.T.Gillies, ed., Stabilizers for Synthetic Resins, Noyes, Park Ridge (NJ) 1981
C.F.Cullis, M.M.Hirschler, eds., The Combustion of Organic Polymers, Oxford University Press, Oxford 1981
A.Davis, D.Sims, Weathering of Polymers, Elsevier, Amsterdam 1983
W.Lincoln Hawkins, Polymer Degradation and Stabilization, Springer, Berlin 1984
V.Ya.Shlyapintokh, Photochemical Conversion and Stabilization of Polymers, Hanser, Munich 1984
R.M.Aseeva, G.E.Zaikov, Combustion of Polymer Materials, Hanser, Munich 1986
C.J.Hilando, Flammability Handbook for Polymers, Technomic, Lancaster (PA), 4th ed. 1990
G.Pal, H.Macskasy, Plastics - Their Behavior in Fires, Elsevier, Amsterdam 1991

10.2.3. Colorants

T.Patton, ed., Pigment Handbook, Wiley, New York 1973-1976 (3 vols.)
T.G.Webber, ed., Coloring of Plastics, Wiley, Somerset (NJ) 1979
M.Ahmed, Coloring of Plastics, Van Nostrand-Reinhold, New York 1979
N.S.Allen, J.F.McKellar, Photochemistry of Dyed and Pigmented Polymers, Appl.Sci.Publ., Barking, Essex 1980
D.H.Solomon, D.G.Hawthorne, Chemistry of Pigments and Fillers, Wiley, New York 1983
P.Lewis, ed., Pigment Handbook, Wiley, New York, 2nd ed. 1988
H.Zollinger, Color Chemistry. Syntheses, Properties and Applications of Organic Dyes and Pigments, VCH, New York, 2nd ed. 1991
W.Herbst, K.Hunger, Industrial Organic Pigments, VCH, Weinheim 1992

10.2.4. Processing Aids

G.Buttlers, ed., Particulate Nature of PVC, Elsevier, Amsterdam 1982

10.3.1. Extending and Reinforcing Fillers

R.D.Deanin, N.R.Schott, eds., Fillers and reinforcements for Plastics, Chem. Soc., London 1975
D.E.Leyden, W.Collins, eds., Silylated Surfaces, Gordon and Breach, New York 1980
S.K.Bhattacharya, ed., Metal-Filled Polymers–Properties and Applications, Dekker, New York 1986
J.V.Milewski, S.Katz, eds., Handbook of Reinforcements for Plastics, Van Nostrand, New York, 2nd ed. 1987
E.P.Plueddeman, Silane Coupling Agents, Plenum, New York, 2nd ed. 1991
J.T.Lutz, Jr., D.L.Dunkelberger, Impact Modifiers for PVC, Wiley, Somerset (NJ) 1992

10.3.2. Fibers

F.Hapey, Ed., Applied Fibre Science, Academic Press, New York 1978 (3 vols.)
M.Lewin, E.Pearce, Eds., Handbook of Fiber Science and Technology, Dekker, New York 1984 (several volumes with various publication dates by various editors)
H.A.Krässig, J.Lenz, H.F.Mark, Fiber Technology. From Film to Fiber, Dekker, New York 1984
J.-B.Donnet, R.C.Bansal, Carbon Fibers, Dekker, New York 1984
H.H.Yang, Aromatic High Strength Fibers, Wiley, New York 1989

10.3.3. Rubbers and Impact Modifiers

D.C.Blackley, Synthetic Rubbers: Chemistry and Technology, Appl.Sci.Publ., London 1983
J.A.Brydson, Rubber Materials and Their Compounds, Elsevier, London 1988
W.Hofmann, Rubber Technology Handbook, Hanser, Munich 1989
J.T.Lutz, Jr., D.L.Dunkelberger, Impact Modifiers for PVC, Wiley, Somerset (NJ) 1992

10.3.4. Plasticizers

J.Kern Sears, J.R.Darby, The Technology of Plasticizers, Wiley, New York 1982

11. Processing

11.1. Introduction

11.1.1. Overview

"Processing" refers in general to all operations on matter that produce a material with the desired microstructure and macroscopic shape. In the manufacturing of ceramics via sol-gel processes, these operations include chemical reactions. In metallurgy, "processing" may include, for example, alloy formation before the shaping and annealing thereafter. In polymer technology, however, "processing" is usually restricted to "shaping" although the preparation of thermosets includes chemical cross-linking.

Plastics processing methods can be subdivided in many ways. Engineers prefer a classification according to the type of machine type (e.g., pressureless vs. pressure-using techniques), process (shaping vs. transforming), or plastics (thermoplastic vs. thermoset). Molecular engineering dictates another classification according to the form stability of the articles, which in turn depends on the rheological characteristics of the starting materials and finished articles.

Two main groups can be distinguished: those that need external supports for the shaping of raw materials into articles (molding processes) and those that do not (extrusion processes). Within each group, many different types of processes exist to convert monomers, prepolymers, or polymers into plastics. In general, four procedures can be distinguished for the processing of raw materials into shaped plastics:

A. From monomer to product by polymerization with simultaneous shaping;
B. From monomer to prepolymer by oligomerization, followed by simultaneous polymerization and shaping to product;
C. From monomer to polymer by polymerization, followed by separate shaping;
D. From monomer to polymer by polymerization, followed by shaping to semi-finished product, and finally to the end product.

In all four procedures, manufactured articles must be sometimes after-treated (degated, polished, surface-coated, etc.).

Procedure A has only one process step and should be the most economical. It is widely used for wire coatings. However, technological problems are often encountered for shaped articles because the process involves the conversion of a low-viscosity liquid into a high-viscosity body. Processing methods include coating, casting, block foaming, extrusion, and injection molding. Resulting polymers ("reaction polymers") may be thermoplastics such as polyamides, polycarbonates) or thermosets such as polyurethanes, unsaturated polyesters, epoxies, phenolics) Since very high polymerization rates are required for short cycle times, few monomers, types of polymerization, and shaping methods are suitable for mass production techniques such as reaction injection molding (RIM).

Procedure B is the typical process for thermosets. An oligomerization is carried out to conversions just before the gel point (sometimes called B-stage). Additives and modifiers are mixed in and the final processing step consists of simultaneous cross-linking polymerization and shaping. Since viscosities remain low at B-stages and are required for mixing, reaction systems lack strength and molds are necessary for most shaping procedures. Processing methods include molding, compression molding, cavity-compression molding (hot pressing), injection compression molding, and injection molding. Extrusion methods may also be used in certain cases, depending on the system. The manufacture of expanded plastics as semifinished goods (slabs etc.) also does not require molds.

Procedure C is generally chosen for thermoplastics. The polymer from the polymerization process is isolated before shaping and stored (e.g., as granulate) for a shorter or longer period of time. The stored polymer must be dried before further processing, otherwise, steam produced at higher processing temperatures may lead to voids and cavities in shaped articles. After compounding, the plastic raw material must be accurately weighed out for each cycle in discontinuous processing procedures. The filling factor = raw material density/bulk density must be known.

The choice of a processing technique is influenced technically by both the rheological properties of the material and the form or shape of the desired article. Economically important are the cost of processing machines and the phase sequence (number of articles produced per unit time). Commodity articles are produced by casting, centrifugal casting, hot pressing, injection compression molding, coating, spraying, roll milling, extrusion, injection molding, cold forming, press forming, stretch forming, blowing, extrusion blowing, sintering, fluidized bed sintering, flame spraying, or hot blast sprinkling; specialty articles are also prepared by molding.

Procedure D is the method of choice for difficult to manufacture articles. Procedures such as welding, stamping, cutting, forging, sawing, boring, turning, milling (in the metal-working sense), and baking are employed for both thermoplastics and thermosets. The surfaces of plastics are sometimes additionally treated for technical (surface hardness, friction) or aesthetic reasons (gloss, color) by polishing, painting, metallizing, coating, etc.

11.1.2. Energy Considerations

Many processes require the heating of materials. Monomers and propolymers must be brought to the appropriate reaction temperatures and prepolymers and plastics to temperatures at which the plastics attains the viscosity needed for the particular fabrication process.

Heat can be supplied externally or internally. *External heat* is taken up by plastics surfaces through contact (hot metal plates, immersion into baths) or by convection (infrared irradiation, blowing hot air); the heat is transported into the interior by conduction. *Internal heat* is generated by vibration of atomic groups through irradiation (ultrasound or high frequency (polar groups only)) and by friction through shearing (kneaders, screw plasticators, etc.). The combination of contact and friction heating is the predominant method (*plastication*), whereas contact heating alone is less often used (*plastifying*). Both terms should not be confused with plasticization by addition of a plasticizer.

Contact heating creates a plastified zone ("melt") near the mold surface to which the melt may or may not adhere. In extruders, part of an adhering melt is shaved off the mold surface by the screw. This melt can enter the voids between pellets and provides contact heating to the unmelted particles but it cannot penetrate cold plugs created by compressed polymer powders. Cold plugs always form from nonadhering melts because these melts cannot be shaved off.

External heating is thus usually supplemented by internal heating through friction. The adiabatic temperature change is given by $\Delta T = \Delta p/(\rho c_p)$, where $\Delta p = \eta\dot{\gamma}$ is the pressure difference, η the kinematic viscosity, $\dot{\gamma}$ the shear rate, ρ the density, and c_p the specific heat capacity. An injection molding with Δp = 100 MPa and $\dot{\gamma} = 10^4$ s^{-1} of a plastic with ρ = 1 g/cm^3, c_p = 2 J/(g K) and $\eta = 10^4$ Pa s would thus increase the temperature by 50 K (50°C).

Heating serves as a means to increase the temperature of plastics that in turn decreases viscosities and thus facilitates processing. Rheological properties can also be reversibly reduced by repeated extrusion called *shear modification* (shear working, shear refining). Such a treatment decreases extensional viscosities (and thus die swells) and increases melt flow indices but hardly modifies steady-state shear viscosities and dynamic properties. It is probably caused by alignments of chain segments or chains because it can be reversed by heating or dissolution.

Cooling of shaped articles is possible only by thermal conductivity, a slow process, especially for thick-walled articles. Cycle times are therefore very often controlled by cooling times.

11.1.3. Processing Steps

All plastics processing methods involve three steps: (1) a softening (melting, fluidizing, etc.) of the starting material (plastic compound, thermosetting resin, reactive monomer) is followed by (2) a shaping of the article (sometimes concurrent with polymerization), and finally by (3) a stabilization of the shape of the article.

Step 1 is always needed in Procedures B-D because of the very high viscosities of solid polymers. It can be circumvented in Procedure A if liquid monomers are used.

Step 2 depends on the melt strength of the material. Polymers subjected to Procedures B, C and D usually have sufficient melt strength and do not need external supports for shaping. Although the polymerizing system becomes more viscous in Step 2 of Procedure A, its melt strength is insufficient and the shaping thus requires the use of external supports (molds, etc.). The oligomers used in Procedure B as precursors for thermosets suffer from the same limitation.

Step 3. The shapes can be stabilized by either cooling the articles below the glass temperature (amorphous thermoplastics) or the melting temperature (semicrystalline thermoplastics) or by cross-linking (thermosets). Semicrystalline thermoplastics require supercooling beyond the crystallization temperature because of the latent heat of crystallization. Solidification by cooling and cross-linking may lead to considerable shrinkage and thus to deformation and stress; pressure must often be applied in order to maintain the desired shape of articles. Low shrinkage, on the other hand, can make unloading difficult.

Thermosets. Raw materials for thermosets (monomers, prepolymers) have low melt strengths. They can be processed only by methods requiring external supports such as molding, winding, compression molding, and injection molding (Table 11-1). The use of raw materials with low viscosities furthermore restricts the processing to those methods that generate only small shear gradients (molding, dipping, winding); examples are diallyl phthalate (DAP) and the raw materials for polyurethanes (PUR). Another monomer processed by molding is methyl methacrylate (MMA).

Table 11-1 Preferred processing methods for various plastics. ++ Extensively used method, + commonly used, (+) processed as monomer, {+} processed as plasticized polymer. Abbreviations: see text and Appendix.

Plastics	Stabilization by external support				No external stabilization needed		
	Molding, dipping	Lamination, winding	Compression molding	Injection molding	Extrusion	Blow forming	Thermo-forming
Thermosets, processed as monomers							
DAP	+						
PUR	++						
EP	+	+	+				
Thermosets, processed as prepolymers							
UP	+	++	++	+			
PF			++	+			
MF			+	+			
Semicrystalline thermoplastics							
POM				++	+		
PA 6, 6.6				++	+		
PET				+	+	++	
PE, PP			+	++	++	++	+
Amorphous thermoplastics							
PC				+	+	+	+
PS				++	+	+	++
PMMA	(+)			+	++		+
PVC	{+}			+	++	++	+
Plastics (%):	?	?	3	32	36	10	?

Other processing: Calendering 6 %, coating 5 %, powder processing 2 %, all other methods 10 %.

Raw materials with higher viscosities can be subjected to methods with greater shear gradients such as compression molding ($\dot{\gamma}$ = (1-10) s^{-1}) and injection molding ($\dot{\gamma}$ = (10^3-10^5) s^{-1}). Examples are stage B resins (prepolymers) of melamine and phenolic resins (MF, PF), unsaturated polyester molecules (UP), and the raw materials for reaction injection molding (not listed in Table 11-1).

Thermoplastics. Melt strengths of thermoplastics are much higher than those of monomers and prepolymers used for thermosets. Thermoplastics can thus be processed by methods with high shear gradients (injection molding) or by methods requiring (at least in principle) no external support for stabilization (extrusion, blow forming, thermoforming) (Table 11-1); however, such supports may be needed to create the desired shapes of articles.

Thermoforming requires plastics that can be deformed without flowing away. The processing temperature must be above the glass temperature (onset of deformation due to cooperative movement of segments). It cannot be very high above the glass temperature, however, because then the melt viscosity would be too low. With the exception of poly(ethylene), crystalline polymers are thus rarely suitable for thermoforming (Table 11-1) because their glass temperatures are far below their softening (Vicat) temperatures, which are near their usually high melting temperatures (Section 5.2.7). Amorphous polymers have temperature dependencies of viscosities that make them suitable for thermoforming.

Thermoforming and blow molding also require high viscosities, which can only be achieved by entangled polymer chains. The ability to form entanglements depends on the mass-average degree of polymerization $\overline{X}_w$; it increases with increasing number of chain atoms $(\overline{N}_c)_w = N_u \overline{X}_w$, where N_u is the average number of chain atoms per monomeric unit. N_u = 2 for styrene units $-CH_2-CH(C_6H_5)-$ and N_u = 7 for ε-caprolactam units $-NH-(CH_2)_5-CO-$.

Polymers by polycondensation typically have chain link numbers $(\overline{N}_c)_w$ than polymers by chain polymerization ("addition polymerization"); the former are usually less suitable for thermoforming and blow molding than the latter. Typical $(\overline{N}_c)_w$ values for injection molding grades of condensation polymers are 1700 for polyamide 6 (PA 6) and 2000 for poly(ethylene terephthalate); PA 6 grades for extrusion require $(\overline{N}_c)_w \approx$ 6000. Poly(ethylene terephthalate)s could only be blow molded after the development of so-called bottle grades with much higher chain link numbers of ca. 7000. Typical values $(\overline{N}_c)_w$ for thermoplastics are 12 000 for poly(styrene) injection molding grades and 30 000 for poly(ethylene)s.

Different processing methods thus require polymer grades with different minimum mass-average molar masses for sufficient shape stabilities. High molar masses lead however not only to high melt strengths but also to high melt viscosities. Low viscosities (low molar masses), on the other hand, allow lower working pressures and reduce operational costs.

Processing is also affected by the molar mass distribution because small polymer molecules act as diluents. Polymer grades are thus selected with those molar masses and molar mass distributions that allow a comprise between sufficient high melt strengths and desirable low operational costs.

From the materials point of view, plastics processing methods can be subdivided according to their rheological requirements (viscous states, elastoviscous states, elasto-plastic states, viscoelastic states, and solid states). In engineering, plastics processing is usually discussed with respect to the machine requirements (pressure-less techniques, processing under pressure, forming processes, machining, bonding, surface treatments).

11.2. Processing of Viscous Materials

Liquid monomers, melted prepolymers, solutions with low polymer concentrations, polymer dispersions, and polymer powders have relatively low viscosities and flow under their own gravity. No pressure is needed to apply the materials but it may be required for the removal of voids during the final processing stage. Processes for viscous materials can be subdivided into casting, coating, and molding methods.

11.2.1. Casting

Monomer Casting. Monomers (methyl methacrylate, styrene, ε-caprolactam, N-vinyl carbazole) or prepolymers (phenolic resins, epoxy resins) are cast into open molds and polymerized (Fig. 11-1 M). The exothermic polymerizations require good temperature controls and long reaction times (slow production rates); unsaturated polyesters are hardly processed by casting because of the strong exothermal copolymerization with the "cross-linkers" styrene or methyl methacrylate. Molds are inexpensive and can even be made from sheet metal, wood, ceramics, or glass. Casting is thus used for the manufacture of articles in small numbers such as dentures, frames for glasses, etc.

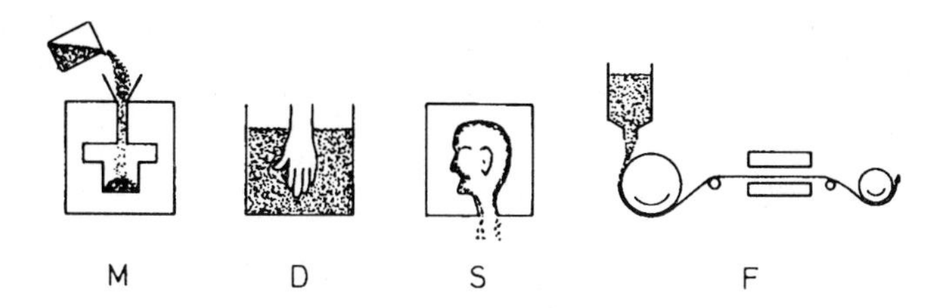

Fig. 11-1 Casting methods: monomer casting M, dipping D, slush casting S, and film casting F.

Film Casting. Polymer solutions, dispersions, or melts are cast on a casting wheel (casting drum) (Fig. 11-1 F) or a continuous metal belt. Solvents evaporate in a drying chamber and the film is wound up. Casting is used mainly for photographic films from cellulose acetate; these films are more homogeneous than those by calendering.

Dipping (dip molding) is used for the casting of very thin hollow articles, for example, surgical gloves and condoms (Fig. 11-1 D). An external mold (dip mold, dipping mandrel) is dipped into a solution (e.g., silicones), latex (e.g., natural rubber), or PVC plastisol for a polymer deposition. The mold is withdrawn and heated or dried to solidify the deposit; plastisols need gelling.

Slush casting (*hollow casting, flow casting*) is used for the casting of thicker hollow articles. Solutions or dispersions of resins, PVC plastes, or hot plastics melts are poured into a two-part hollow mold where a skin of the plastic is formed at the inner wall. The mold is reversed to let excess resin flow out (Fig. 11-1 S). The deposited plastic is hardened by cooling (melts) or heating (solutions). Hollow casting from dry, sinterable powders is known as *slush molding*; however, the terms slush molding and slush casting are also used interchangeably.

Centrifugal casting produces thick-walled rotationally symmetrical articles or inner coatings by centrifuging a partially filled mold (rotation around *one* axis). The method is used for the manufacture of pipes, bushings, sockets, etc., from polymerizing ε-caprolactam and for inner coatings of hollow articles by meltable powders.

Rotational casting is a method for the automatic casting of hollow articles. Powdered thermoplastic resins (ca. 500 μm diameter) or PVC plastisols are metered into forms (Fig. 11-2). The forms are closed, rotated around *two* perpendicular axes for a homogeneous distribution of the material at the walls, and heated to melt the thermoplastics or to gelatinate (gelify, gelatinize, gel, fuse, flux) the plastisol. The thickness of articles is determined by the amount of resin and the rotational speed. The form is opened, and the article is removed after cooling.

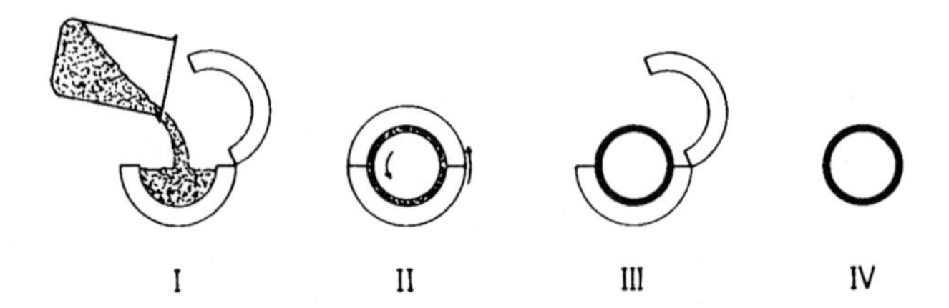

Fig. 11-2 Rotational casting: the mold is filled with resin (I), closed, rotated and heated (II), and opened (III), whereupon the article can be removed (IV).

11.2.2. Coating

Coating processes are used to deposit plastics permanently on other materials, either to modify surfaces (Section 11.7) or to manufacture composite materials. **Spread coating** is used to coat sheet materials (paper, fabric, etc.) with fluid materials such as polymer solutions, dispersions, pastes or melts; methods include brush, knife, roll, and spray coating. Three-dimensional articles are coated by flame spraying of plastics pellets and by fluidized bead coating or electrostatic spraying of polymer powders. Direct dipping of articles into polymer melts is not a commercial process.

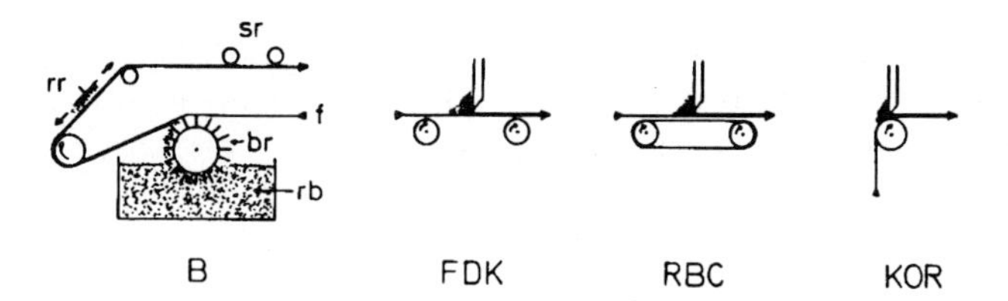

Fig. 11-3 Spread coating processes: B = brush coating with impregnation of fabric f by brush roll br, reciprocating brush rr, and smoothing rolls sr; FDK = coating with floating doctor knife, RBC = rubber blanket coating, KOR = knife-over-roll coating.

In **brush coating**, the fluid coating compound is transferred from a coating pan to the fabric by a brush roll (Fig. 11-3 B). The coating is spread by a reciprocating brush and finally smoothed by smoothing rolls.

Knife coating employs doctor knives (doctor blades) instead of brushes. *Floating doctor knives* (not to be confused with air knives; see below) press against the fabric without support and remove excess fluid coating resin; the fabric is held under tension by two supporting rolls (Fig. 11-3 FDK). The doctor knife may be supported by an endless rubber belt in a rubber blanket coater (Fig. 11-3 RBC), or a backing roll in a knife-over-roll coater (Fig. 11-3 KOR), which carry the fabric past the doctor knife.

Roll Coating. A roll dips into the coating pan and transfers the fluid coating resin to the underside of the fabric which is supported by a roll rotating in counter direction; excess coating resin is squeezed off (Fig. 11-4 RCS). In *reverse roll coating*, fabric is running between two reverse rolls. A center coating roll is continuously coating the fabric; excess coating resin is removed by a doctor knife and collected in a catch pan (Fig. 11-4 RRC). Very many different configurations of roll coaters are known. For example, excess coating resin may be removed in roll coating by using an air-jet (*air knife*) instead of a doctor knife (Fig. 11-4 RAK).

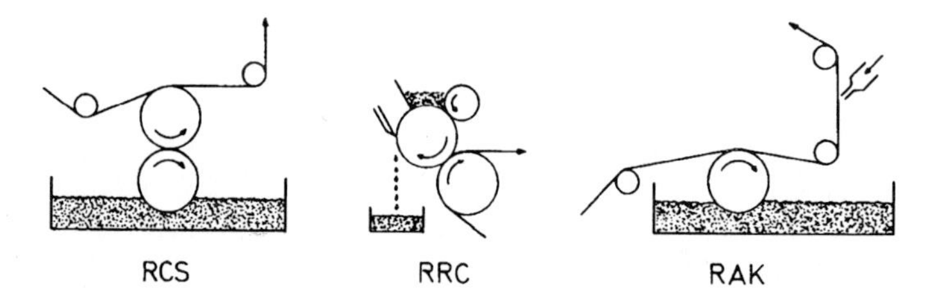

Fig. 11-4 Some roll coating methods. RCS = Roll coating with squeezing-off excess resin; RRC = reverse roll coating with doctor knife; RAK = roll coating with air knife ✔.

Spray coating is used to spray materials (especially foams) continuously through spray nozzles unto moving webs (Fig. 11-5, SC).

Flame Spraying. Plastics pellets are heated in a flame-spraying gun by combustion of a combustible gas with air or oxygen and sprayed on heated metal articles (Fig. 11-5, FS). Suitable polymers are poly(ethylene), poly(vinyl chloride), cellulose esters, and epoxy resins.

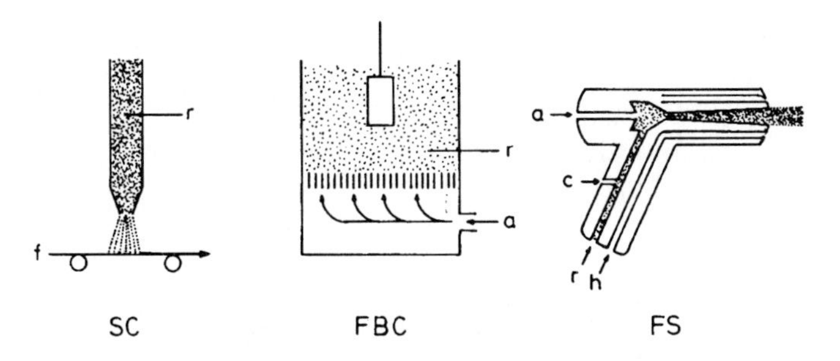

Fig. 11-5 Coating processes: SC = spray coating; FBC = fluidized bed coating; FS =flame spraying. a = Air, c = control of powder flow, f = fabric or film, h = heating gas, r = resin powder.

Hot spraying does not require the heating of metal articles. Polymer powders (polyamides, epoxy resins) are rather sprayed on the metal parts under a blanket of an inert gas (Ar, He, N_2) while being heated by a 1600°C arc. Metal temperatures do not exceed 50-60°C; metal structures thus does not change as in flame spraying.

Fluidized bed coating is often called *fluidized bed sintering* although it is not a sintering process. Metal parts are degreased and roughened or treated with primers for better adhesion. The heated parts are immersed in a fluidized bed containing polymer particles of ca. 200 μm diameter (Fig. 11-5, FB), which are agitated by inert gases to avoid oxidation. The polymer particles adhere to the metal surface where they melt and form a dense film of ca. 200-400 μm. Suitable polymers must possess unique combinations of melt viscosities, heat stabilities, and mechanical properties. Used are mainly polyamides, poly(ethylene)s, and poly(vinyl chloride), to a lesser extent also cellulose acetobutyrate.

11.2.3. Molding

The term *molding* is applied to most (but not all) processes requiring a mold; it is used as a generic term for molding operations on fiber mats that are impregnated by thermosetting resins. Molds can be negative (female molds, cavities) or positive (male molds, plugs) or can have both female and male parts. A female mold was formerly called a matrix (Lat.: *matrix* = womb; from *mater* = mother); the corresponding term *patrix* (from Lat.: *pater* = father) was never used for male molds.

Hand lay-up molding (*contact molding, impression molding*) is the simplest discontinuous, pressure-less processing method (Fig. 11-6, HL) Molds are treated first with a release agent or a sealing film. A (300-600) μm thick layer of fiber-free resin (gel-coat) is applied with a brush or spray gun and partially cured; the gel-coat may be substituted by a resin-impregnated surfacing mat. One ply (or more plies) of fiber mats is pressed against the first layer and is impregnated with the resin-catalyst mixture (unsaturated polyester, epoxy resin), usually followed by another gel-coat. The composite is cured in place, e.g., by infrared radiation. Hand lay-up is especially suited for small numbers of large parts (e.g., hulks of sail boats, rotor blades).

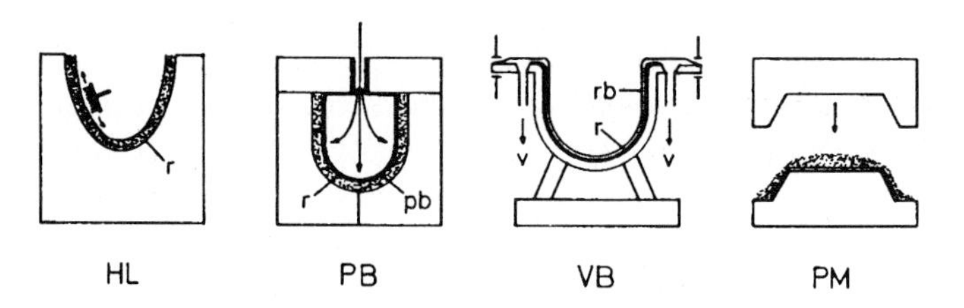

Fig. 11-6 Molding processes for reinforced plastics.
HL = Hand lay-up of resin-impregnated mat r on gel coat (dotted);
PB = pressure bag process with rubber bag pb (thick black line) over resin-impregnated fabric r (dotted area);
VB = vacuum bag process: a rubber bag rb (thin black line) is pressed against a resin-impregnated fabric r (thick black line) by application ↓ of vacuum v;
PM = press molding: a mat or fabric is placed over a male mold, resin is applied to the center of the mold where it runs off, and is evenly impregnating the mat upon closure of the mold.

Hand lay-up spray molding can be simplified by use of preimpregnated fibers (*prepregs*): fibers are continuously impregnated with resin and partially cured in an oven before they are applied to the mold:

Preimpregnated fabrics are fabrics of parallel glass, carbon, or aramide fibers impregnated with phenolic or epoxy resins; they are mainly used for high-load, thin-walled articles.

Preimpregnated rovings contain parallel strands or filaments of textile glass fibers, which were assembled without intentional twist.

Sheet-molding compounds (SMCs) are glass-fiber mats that are impregnated with a mixture of UP resins and mineral fillers (chalk, talc, clay).

Fiber-containing molding compounds (*bulk molding compounds*, BMCs) are compounds of UP resins, chopped glass fibers, and mineral fillers.

Unfilled UP resins shrink 6-8 % and BMCs still 0.2-0.4 % during polymerization (Section 3.2.4), which leads to rugged surfaces (profiles). The shrinkage can be reduced by the addition of syrupy (up to 25 %) solutions of thermoplastics (poly(styrene), poly(methyl methacrylate), cellulose acetobutyrate) to the UP resins (*low-profile resins*). Upon heating for curing, part of the styrene evaporates prior to polymerization and "foams" the resin, which compensates for shrinkage. Furthermore, parts of the thermoplastics demix and move to the surface. Foaming and demixing generate smoother surfaces.

Fiber-spray gun molding (*spray-up molding*) is a semimechanized hand lay-up process. Unsaturated polyester resins, catalysts, and chopped fibers are sprayed through separate nozzles by pressurized air. They mix in the air before they contact the mold. Advantages are long shelf-lives of resin and catalyst (which are stored separately) but short cure times (mixing of resin and catalyst in the spray gun), and no loss of fibers whereas in hand lay-up mats, rovings, etc., have to be cut to size. Fiber-spray gun molding needs however more skill for an even application of the fiber-filled resin and extensive starting-up and clean-up operations for the equipment.

More compact thermosets can be obtained if pressure is applied to molds during curing. Single-component rigid molds and elastic sheets as countermolds are used in:

Bag Molding Processes. In the *pressure bag process* (Fig. 11-6, PB), molds are filled with the resin mixture. Fabrics or mats are added and covered with a rubber bag. The mold is closed; on application of air pressure, the bag is pressed against the reinforcing material, which is impregnated by the resin. Vacuum is used for impregnation in the *vacuum bag process* (Fig. 11-6, VB). Complex high-quality articles are manufactured by *autoclave processes*, which employ heated, pressurized autoclaves.

Press molding of glass fiber (fabric or mat) reinforced resins requires greater pressure. In *liquid resin press molding* (Fig. 11-6, PM), the mat is placed in the mold and resin is poured into the center of the mold where it is evenly distributed upon mold closure. *Cold liquid resin press molding* of epoxy resins uses pressures of 0.1-1 MPa at (20-60)°C. The slow reaction time of the cold liquid resin process requires the use of accelerators whereas none are needed in the *hot liquid resin press molding* where higher pressures and higher temperatures are common (Table 11-2). Unsaturated polyesters harden by free radical polymerization; no volatile leaving molecules are formed and less pressure is needed than in the polycondensation of phenol, urea, and melamine resins. Press molding leads to edges at seals; articles must be degated.

Table 11-2 Processing of thermosetting resins by press molding, transfer molding, and injection molding of PF = phenol-formaldehyde, UF = urea-formaldehyde, MF = melamine-formaldehyde, and UP = unsaturated polyester. [a] Applied external pressure; [b] per mm wall thickness; [c] approximately independent of wall thicknesses up to ca. 20 mm. Data of [1].

	Phys.unit	PF	UF	MF	UP
Pressure					
press molding	MPa	15 - 80	15 - 80	15 - 80	10 - 40
transfer molding[a]	MPa	50 - 200	50 - 200	50 - 200	30 - 80
injection molding[a]	MPa	80 - 250	80 - 250	100 - 250	30 - 100
Mold temperature					
press molding	°C	160 - 190	130 - 160	150 - 170	130 - 190
transfer molding	°C	160 - 190	130 - 160	150 - 170	140 - 190
injection molding	°C	170 - 190	140 - 160	160 - 180	150 - 190
Curing time					
press molding[b]	s	30 - 60	20 - 40	20 - 40	10 - 20
transfer molding[c]	s	40 - 120	30 - 120	30 - 120	10 - 60
injection molding[c]	s	20 - 80	15 - 80	15 - 80	10 - 60

Resin transfer molding (RTM; *transfer molding*) uses multicomponent rigid molds. The cut-to size braided or woven preform structure is placed in the mold (Fig. 11-7). After hydraulic closing, low viscosity UP or EP resin is injected under pressure; the impregnation of the reinforcing material may be assisted by vacuum. Heat is applied for cross-linking polymerization and the cured composite is demolded. Transfer molding is thus an intermediate between compression molding and injection molding.

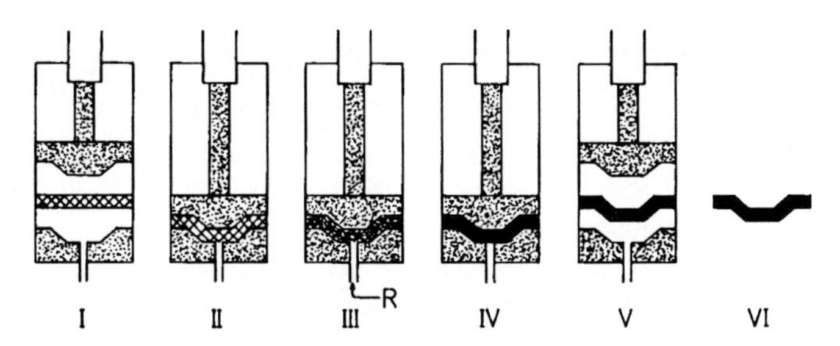

Fig. 11-7 Resin transfer molding. A mat is placed in the mold (I) and the mold is closed (II). The resin R is injected under pressure (III) and impregnates the mat. Applied heat causes cross-linking of the resin (IV). The cured article (VI) is demolded (V).

Filament winding is a semicontinuous process for the manufacture of large hollow articles (Fig. 11-8). Glass-fiber rovings are impregnated with the resin (+ initiator or catalyst) and wound under tension around a rotating mandrel in exact geometric patterns. Smaller moldings are cured in ovens; larger ones in the heated mold or by ultraviolet radiation. The cured moldings shrink on the mandrel, which is therefore made collapsible.

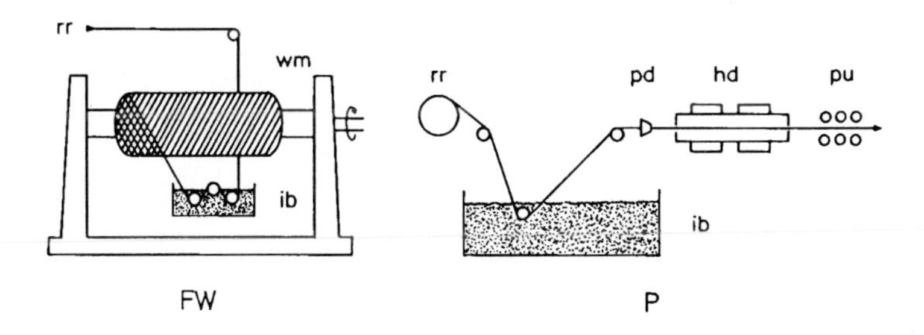

Fig. 11-8 Impregnation of rovings by filament winding F and pultrusion P.

Filament winding F: Rovings from roving reels rr are impregnated by resin in an impregnating bath ib, and led by a reciprocating fiber guide unto a winding mandrel wm. The collapsible mandrel is removed after curing (cross-linking of the resin).

Pultrusion P: Rovings from roving reels rr are pulled through a roving guide, an impregnating bath ib, and a preforming die pd (inlet nozzle) into a heated matched metal die hd where it receives the final shape. A gap between the die hd and the puller unit pu allows the cured laminate to cool and to develop strength before it is gripped by the clamps of the puller unit.

Pultrusion is the name for a continuous process for the mass production of uniaxial, "endless" semifinished products from glass fiber-reinforced resins (Fig. 11-8). Impregnated fibers pass through an inlet nozzle where excess resin and entrapped air are removed. The fiber bundle is then pulled into a heated mold where it polymerizes partially and receives a profile; despite the name "pultrusion", the material is not extruded by pressure.

Careful adjustment of resin reactivity, heating temperature, and pulling speed is essential in order to prevent premature gelation (causes adhesion to the mold) and dimensionally unstable materials (produces deformation of profiles). The profile is cured completely after it has passed through a heating channel.

11.3. Processing of Viscoelastic Materials

Thermoplastic resins have, in general. higher molar masses than thermosetting resins. Their polymer molecules are therefore often entangled. The resins thus behave as temporary physically cross-linked materials that show viscoelastic effects on processing. The compacting of these elastic networks requires application of pressure, which in turn may lead to high shear gradients.

11.3.1. Plastication

Thermoplastics often need to be plasticated before shaping. In some processes, plastication is done concurrently with shaping; in others, special equipment is used.

Kneading is used for the formulation and blending of rubbers, the compounding of plastics with additives and fillers, and the plastication required for machines without incorporated plastication units (presses, calenders). The high viscosity material is kneaded between a moving screw and stationary teeth, by overlapping kneading blades, kneading disks, sigma blades (Fig. 11-9, K), etc.

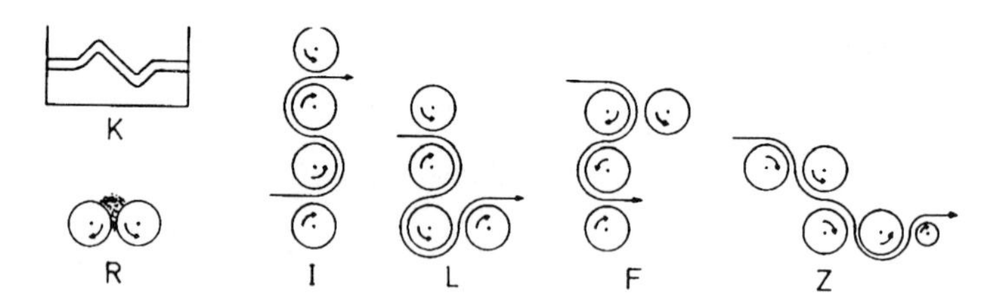

Fig. 11-9 Kneader K (here with sigma blade undergoing a crankshaft motion), rolls R, and I, L, F, and Z type calenders.

Rolling is employed for compounding, plastication, and homogenization. Two heated rolls rotate with different speed, which forces the kneading stock to circulate around one roll only (Fig. 11-9, R). The kneading stock is assembled in the roll gap where it is plasticated by the combined action of shearing and kneading, often under degradation to smaller molar masses.

Calendering. Three to seven rolls are required for the formation of endless sheets and films; most common are calenders with four rolls (Greek: kylindros = cylinder). The configurations of these rolls resemble capital letters; one thus distinguishes I, L, F, and Z calenders (Fig. 11-9), S calenders, inverted L calenders, etc. The rolls may be arranged vertically or in part at 90° to each other (see Fig. 11-9); they are sometimes placed at angles of 45°. Rolls are thicker in the middle than at the ends which compensates for roll deflection and prevents the formation of films of uneven thickness. They are heated by steam, hot water, or electricity. Residence times of plastics decrease in the order L calender > F calender > Z calender.

11.3.2. Extrusion

Extrusion is one of the most important processing methods for rubbers and powdered or pelletized plastics (Table 11-1). Extruders are built in many different configurations. They are universal machines in which the polymers are transported, degassed, compacted, melted, mixed, shaped, and finally extruded as a melt (Fig. 11-10). The parison is not externally supported; it must maintain its shape until it is solidified by cooling. The self-support is provided by highly entangled chains; sometimes also by slightly cross-linked polymers. Extrusion thus requires high molar mass polymers.

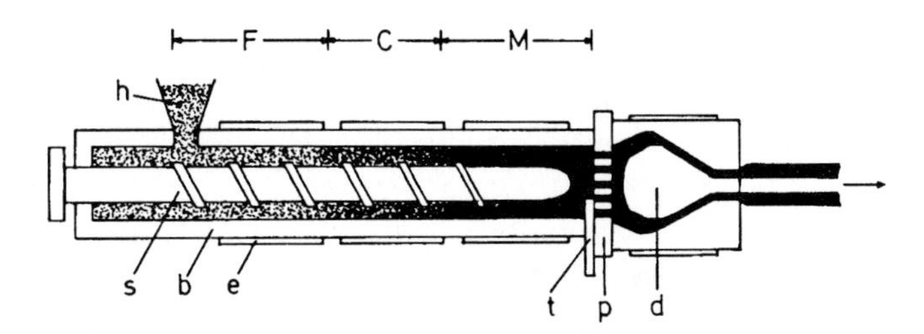

Fig. 11-10 Standard extruder. Resin powder from a feed hopper is transported by a screw s in a barrel b with heating by electrical heaters e (and friction!) from the feed zone F to the compression zone C and further into the metering zone M. The melt passes a pressure transducer t, a braker plate and screen pack p, and a die mandrel d (for pipes) before it is extruded from the die body.

Extrusion of melts through an orifice deforms polymer coils (Fig. 11-11, top left). Segments of entangled chains can no longer slip from each other at high stresses and short times because entanglements act as physical cross-links. A normal stress builds up perpendicular ("normal") to the stress direction. At the die exit, this stress is relieved and the coils return to the thermodynamically more favorable shape of unperturbed coils: the melt expands perpendicular to the flow direction. This phenomenon is known as Barus effect or memory effect (melts), Weissenberg effect or rock climbing (solutions), parison swell (extrusion), swelling (blow molding), etc. Because normal stresses are different in the various sections of noncircular dies, shapes of parisons and die orefices differ, too (Fig. 11-11, bottom left). The Barus effect is especially strong for entangled polymers with high molar mass tails because of the 3.4 power dependence of η on $\overline{M}_w$.

Negative Barus effects are known for lyotropic solutions of rod-like molecules or thermotropic LCPs with rod-like mesogens (diameter of strand smaller than diameter of orifice). If such molecules crystallize after exiting from the die, the strand contracts perpendicular to the extrusion direction and the strand diameter becomes smaller than the diameter of the orifice.

The extent of normal stress can be determined by the Bagley equation. A force $F_f = \pi R^2 p$ is exerted on a liquid during the flow through a capillary with radius R and length L under a pressure p. It is counteracted by a frictional force $F_r = 2\,\pi R L \sigma_{21}$. In steady state, $F_f = F_r$; since $\sigma_{21} = \eta\dot{\gamma}$, one arrives at $p = 2\,\sigma_{21}(L/R) = 2\eta\dot{\gamma}(L/R) = K'(L/R)$. Pressure p is accordingly plotted at constant shear rate against die geometry

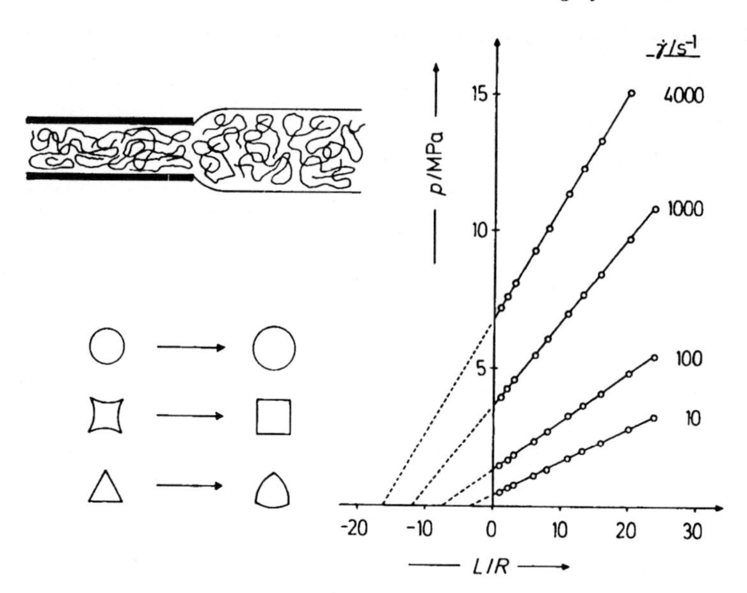

Fig. 11-11 Parison swell.
Left: Barus effect upon extrusion (top left) and effect of cross-sectional shapes of dies on swell of resulting parison shapes (bottom left).
Right: Bagley diagram of a molten high-impact poly(styrene) at 190°C at various shear rates. The intercept p_o at $L/R \rightarrow 0$ is the pressure correction; the intercept at $p \rightarrow 0$, the Bagley correction. Bagley diagram reproduced with permission by Hanser Publishers, Munich, Germany [2].

L/R in the Bagley diagram (Fig. 11-11, right). For non-Newtonian liquids, a relationship $p = p_o + K(L/R)$ is found. The intercept p_o at $L/R \rightarrow 0$ is identified with the pressure loss caused by the elastically stored energy of the flowing melt and the formation of a steady-state flow profile at both ends of the capillary (die). The higher L/R, the higher must the applied pressure be. Die lengths should thus be as short as possible.

High extrusion speeds can be achieved by thermoplastics with broad molar mass distributions. The low molar mass fractions act as lubricants which allows higher shear rates; the greater extrusion speeds lead also to greater parison swells, however.

Extruders now use almost exclusively single or double screws for the transport and plastication (Fig. 11-10); torpedos are still employed in special cases. The ratio of screw length L to screw diameter D is usually $L/D \approx$ 20:1 to 30:1; the compression ratio, 2-4. Many different extruder designs exist for different polymers and articles. A special case is for example the extrusion of films (*cast film extrusion*) through a broad-slit die (*flat sheet die, slot die*). The 20-100 μm thick films are then solidified by cooled rolls (*chill roll extrusion*) or cold water baths. Slot dies are also employed for the extrusion coating of paper and card board by poly(ethylene). *Coextrusion* of two or more polymers is used extensively for packaging films and semifinished goods (e.g., sheets that are later deep-drawn into refrigerator doors).

The Netlon® process manufactures nets by extrusion of thermoplastic polymers through two counterrotating dies whose slits are arrranged in circles. A strand is produced if slits from both dies overlap. On further rotation, the strand is split into two substrands and then reunited. Warps and wefts are thus not connected by knots as in weaves. Diamond-like openings result if both dies rotate with the same speed; other structures can be created by changing speeds, types of slits, etc. The process delivers tubes which can be cut open to give planar nets.

Extrusion blow forming (*extrusion blowing*) is a special kind of extrusion for the manufacture of hollow bodies (bottles, containers) from poly(ethylene), poly(vinyl chloride), polycarbonates, polyamides, and high-impact poly(styrenes). One-step extrusion blowing requires polymers with broad melting ranges.

The many known machine designs differ in the type of blowing (from the top, the bottom, or the side), and the taking and transporting of the parison, mold, and mandrel (shot-in of mandrel, sliding mold, rising table, etc.). In the example of Fig. 11-12, the extruder head points downwards. The extruded endless tubes are "blown" into a mold by air pressure; upon mold closure, they form hollow bodies with seams at the bottom. Flash may amount to 20 % of the material in bottle manufacturing.

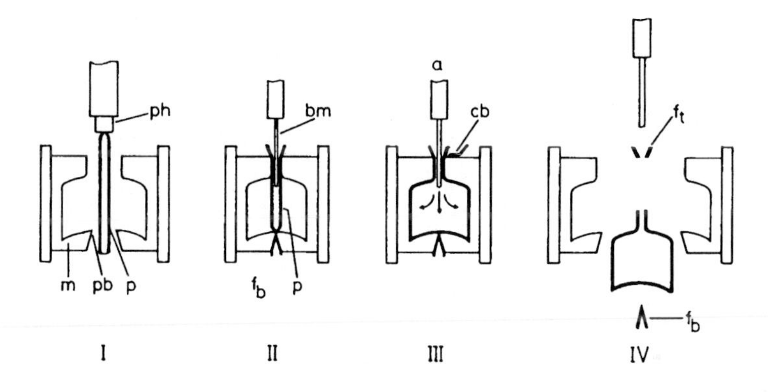

Fig. 11-12 Extrusion blow molding.
I: The parison p is extruded from the parison head ph into the open mold m; the blow mandrel bm is lifted (not shown).
II: The pinching blade pb generates bottom flash f_b upon closing of the mold; the blow mandrel is inserted.
III: Pressurized air a causes the parison to adopt the shape of the mold. After the blow mandrel is lifted, a cutting blade cb cuts off the top flash f_t.
IV: The mold is opened.

Such hollow bodies can also be prepared by other processes. Filament winding and rotational casting also require only one step; they are, however, much more expensive. Injection molding needs two steps: separetely molded halves have to be welded or glued together. It is difficult, however, to achieve constant wall thicknesses by extrusion blow forming.

Film blowing is the extrusion of tubular film from a circular die (Fig. 11-13). A blow head inflates the film tubing. The blown film tubing is cooled by an air cooling ring. The resulting bubble is collapsed by rolls or boards and laid flat by pinch rolls (squeeze rollers). Tubular film can be converted into flat film by slitting it open or by trimming both sides.

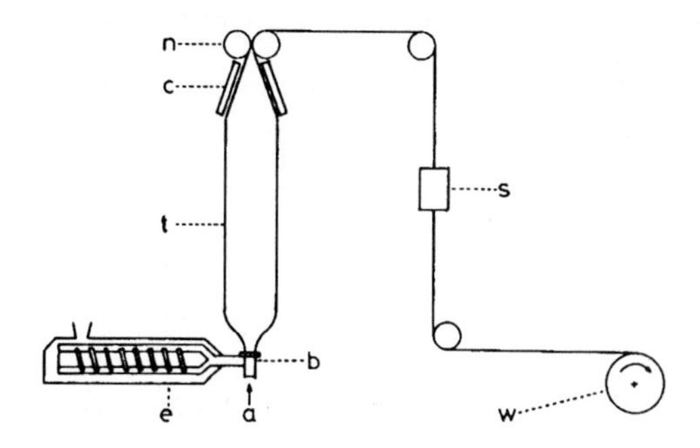

Fig. 11-13 Film blowing. A tube is extruded from an extruder e with a blow head b directed upwards. Air a expands the tube to a tubular film t ("bubble"), which is collapsed by bubble collapsing boards (or rolls) c. The film is flattened further by nip rollers n (squeeze rollers) and slit by a slitter s before it is wound by a winder w.

11.3.3. Injection Molding

Injection molding is an extrusion of hot, plasticated polymer into a "cold" mold under pressure (Table 11-1). This most important process for thermoplastics is also used for some thermosets (Table 11-1). Compact articles are mostly produced by single-screw machines with shot weights up to 30 kg and high clamping forces (ca. 3000 t for parts with 1 m^2 cross section) (Fig. 11-14).

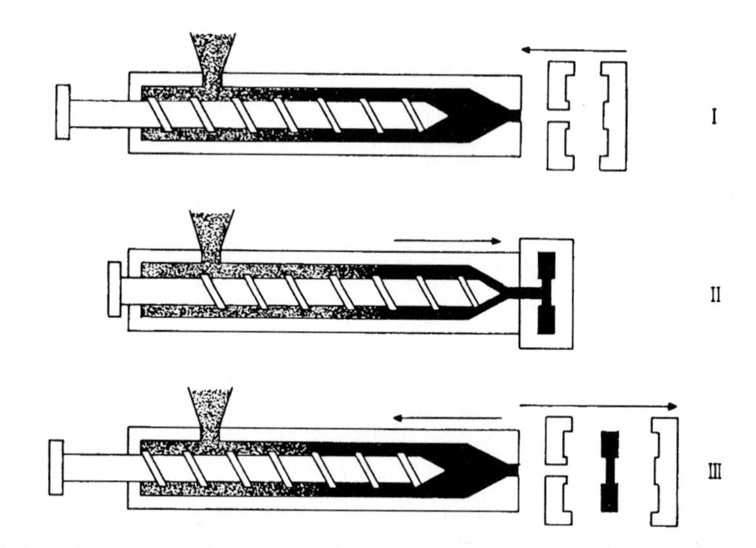

Fig. 11-14 Injection molding (here with reciprocating screw). (I) Pellets are fed into the machine with a feed hopper. They are transported by the rotating screw through the feed, compression, and metering zones where they are heated externally by heating elements and internally by friction. (II) The plasticated material is injected into the cold mold where it solidifies almost instantaneously. (III) The specimen is demolded; the reciprocation of the screw plasticates new material.

Table 11-3 Injection molding of thermoplastics with glass temperatures T_G, melting temperatures T_M, and heat distortion temperatures T_{HDT} (method A). T_{polym} = melt temperature, T_{mold} = mold temperature, p = injection pressure.

Polymer	T_G/°C	T_M/°C	T_{polym}/°C	T_{mold}/°C	T_{HDT}/°C	p/MPa
Amorphous polymers						
PC	150	-	280-320	85 - 120	132	80 - 150
SAN	120	-	200-260	30 - 85	103	
ABS	100	-	200-280	40 - 80	98	60 - 180
PS	100	-	170-280	5 - 70	104	
PMMA	105	-	150-200	50 - 90	90	70 - 120
PVC-u	82	-	180-210	20 - 60	68	100 - 180
Semicrystalline polymers						
PET	70	265	270-280	120 - 140	85	120 - 140
PCTFE	40	220	220-280	80 - 130	66	150
PA 6	50	215	230-290	40 - 60	65	90 - 140
POM	- 82	181	180-230	60 - 120	136	80 - 170
PP	- 15	176	200-300	20 - 60	56	80 - 180
PE-HD	- 80	135	240-300	20 - 60	49	60 - 150
PE-LD	- 80	115	180-260	20 - 60	37	60 - 150

Torpedoes are no longer used in standard injection molding machines: screws permit higher production speeds and create new surfaces by shearing; this allows a better degassing and an increased flow of the melt into the interstitial zones between pellets. Large foamed articles are manufactured by special injection molding units using torpedoes instead of screws.

Processing temperatures of plasticated polymers must be high enough to allow a good flow into the mold; they are considerably above the glass temperature of amorphous polymers and at least just above the melting temperature of semicrystalline ones (Table 11-3), yet lower than the decomposition temperatures. The injection pressure depends on the viscosity of the polymer. In order to achieve shape stability, mold temperatures need to be lower than the softening temperatures T_{HDT} of amorphous polymers; semicrystalline polymers attain often shape stability above T_{HDT}.

Thermoplastics are used as pellets because powders are compacted and form an undesirable plug in the center of the machine (see Section 11.1.2.). The plasticated polymer is injected into the mold where it forms a ca. 0.05 mm thick layer at the cold walls. The center material flows and solidifies more slowly; after-pressure (dwell pressure, holding pressure) is subsequently applied in order to prevent void formation. Injection-molded articles thus have skin-core structures: a (100-600) μm thick skin surrounds a lower-density core with different morphology. The flow also creates radial orientations of polymer segments and filler particles, especially near sharp edges and bends. Injection-molded articles are thus always anisotropic; molds need careful design to avoid stress zones in injection-molded plastics.

Crystallizable polymers must crystallize fast, which can be achieved by addition of nucleating agents. Slow crystallization leads to spherulite formation and/or secondary crystallizations that can lead to shrinkage and warping.

Sandwich molding (short for "sandwich injection molding") allows the subsequent injection of two polymers from two different injection units into the same mold. The injected second polymer balloons the first one and presses it against the wall. The process is used for the sheathing of inexpensive polymers by more expensive ones or the forming of strong skins around foamed polymers (*foam sandwich molding*).

Reaction injection molding (RIM) is a fast simultaneous injection molding and polymerization with cycle times of a few minutes. It is used for the manufacture of articles from cross-linked polyurethanes and from interpenetrating networks of polyurethanes and polyacrylates. Reaction injection molding of other fast polymerizing monomers such as ε-caprolactam or dicyclopentadiene was not a commercial success (see Chapter 12.6.7.).

Injection blow molding is a blow molding into closed molds where the parison is blown. The core of the injection mold acts as the blow mandrel. See "extrusion blow molding".

11.4. Solid State Processing

Forming is the general term for the changing of solid semifinished goods by external forces into the desired shapes. These processes require ductile plastics; they are performed near in the ductile region of the stress-strain curve (Fig. 7-2). Plastics are not "melted" at the applied temperatures; they remain "solid" at these "cold" processing conditions. Applied tensile forces are generally assisted by vacuum or pressure.

Two groups of forming processes are distinguished: *strengthening* for one-dimensional and two-dimensional semifinished goods (fibers, rods, tapes, films, sheets) and *transformation* for the manufacture of three-dimensional articles from sheets.

11.4.1. Strengthening

Drawing, rolling, and extrusion are processes that are performed between room temperature and up to temperatures near the glass temperature of amorphous polymers or just below the melting temperature of semicrystalline polymers. The applied tensile forces partially orient mobilized chain segments and crystalline lamellae. The resulting orientation is frozen-in by cooling under tension. The finished articles are thus under considerable internal stress because molecule segments are not in their thermodynamically preferred random coil conformations. Tempering (annealing) below the softening temperature increases segmental mobilities. The molecules return to their random coil states and unwanted stresses are relieved (*thermofixing*).

Oriented polymer films with internal stresses are utilized as *shrink films*. The films shrink upon fast heating just above the softening temperature and cover packaged goods tightly.

Cold drawing of fibers, rods, and tapes may be performed without auxiliaries or may be assisted by mandrels or dies (Fig. 11-15, top); drawing of tapes and sheets with the help of rolls is called *rolling* (Fig. 11-15, bottom, left). The drawing of cables, pipes, and sheets is also named *stretching* whereas the term *orientation* is used for the drawing of sheets and films. These terms refer to the macroscopic processes; molecular orientation on the microscopic level occurs with all processes. Unidimensional drawing of films leads to monoaxial orientation; such films are used for the manufacture of fibrillated fibers.

Cold extrusion of rods and tapes is a pushing process whereas drawing refers to pulling. **Hydrostatic extrusion** applies an all-sided pressure to the specimen.

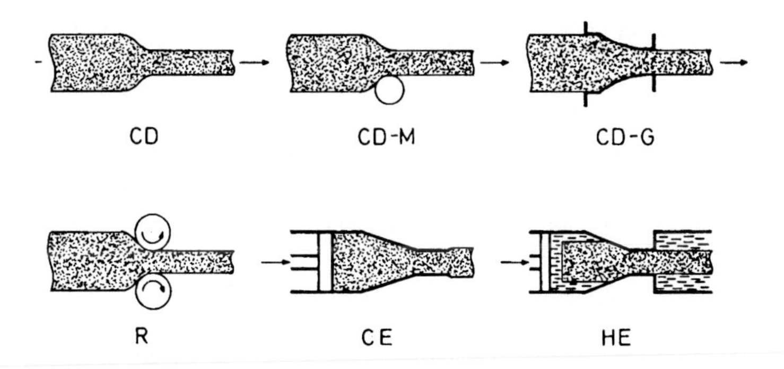

Fig. 11-15 Solid-state forming of one-dimensional and two-dimensional structures.
Top row: Cold drawing (fibers, rods, tapes) without assistance (CD), with a mandrel (CD-M), or with a guide (CD-G).
Bottom row: Rolling R (tapes, sheets), cold extrusion CE (rods, tapes), and hydrostatic extrusion HE (rods, tapes).

11.4.2. Transformation

Transformations of two-dimensional semifinished goods (sheets, films) into three-dimensional articles are carried out near the maximum of the ductile region of the stress-strain curve (Fig. 7-2). In order to avoid orientations and to obtain stress relaxations, they are performed with heated sheets or films at higher temperatures than those in strengthening processes. Processing temperatures are above the glass temperature of amorphous polymers and at the melting temperature of semicrystalline polymers; upper processing temperatures are limited by the onset of flow. These transformations are thus called *thermoforming*. Many different thermoforming processes are known (Fig. 11-16):

Plug-and-Ring Forming. Simple shapes can be prepared from preheated sheet stock with a clamping ring and a plug (forming die) and without a female mold (*deep drawing*) (Fig. 11-16, PRF). Excess material has to be cut off from the shaped article.

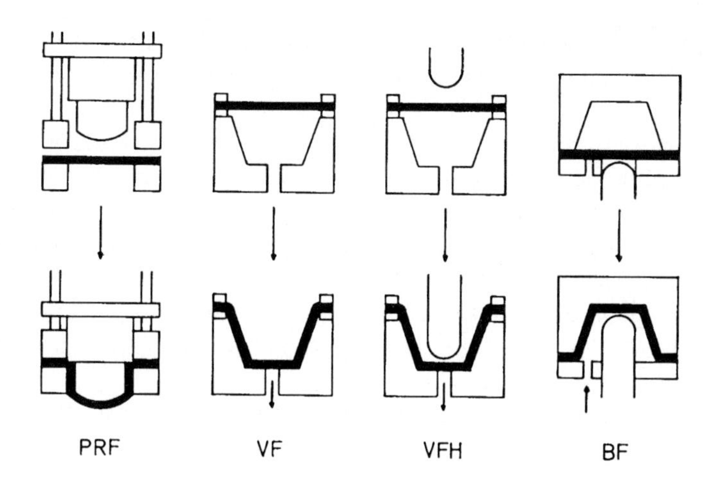

Fig. 11-16 Some thermoforming methods. PRF = plug-and-ring forming (deep drawing); VF = vacuum forming; VFH = vacuum forming with helper (plug-assisted vacuum forming; BF = blow forming (blowing).

Stretch Forming. A cut-to-size, preheated billet is clamped in a blankholder on top of a heated female mold. The plug (punch) stretches the billet; the clamped rim becomes the flange of the container (Fig. 11-16, SF).

Vacuum forming is used in many variations. For simple parts, a heated, clamped sheet is drawn into a female mold (Fig. 11-16, VF); the forming may be assisted by plugs. The resulting articles have uneven wall thicknesses in straight-vacuum forming; they are thinnest at the lowest positions. More complicated shapes are thus prepared by first prestretching the heated film by air pressure and then applying vacuum to suck in and form the sheet in the same mold.

Male molds often allow more uniform thicknesses and dimensional stabilities. In *drape and vacuum forming*, the sheet is clamped into a movable clamping frame and heated before it is pulled over the male mold where it is vacuum molded. In *deep drawn vacuum forming* (*vacuum deep drawing*), the plug is coming from above and combines vacuum forming and deep drawing.

Blow forming (*blowing*) is the forming of sheets with the help of pressure. Blowing may be with air pressure into a female mold (Fig. 11-16, BF), assisted with a plug, with a rubber blanket, or into the open. It differs from *film blowing*, which is an extrusion process (Fig. 11-13).

11.5. Fabricating

Fabricating is the process of modifying preformed articles or their assemblies to final products; this term should not be used for the manufacture of the preformed articles themselves. Three main groups of operations can be distinguished: transforming, machining, and bonding.

11.5.1. Transforming

Transforming is the *conversion* of semifinished goods into final goods without formation of shavings or addition of other materials. *Forging* is used for the transformation of poly(ethylene)s PE-UHMW with ultrahigh molecular masses of several millions and poly(*p*-hydroxybenzoate) PHB. These polymers cannot be injection molded, etc., because their high entanglement densities (PE-UHMW) or their frozen-in liquid crystalline structures (PHB) cause very high viscosities.

11.5.2. Machining

Plastics articles sometimes need to be detached, either to remove gates, sprues, etc., or to receive their final shapes. Detaching operations are always performed below the melting temperature (semicrystalline polymers) or below the glass temperature (amorphous polymers). They may occur without or with formation of shavings. No shavings are produced by cutting, shearing, or punching. Mechanical *cutting* is used for films of celluloid and poly(tetrafluoroethylene); newer systems employ high pressure air jets (*waterjet cutting*) or lasers. *Shearing* is a cutting by knives with lateral motion. *Punching* is used for special parts such as O-rings.

Detachment operations with formation of shavings are known as *machining*; they include *sawing, drilling, turning, milling*, and *threading*. In these operations, high speeds have to be avoided because the low heat conductivity may cause the plastic to heat up, become viscoelastic, and start smearing. High speeds thus require small cross-sections of shavings.

11.5.3. Bonding

Bonding is the joining of parts or semifinished goods; it includes the joining by mechanical devices, adhesive joining, heat sealing, and welding.

Mechanical bonding of plastics can be achieved by screws, rivets, or bolts of the same plastics or of different materials; mechanical strengths are usually sufficient for small loads or stresses. The deformability of plastics is utilized in snap-in or clamp connectors, which consist of hooks or other protrusions that interlock on contact with undercuts or other indentations.

Welding is the physical joining of top layers of two parts by heat and pressure. The materials of the two parts must be compatible. Polymer welding is thus restricted to parts of the same polymer; compatible polymers of different structures are rarely welded. On welding, molecule coils from the two parts interpenetrate. The welding of polymers thus requires a certain mobility of polymer segments, which is provided by heating (thermal conduction, convection, radiation, friction, induction) above the softening temperature. Polymers cannot be welded if they are chemically cross-linked (thermosets) or highly entangled polymers (e.g., ultrahigh molar mass poly(ethylene)), or if they decompose under welding conditions.

The softened top layers are then pressed together. Mutual diffusion of segments across the interface leads to interpenetration of molecule coils. This diffusion is *not* effected by pressure whose rôle is rather to provide good surface contact and to smoothen surface irregularities.

Solvent adhesion is also a welding process and not an adhesive bonding. The top layers of parts are swollen by solvents that have approximately the same solubility parameter as the polymer parts to be joined. The swelling provides polymer segments with the mobility necessary for mutual diffusion. The joints are initially weak because of the large times required for the solvent to diffuse out of the joint through the polymer parts into the atmosphere. This problem can be circumvented if the solvent is a polymerizable monomer.

Adhesive joining, on the other hand, can be used to combine parts from the same *or* from different materials because only surfaces are involved and not top layers. The two parts need not be in close contact; bonding is provided by an intermediate adhesive layer of a different material. The adhesive may be a melt, a solution, or a dispersion of polymers. It must wet the surfaces, that is, its surface tension must be equal or lower than the critical surface tension of the polymer. Surfaces must often be cleaned to remove surface contaminants, which prevent good contact between adhesive and surface. Contact areas can be increased by roughening surfaces. Inert surfaces may be etched, oxidized, etc., in order to generate groups, which interact more easily with the groups of the adhesive. Examples are hydroxyl groups, that may form hydrogen bonds or react with other reactive groups, or free radicals from the treatment of poly(tetrafluoroethylene) with sodium.

11.6. Surface Treatment

Surfaces of plastics parts need sometimes be modified by other materials for aesthetic (color, gloss) or technical reasons (hardness, abrasion, corrosion). These materials are either deposited as such on the surface by physical or chemical means or they react chemically with the surface; one can thus distinguish between surface-deposition and surface-modification processes. Surfaces should be clean and dry; they sometimes need to be pretreated (see above).

Painting. Plastics are usually colored in bulk by pigments. Some plastics part require painting, however, since their color and gloss must match those of, for example, metal parts such as automobile fascia. The paint must wet the surface but the vehicle should not modify the physical structure of the plastic (no swelling, recrystallization, stress corrosion etc.). The plastics part must be carefully designed because paints recede from sharp edges, leaving only a thin coating.

Printing. All printing techniques can be applied to plastics. Hard and nonabsorbent plastics surfaces require indirect techniques, which use an elastic intermediate to transfer the printing ink to the substrate.

High vacuum metal deposition is used to provide plastics parts and films with high gloss coatings. The process requires a thorough outgassing, degreasing, and drying of plastics surfaces. An example is the deposition of aluminum on poly(ethylene terephthalate) for packaging films. A disadvantage is the rather low adhesion of the metal to the plastic, which can be improved be a chemical pretreatment of the polymer surface. The surface receives a coating that contains oxides of cadmium, zinc, or lead. The metal oxides are chemically reduced. The resulting well adhering metal layers can then be coated by aluminum vapor.

Chemical deposition by the chemical reaction of a metal compound on the surface of a part or film is used directly for the deposition of silver by treating an aqueous silver salt solution with a chemical reducing agent. Chemical deposition is also employed for the generation of primers, that allow a better electroplating of ABS polymers.

Polymer electroplating. Etching ABS polymers by a chromic acid-sulfuric acid bath oxidizes dispersed rubber particles, leaving submicroscopic pores at the surface of the plastic. These pores are the anchoring sites for chemically deposited metals such as copper or nickel. The resulting metal film can be reinforced by electroplating with chromium.

Glazing is used both as a general term for the generation of high gloss surfaces by infrared radiation or polishing rolls immediately after processing or for the specific process of depositing an inorganic glass on a surface by electron beams. Glasses may consist of borosilicate or SiO_{1-2}; quartz cannot be used because its thermal expansion coefficient is very different from those of plastics.

Sol-gel processes utilize the evaporation of alcohol from alcoholic solutions of hydrolyzable alcoholates of multivalent metals such as Ti, Si, or Al to form networks. Coatings generated at low temperatures contain many metal hydroxide groups; the resulting surfaces are hydrophilic and antistatic. At higher temperatures, metal oxides result, which form scratch-resistant surface coatings.

Fluorination provides the plastics surface with a thin layer of fluorinated material by treating the plastics with a gas consisting of 10 % fluorine and 90 % nitrogen at temperatures up to 110°C, sometimes with addition of oxygen. Fluorinated surfaces are less wettable and less permeable to organic solvents.

Plasma polymerization generates films on plastics surfaces. The term refers to the vacuum polymerization of low molar mass compounds by gas plasmas (partially ionized gases consisting of ions, electrons, and neutral species). It is not a polymerization of monomers to polymers with the same monomeric units but rather a molecular polymerization of molecule fragments. For example, the plasma polymerization of ethane C_2H_6 leads to $(C_2H_3)_n$ and that of ethylene C_2H_4 in air to $(C_2H_{2.6}O_{0.4})_n$. Both ethylene and hexamethyltrisiloxane are used industrially as "monomers".

Literature

11.1. Introduction: General Texts

W.A.Holmes-Walker, Polymer Conversion, Wiley, New York 1975
S.Middleman, Fundamentals of Polymer Processing, McGraw-Hill, New York 1977
D.C.Miles, J.H.Briston, Polymer Technology, Chem.Publ.Co., New York 1979
J.L.Throne, Plastics Process Engineering, Dekker, New York 1979
Z.Tadmor, C.G.Gogos, Principles of Polymer Processing, Wiley, New York 1979
R.J.Crawford, Plastics Engineering, Pergamon, Oxford 1981 (= Progr.Polym.Sci. **7**)
S.S.Schwartz, S.H.Goodman, Plastics Materials and Processes, Van Nostrand Reinhold, New York 1982
J.R.A.Pearson, S.M.Richardson, Computational Analysis of Polymer Processing, Appl.Sci.Publ., London 1983
G.R.Moore, D.E.Kline, Properties and Processing of Polymers for Engineers, Prentice-Hall, Englewood Cliffs 1984
G.Astarita, L.Nicolais, ed., Polymer Processing and Properties, Plenum, New York 1984
J.R.A.Pearson, Mechanics of Polymer Processing, Elsevier, New York 1985
M.J.Folkes, Processing, Structure and Properties of Block Copolymers, Elsevier Appl.Sci., London 1985
Radian Corp., Plastics Processing. Technology and Health Effects, Noyes Publ., Park Ridge 1986
N.G.McCrum, C.P.Buckley, C.B.Bucknall, Principles of Polymer Engineering, Wiley, New York 1988
D.H.Morton-Jones, Polymer Processing, Chapman and Hall, New York 1989
C.L.Tucker III, Fundamentals of Computer Modeling for Polymer Processing, Hanser, Munich 1990
C.Rauwendaal, Ed., Mixing in Polymer Processing, Dekker, New York 1991
J.-F.Agassant, P.Avenas, J.-Ph.Sergent, P.J.Carreau, Polymer Processing. Principles and Modeling, Hanser, München 1991
J.-M.Carrier, Polymeric Materials and Processing, Hanser, Munich 1991
A.I.Isayev, ed., Modeling of Polymer Processing, Hanser, Munich 1991
P.J.Corish, ed., Concise Encyclopedia of Polymer Processing and Applications, Pergamon Press, Tarrytown (NY) 1991, Oxford 1991
J.M.Vergnaud, Drying of Polymeric and Solid Materials, Springer, London 1992
J.F.Chabot, The Development of Plastics Processing - Machinery and Methods, Wiley, New York 1992

11.1. Introduction: Rheology in Processing

C.D.Han, Rheology in Polymer Processing, Academic Press, New York 1976
K.Murakami, K.Ono, Chemorheology of Polymers, Elsevier, Amsterdam 1979
F.H.Cogswell, Polymer Melt Rheology, Godwin, London 1981
C.D.Han, Multiphase Flow in Polymer Processing, Academic Press, New York 1981
H.Janeschitz-Kriegl, Polymer Melt Rheology and Flow Birefringence, Springer, Berlin 1983
J.Ferguson, N.E.Hudson, Extensional Flow of Polymers, in R.A.Pethrick, ed., Polymer Yearbook **2**, Harwood Academic Publ., Chur 1985, p. 155
S.W.Churchill, Viscous Flows: The Practical Use of Theory, Butterworths, Stoneham, MA 1988

J.L.White, Principles of Polymer Engineering Rheology, Wiley, New York 1989
H.A.Barnes, J.F.Hutton, K.Walters, An Introduction to Rheology, Elsevier, Amsterdam 1989
D.H.Morton-Jones, Polymer Processing, Chapman and Hall, New York 1989
J.M.Dealy, K.F.Wissbrun, Melt Rheology and Its Role in Plastics Processing, Van Nostrand-Reinhold, New York 1990
N.P.Cheremisinoff, An Introduction to Polymer Rheology & Processing, CRC Press, Boca Raton (FL) 1992

11.2.2. Coating

C.I.Hester, R.L.Nicholson, M.A.Cassidy, Powder Coating Technology, Noyes Publ., Park Ridge (OH) 1990
D.R.Randall, Radiation Curing of Polymers. II, CRC Press, Boca Raton (FL) 1991

11.2.3. Molding

D.V.Rosato, C.S.Grove, Jr., Filament Winding, Interscience, New York 1968
R.W.Meyer, Handbook of Pultrusion Technology, Chapman and Hall, London 1985
R.E.Wright, Molded Thermosets, Hanser, Munich 1992
R.J.Crawford, Rotational Molding of Plastics, Wiley, New York 1992

11.3.1. Plastication

R.E.Elden, A.D.Swan, Calendering of Plastics, Iliffe, London 1971
H.Kopsch, Kalandertechnik, Hanser, München 1978

11.3.2. Extrusion

M.J.Stevens, Extruder Principles and Operation, Elsevier, New York 1985
N.P.Cheremisinoff, Polymer Mixing and Extrusion Technology, Dekker, New York 1987
F.Hensen, W.Knappe, H.Potente, Kunststoff-Extrusionstechnik, Bd. I, Grundlagen, Hanser, München 1989
F.Henson, ed., Plastics Extrusion Technology, Hanser, München 1988
J.L.White Twin-Screw Extrusion - Technology and Principles, Hanser, Munich 1990
C.Rauwendaal, Polymer Extrusion, Hanser, Munich 1990
M.Xanthos, Reactive Extrusion, Hanser, Munich 1992
K.T.O'Brien, ed., Application of Computer Modeling for Extrusion and Other Continuous Polymer Processes, Hanser, Munich 1992

11.3.3. Injection Molding

I.I.Rubin, Injection Molding: Theory and Practice, Wiley, New York 1973
F.Johannaber, Injection Molding Machines, Hanser, München 1983
D.V.Rosato, D.V.Rosato, eds., Injection Molding Handbook, Van Nostrand Reinhold, New York 1986
A.I.Isayev, ed., Injection and Compression Molding Fundamentals, Dekker, New York 1987

L.T.Manzione, ed., Applications of Computer Aided Engineering in Injection Molding, Hanser, München 1987

11.3.3.a. Blow Molding

E.G.Fisher, Blow Moulding of Plastics, Butterworth, London 1971
D.V.Rosato, D.V.Rosato, eds., Blow Molding Handbook; Technology, Performance, Markets, Economics, Oxford University Press, New York 1989

11.3.3.b. Reaction Injection Molding

W.E.Becker, ed., Reaction Injection Molding, Van Nostrand Reinhold, New York 1979
F.M.Sweeney, Introduction to Reaction Injection Molding, Technomic, Lancaster (PA) 1979
J.E.Kresta, ed., Reaction Injection Molding and Fast Polymerization Reactions, Plenum, New York 1982
F.M.Sweeney, Reaction Injection Molding Machinery and Processes, Dekker, New York 1987
C.Macosko, Fundamentals of Reaction Injection Molding, Hanser, München 1989

11.4.2. Transformation

A.Höger, Warmformen von Kunststoffen, Hanser, München 1971
J.L.Throne, Principles of Thermoforming, Hanser, München 1986
J.L.Throne, Thermoforming, Hanser, Munich 1987
J.Florian, Practical Thermoforming. Principles and Applications, Dekker, New York 1987

11.5.2. Machining

A.Kobayashi, Machining of Plastics, McGraw-Hill, New York 1981

11.5.3. Bonding

M.N.Watson, Joining Plastics in Production, The Welding Institute, Cambridge, United Kingdom, 1989

11.6. Surface Treatment

J.M.Margolis, ed., Decorating Plastics, Hanser, München 1986

References

[1] S.Artmeyer, K.-H.Seemann, G.Zieschank, G.Spur, W.Brockmann, P.Berns, K.Wiebusch, Kunststoffe, Verarbeitung, in Ullmann's Enzyklopädie der technischen Chemie, Verlag Chemie, Weinheim, 4th ed. 1978, Vol. 15, p. 281 ff., Table 5
[2] J.Meissner, in R.Vieweg, G.Daumiller, eds., Kunststoff-Handbuch **5** (1969) 162, Fig. 10

12. Types of Plastics

12.1. Introduction

12.1.1. Classification

Plastics are usually divided into four groups according to their mechanical properties: commodity plastics, engineering plastics, high-performance plastics, and functional plastics (Chap. 1.2.1). Thermosets, and sometimes also fluoroplastics, are often considered separate groups.

Each of these groups can be subdivided into "clans" of polymers. Members of a clan share common monomeric units. For example, all members of the clan of poly(styrene) plastics ("styrenics") possess significant proportions of styrene units.

Clans can be subdivided further into "families". A family consists of plastics based on polymers with similar chemical and physical structures (Table 12-1). For example, the clan of styrenics comprises the families of styrene homopolymers (PS), copolymers of styrene and butadiene (SB plastics), copolymers of styrene and acrylonitrile (SAN), ABS polymers from acrylonitrile-butadiene rubbers and styrene, and ASA polymers from acryl esters, acrylonitrile, and styrene.

Table 12-1 Chemical and physical structure of plastics families.
L = linear, X = cross-linked, A = amorphous,
C = semicrystalline, S = single-phase, M = multiphase,
HI = high-impact, * = slightly crystalline.

Type	Constitution		Phase		Morphology		Families
I	L		S		A		PS, SAN, PMMA, PC, PVC*, amorphous types of PA's
II	L		S			C	PE (HD, LD, LLD), PP, PET, PA (6, 11, 12, 6.6, 6.12)
III	L			M	A		ABS, TPE's
IV	L			M		C	PUR, E-*block*-P
V		X	S		A		PF, MF, UF, EP
VI		X	S			C	PE-X
VII		X		M	A		EP-HI, PF-HI, UP-HI
VIII		X		M		C	PUR

Plastics based on polymers from only one monomer do not necessarily belong to the same family. Ethylene is, for example, the monomer for the plastics families of low density poly(ethylene)s (PE-LD = LDPE), high density poly(ethylene)s (PE-HD = HDPE), linear low density poly(ethylene)s (LLDPE), very low density poly(ethylene)s VLDPE, ultrahigh molecular weight poly(ethylene)s UHMWPEs, etc. LLDPEs are copolymers. PE-LDs and PE-HDs are homopolymers but they differ so much in their branching and thus in physical structure and mechanical properties that PE-LDs and PE-HDs are considered different families.

Families are further subdivided into grades that differ in molar masses, molar mass distributions, configuration of their polymers, and/or in the presence/absence and proportion of additives and modifiers.

Reported properties of plastics grades, families, clans, and groups are in general difficult to compare because they were obtained by different national or international standards, specimen preparations, and often also under "optimization" of test results. Whenever possible, properties discussed in Chapter 12 are those reported by companies adhering to the CAMPUS® system. CAMPUS stands for "Computer Aided Material Preselection by Uniform Standards", a system of up to ca. 50 properties per grade, that were obtained by uniform preparation and testing of specimens. At present, 22 European companies and European subsidiaries of American companies adhere to the rules of the system. Each company reports its test data on computer diskettes; mainly for PCs (IBM and IBM clones), but sometimes also for Macintosh; newer versions contain graphs of e.g., property variations with load or time. The diskettes describe also typical applications and other relevant information.

12.1.2. Production

The world production of plastics was ca. 100 million tons per year in 1991; this figure includes $6.1 \cdot 10^6$ t/a engineering plastics, and ca. 37 000 t/a high-performance plastics (PPS, PSU, PES, PAR, PEI, PEEK, LCP). Rubber-modified plastics amounted to ca. $4 \cdot 10^6$ t/a and blends of engineering thermoplastics to 340 000 t/a.

Commodity resins are based on inexpensive, simple monomers, that are obtained in total or in part from low-cost intermediates such as ethylene (0.53 $/kg), benzene (0.45 $/kg), propylene (0.33 $/kg), and chlorine (0.19 $/kg). The relative cost of raw materials and polymerization processes is reflected in the ratio of polymer to monomer prices: 2.17 for high density poly(ethylene), 2.34 for poly(styrene), 2.59 for poly(vinyl chloride), and 3.33 for poly(propylene) (1990 data).

The price development of commodity resins depends largely on the cost of catalysts, polymer work-up, and plants. It is thus based on the economy of scale. Because equipment costs usually increase with the 0.6th power of name plate capacity, polymer prices decrease in with the -0.4th power of annual production (Fig. 12-1).

High-performance resins, on the other hand, are driven by performance, not by economy. They can be based on more complex monomer structures; their price is often higher than prescribed by the experience curve. Engineering plastics occupy a middle ground between commodity and high-performance resins: those from inexpensive monomers fall below the experience curve, for example, poly(oxymethylene) POM; those from expensive ones, above, for example, poly(tetrafluoroethylene) PTFE.

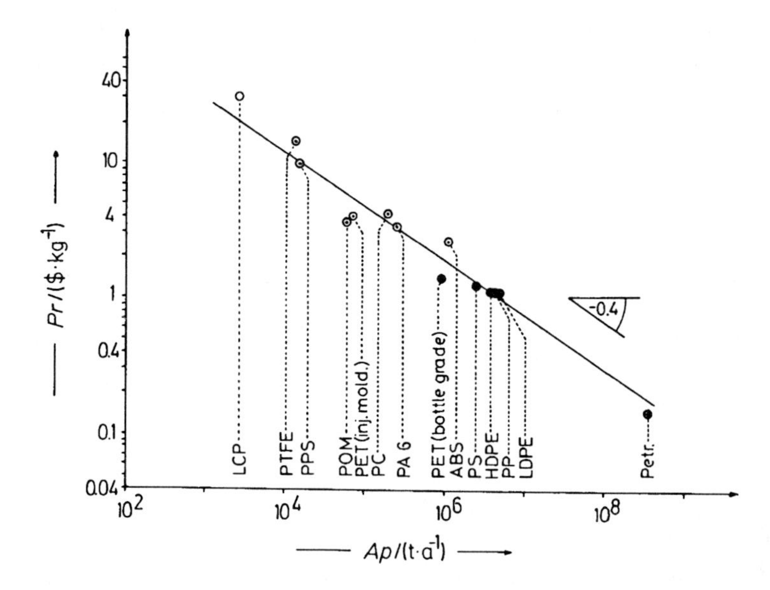

Fig. 12-1 Experience diagram (Boston diagram) of the price *Pr* of resins as function of the annual U.S. production (data for early 1990). ● Commodity plastics, ⊙ engineering plastics, O high-performance resins, ⊕ Petr. = petroleum (Texas light).

12.2. Carbon Chain Thermoplastics

12.2.1. Olefin Polymers

Poly(olefin)s comprise polymers based on ethylene (ethene) $CH_2=CH_2$, propylene (propene) $CH_2=CHCH_3$, butylene-1 (1-butene) $CH_2=CH(CH_2CH_3)$, isobutylene (isobutene) $CH_2=C(CH_3)_2$, and 4-methylpentene-1 $CH_2=CH(CH_2C(CH_3)_2)$. The term "poly(olefin)s" is sometimes used for these polymers *and* styrenics, vinyls, polyvinyls, fluorinated polymers, and polydienes. Despite the names poly(olefin) or olefin polymer, these polymers are not unsaturated but have saturated main chains with the idealized structure $\sim[CH_2-CR_2]_n\sim$, where R may be H and/or alkyl substituents.

Poly(ethylene) PE is the industry term for ethylene homopolymers and for ethylene copolymers with small proportions of olefinic comonomers. All commercial poly(ethylene)s are more or less branched (Fig. 12-2). The extent of branching determines crystallinity, which in turn effects density and other properties. Poly(ethylene)s are thus classified according to their density (Table 12-2). High, medium, and low density poly(ethylene)s are homopolymers of ethylene, whereas linear low density and very low density poly(ethylene)s are copolymers of ethylene with usually 8 % butene-1, hexene-1, or octene-1. Copolymers with 10-20 % octene have been called "plastomers"; this term has also been used in Germany for *all* thermoplastics.

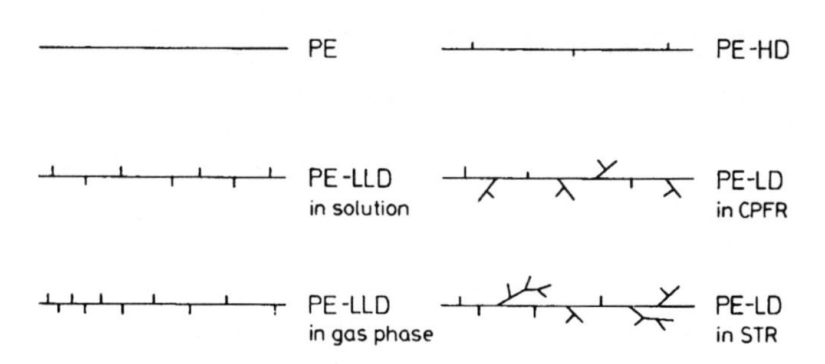

Fig. 12-2 Schematic representation of the branching of various poly(ethylene)s (see text). PE-LLD can be prepared in solution or in the gas-phase, PE-LD in continuous plug-flow reactors CPFR or in stirred tank reactors STR.

Low density poly(ethylene) was discovered by ICI in England; it went into production in 1939, High density poly(ethylene) was first observed 1954 by Karl Ziegler in Germany and by Standard Oil and Phillips Petroleum in the United States; production started in 1957 in Germany, Italy, and the United States.

Low and medium density poly(ethylene)s are manufactured by free radical polymerization; high density, low density, and very low density poly(ethylene)s, with transition metal catalysts. The type of branching and the molar mass distribution also depends on the medium (solution, gas phase) and the distribution of residence times in the reactor (Fig. 12-2). Conventional poly(ethylene)s have relative molecular masses of M_r < 300 000. High density poly(ethylene)s with higher molar masses are known as high molecular weight (HMW-HDPE; $3 \cdot 10^5 < M_r < 4°10^5$), extra high molecular weight ($5 \cdot 10^5 < M_r < 1.5 \cdot 10^6$), and ultrahigh molecular weight ($M_r > 3 \cdot 10^6$).

Tab. 12-2 Classification of poly(ethylene)s.

Type Name	Abbreviations ASTM	ISO	Conventional	Density in g/cm³
-	IV	-	-	> 0.96
High density	III	PE-HD	HDPE	0.941-0.960
Medium density	-	-	MDPE	0.926-0.940
Low density	I	PE-LD	LDPE	0.910-0.925
Very low density	-	-	VLDPE	< 0.910
Linear low density	II	-	LLDPE	0.925-0.940

Poly(ethylene)s are semicrystalline polymers. Density, crystallinity, and melting temperatures increase with decreasing branching: melting temperatures are ca. 115°C for PE-LD and PE-LLD, 135°C for PE-HD, and 144°C for ideal crystals of unbranched poly(ethylene). Glass temperatures are controversial: some authors claim -80°C, others -30°C as true glass temperatures.

The variation of type of initiator, catalyst, reactor, medium, process control, etc., leads to a great number of very different poly(ethylene)s. Table 12-3 indicates the spread of properties for some of the 17 poly(ethylene) grades offered by Neste; for each family, grades with the lowest and the highest densities were selected from the CAMPUS™ data bank. Increasing crystallinity (density) leads to higher Young's moduli, tensile strengths, heat distortion and Vicat temperatures, hardness, opacity, barrier properties, and heat and chemical resistance, but decreases impact strengths and resistance against stress corrosion.

High density PEs are processed into containers, corrugated pipe, monofils, split fibers, wire and cable sheatings, household wares, and toys. Low and linear low density PEs are used mainly as packaging films, coatings, and electrical insulations.

Tab. 12-3 Unfilled injection molding grades poly(ethylene)s from Neste. MVR = Melt volume rate, NB = no break. All grades: relative permittivity of $\varepsilon_r = 2.3$, volume resistivity of 10^{15} Ω cm, surface resistivity of 10^{14} Ω, water absorption of 0.01 % (23°C, saturation), moisture absorption of 0.01 % (23°C, 50 % RH).

Physical property		Polymer types and grades					
Name	Unit	LD NCPE 1515	LLD NCPE 8030	LLD NCPE 8644	LD NCPE 3416	HD NCPE 7003	HD NCPE 7007
Density	g/cm³	0.915	0.919	0.935	0.958	0.958	0.964
Heat distortion temp. B	°C	41	46			84	80
Heat distortion temp. A	°C			39	58	50	54
Vicat temperature A	°C	75	86	110		130	130
Vicat temperature B	°C		48	65	75	85	80
Young's modulus (1 mm/min)	MPa	150	830	650	1350	1350	1600
Yield strength (50 mm/min)	MPa	8	12	17	30	30	31
Strain at yield (50 mm/min)	%		12	12	8.1	8.2	7.1
Strain at break (50 mm/min)		> 50	> 50	> 50	> 50	> 50	> 50
Impact strength (23°C)	kJ/m²	NB	NB	NB	NB	NB	NB
Impact strength (-30°C)	kJ/m²	NB	NB	NB	NB	NB	NB
Notched impact str. (23°C)	kJ/m²	NB	NB	6.5	NB	5.2	5.6
Notched impact str. (-30°C)	kJ/m²	NB	11	5.9	5	5.4	4.6
MVR (190°C, 2.16 kg)	mL/10 min	18	36	9.5	0.24	3.6	8.3
Melt density (225°C)	g/mL		0.787	0.794		0.814	0.819
Melt thermal conductivity	$W\ m^{-1}\ K^{-1}$		0.287	0.303		0.326	0.353
Melt specific heat capacity	$J\ kg^{-1}\ K^{-1}$		3070	3150		3350	3420

Poly(ethylene) copolymers are commercially produced by free radical initiation. Copolymers with 1-40 % vinyl acetate (E/VAC; EVA) find application as films; those with higher VAC contents as elastomers, melt adhesives, or PVC modifiers. Ethylene-vinyl acetate copolymers with 20-50 % VAC are saponified to ethylene-vinyl alcohol copolymers (EVAL®, E/VAL, EVOH); those with lower VAL contents are used for fluidized bed coatings and those with higher VAL contents as barrier polymers.

Radical copolymerization of ethylene with up to 20 % acrylic acid leads to highly branched EAA copolymers. These polymers are tough (intermolecular hydrogen bonds between COOH groups) and have excellent adhesive properties (hydrogen bonds to metals and glass; dispersion forces between CH_2 groups of EAA and poly(olefin)s); they are used in laminates with aluminum films and as packaging films.

Copolymers of ethylene with up to 15 % methacrylic acid are partially neutralized with Na^+ or Zn^{2+}. These *ionomers* contain ion clusters and domains that act as physical cross-linking agents at low temperatures but dissociate at processing temperatures. They form excellent extrusion coatings.

Poly(propylene)s PP are linear polymers that are produced from propylene by Ziegler-Natta polymerizations. The poly(propylene)s of commerce are either isotactic homopolymers of propylene, random copolymers with mainly ethylene or butene-1, or multiphase block copolymers poly(propylene)-*block*-poly(ethylene-*co*-propylene) (Table 12-4). Early homopolymerizations delivered considerable amounts of "atactic polymers" as byproducts of the desired isotactic PPs; these noncrystallizable polymers are actually highly branched. Atactic PPs were initially incinerated or deposited but later found use as melt adhesives, in carpet backing, paper lamination, etc. Because newer propylene polymerizations to isotactic polymers no longer deliver atactic byproducts, "atactic poly(propylene)s" are now produced by special syntheses.

Table 12-4 Six of the 93 poly(propylene) grades offered by Hüls AG (Vestolen P®). All grades have at 23°C the same thermal expansion coefficients ($1.5 \cdot 10^{-4}$ K^{-1}), relative permittivities (2.3), volume resistivities (10^{15} Ω cm), surface resistivities (10^{13} Ω), dissipation factors ($5 \cdot 10^{-4}$ at 50 Hz), comparative tracking indices (600 steps), and water absorptions (0.01 %). MVR = Melt volume rate.

Physical property		Poly(propylene) types and grades					
Name	Unit	Homopolymers			Blockcopolym.		Random
		P 2000	P 5000	P 7000	P 9500	P 5300	P 9421
Density	g/cm³	0.903	0.902	0.901	0.901	0.898	0.901
Isotaxie index	-	95	95	95	95	85	85
Viscosity coefficient	mL/g	150	220	310	450	190	450
MVR (230°C, 2.16 kg)	mL/10 min	55	15	3.2	0.5	18	0.5
Heat distortion temp. A (1.8)	°C	60	55	55	55	50	45
Heat distortion temp. B (0.45)	°C	105	100	100	90	80	75
Vicat temperature B	°C	105	100	100	90	70	65
Stress at yield (50 mm/min)	MPa	38	37	35	32	25	26
Strain at yield (50 mm/min)	%	8	8	8	10	12	12
Strain at break (50 mm/min)	%	25	> 50	> 50	> 50	> 50	> 50
Young's modulus (1 mm/min)	MPa	1800	1600	1500	1300	800	700
Creep modulus (1 h)	MPa			1150	1200		700
Creep modulus (1000 h)	MPa			500	400		300
Impact strength (Izod, 23°C)	kJ/m²	50	65	130	NB	NB	NB
Impact strength (Izod, -30°C)	kJ/m²	12	15	17	40	15	30
Notched impact strength (23°C)	kJ/m²	2	2.5	5	20	3.5	25
Notched impact strength (-30°C)	kJ/m²	1.5	1.6	2	3	1.5	2.5
Dielectric strength	kV/mm	40	40	40	40	35	35

Isotactic homopolymers crystallize in the macroconformation of a 3_1 helix. The resulting compact chain conformation leads to a higher melting temperature (ca. 176°C) than that of the zigzag chains of poly(ethylene)-HD (ca. 135°C). High-impact poly(propylene)s are blends of it-poly(propylene)s with small proportions of EPDM rubbers (copolymers of ethylene, propylene, and a small proportion of a nonconjugated diene). Many PPs are reinforced with short glass fibers, talc, or chalc.

Poly(propylene)s exhibit good mechanical properties and heat and chemical resistance at moderate cost. They have found many applications as pipes, electrical and automotive parts, household goods, films, and fibers.

Poly(butene-1) PB is an isotactic polymer with the structure ~$[CH_2–CH(C_2H_5)]_n$~, a glass temperature of -24°C, a melting temperature of 132°C, and a continuous service temperature of ca. 100°C. It has a good creep resistance at temperatures above 80°C and is mainly used for hot water pipes, fittings, large containers, etc.

Poly(4-methylpentene-1) PMP with the structure ~$[CH_2–CH(CH_2CH(CH_3)_2]_n$~ has the lowest density of all thermoplastics (0.83 g/cm^3). The isotactic polymer is glass clear despite its crystallinity (melting temperature 242°C; glass temperature 29°C). Its high continuous service temperature of 120°C allows its use for sterilizable medical instruments, films for ready-to-eat meals, etc.

12.2.2. Styrenics

Styrenics are polymers of styrene $CH_2=CH(C_6H_5)$ (phenyl ethylene, vinyl benzene). First produced industrially in 1931 in Germany, they became major commercial materials in 1936. The clan of styrenics comprises homopolymers, impact-modified polymers, and various copolymers. All these materials are sometimes classified as poly(styrene)s although that name should be used only for homopolymers.

Homopolymers of styrene are practically linear, atactic polymers with constitutional repeating units ~$[CH_2–CH(C_6H_5)]_n$~. Commercial products are manufactured by free radical polymerization in bulk, suspension, emulsion, and solution (Table 3-1); molding grades have molar masses of $(2\text{-}3)\cdot 10^5$ g/mol, i.e., well above the critical molar mass for entanglements. Syndiotactic poly(styrene)s are obtained with special Ziegler-Natta catalysts; these crystalline polymers are presently evaluated.

All available commercial poly(styrene)s are atactic and amorphous; they do not crystallize, even when stretched. So-called crystal poly(styrene)s are noncrystalline homopolymers that have been polymerized thermally in bulk; their name results from their high clarity and luster due to a high refractive index of ca. 1.59 and the absence of haze (no dispersed impurities). Crystal and standard poly(styrene)s have densities of ca. 1.05 g/cm^3; expandable grades are also available.

Properties of styrene homopolymers vary according to grade; Table 12-5 shows a few grades available from one manufacturer. Some physical properties of poly(styrene)s depend on molar masses, which are usually characterized by their viscosity numbers (c = const.). Low molar masses lead to low melt viscosities (Fig. 5-5). The low molar mass Grade 144 C is thus characterized by BASF as "very easy flow grade". Heat distortion and Vicat temperatures increase accordingly with viscosity numbers in the series 144 C - 143 E - 165 E since these properties are affected by the resistance of specimens against flow under load. The creep modulus increases similarly whereas Young's modulus remains constant because the latter is obtained as secant modulus

Table 12-5 Properties of different grades of BASF poly(styrene)s.

144 C = Very easy flow; 143 E = easy flow; 165 H high molecular weight;
158 K = heat resistant; 168 N = heat resistant, high molecular weight.

Physical property Name	Test conditions	Unit	Poly(styrene) grades 144 C	143 E	165 H	158 K	168 N
Viscosity number	?	mL/g	74	96	119	96	119
Heat distortion temp. B	0.45 MPa	°C	80	82	84	98	98
Heat distortion temp. A	1.8 MPa	°C	70	72	76	86	86
Vicat temperature A	10 N	°C	88	88	92	106	106
Vicat temperature B	50 N	°C	84	84	89	101	101
Young's modulus	1 mm/min	MPa	3150	3200	3150	3200	3250
Creep modulus	1000 h	MPa		2300	2830	2700	2850
Tensile strength	5 mm/min	MPa	46	50	56	50	63
Strain at break	5 mm/min	%	1.5	2	2	2	3
Impact strength	-30 to +23°C	kJ/m^2	6	9	11	10	13
Notched impact strength	-30 to +23°C	kJ/m^2	2	2	2	2	2

from the slope of stress-strain curves at (ideally) infinitely small loads. Increased entanglements are responsible for the higher tensile and impact strengths of grades with higher viscosity numbers.

Grades 158 K and 168 N are described by BASF as poly(styrene)s with high heat resistance that manifests itself by higher heat distortion and Vicat temperatures. These polymers have also higher tensile and impact strengths. The improvements cannot be correlated with structure variations since BASF did not disclose the way by which the higher heat resistance was achieved.

Poly(styrene)s are polymers with polarizable phenyl rings. Their relative permittivities are thus slightly higher ($\varepsilon_r \approx 2.5$) then those of poly(ethylene) ($\varepsilon_r \approx 2.25$) but still independent of frequency in the range 50 Hz to 1 MHz. The volume resistivity is $10^{16}\ \Omega$ cm; the surface resistivity, $10^{14}\ \Omega$; and the dielectric strength (K 20/P 50), 135 kV/mm. The dissipation factor (at 1 MHz) is 0.7 for grades 144 C, 143 E, and 165 H, but 0.5 for the modified grades 158 K and 168 N.

Poly(styrene)s are easy to color and process. Compact grades are used mainly as containers, in household goods, and in toys; expandable grades in packaging and in building and construction.

Copolymers and Impact-Resistant Grades. Ca. 60 % of styrene is used in copolymers and impact-modified grades. Commercial copolymers include free-radically polymerized bipolymers with maleic anhydride (SMA), maleic imide, or acrylonitrile (SAN). These polymers have higher heat distortion temperatures than homo-poly-(styrene)s (cf. Tables 12-5 and 12-6); they are used in appliances and furniture. SANs with 60-70 % acrylonitrile units are barrier polymers.

Impact-modified styrenics are obtained by grafting monomers free-radically on rubbers. Commercially available are graft copolymers of styrene on EPDM elastomers (AES), and graft copolymers of styrene + acrylonitrile on acrylonitrile-butadiene rubbers (ABS), acryl rubbers (ASA), or rubbers from chlorinated poly(ethylene)s (ACS).

The rubbers are dispersed in the continuous poly(styrene) matrix (see Section 13.7.3.). The diameters of the rubber domains usually exceed the wave length of the incident light; these impact-modified styrenics are thus nontransparent.

Anionic copolymerization serves for the synthesis of diblock copolymers of styrene and butadiene (SB) and triblock copolymers of styrene and butadiene (SBS) or isoprene (SIS). The poly(butadiene) blocks of SB thermoplastics form very small domains in the poly(styrene) matrix. In contrast to ABS and ACS polymers, they are transparent and are thus called "glass-clear impact-resistant poly(styrene)s". Triblock copolymers with more than ca. 70 % diene units are thermoplastic elastomers.

Table 12-6 Properties of some BASF styrene copolymers.

Physical property Name	Unit	Copolymer types and grades					
		SAN Luran 368 R	ASA Luran S 776S	SB Polystyr. 2711	SB Polystyr. 466 I	SB-*b* Styrolux 637 D	ABS Terluran 2877
Density	g/cm³	1.08	1.07	1.05	1.05	1.02	1.07
Heat distortion temp. B	°C	101	101	77	90	78	104
Heat distortion temp. A	°C	98	96	67	82	68	99
Vicat temperature B	°C	101	93	79	92	68	101
Young's modulus (1 mm/min)	MPa	3800	2200	1300	2250	2000	3000
Creep modulus (1000 h)	MPa	2800	1170				1870
Yield strength (50 mm/min)	MPa		47	15	31	36	60
Tensile strength (5 mm/min)	MPa	75	47				
Strain at yield (50 mm/min)		5	3.3	1.4	1.7	2.2	3
Strain at break (50 mm/min)			30	> 50	40	35	15
Impact strength (23°C)	kJ/m²	19		94	72		60
Impact strength (-30°C)	kJ/m²	19		54	47		30
Notched impact strgth. (23°C)	kJ/m²	2		8	10		5
Notched impact strgth. (-30°C)	kJ/m²	2		6	6		3

12.2.3. Vinyls

"Vinyls" is the industry term for thermoplastics containing vinyl chloride units $-CH_2-CHCl-$, although the "vinyl" of the chemical literature refers to any chemical compound with the group $CH_2=CH-$ (IUPAC: ethenyl). About 90 % of the vinyls are in form of the homopolymer, poly(vinyl chloride) $\sim[CH_2-CHCl]_n\sim$, which was produced first in the United States in 1928. Ca. 10 % are vinyl chloride copolymers with (3-20) % vinyl acetate, (3-10 %) propylene, or 87 % vinylidene chloride $CH_2=CCl_2$. Some "vinyls" are graft copolymers on butyl acrylate or ethylene-vinyl acetate copolymers. Several vinyls contain impact-modifiers.

All vinyls are produced by free radical initiated polymerization, most in suspension, others in emulsion or in bulk. The propagation step $\sim[CH_2-CHCl]_{n-1}CH_2-{}^{\bullet}CHCl + CH_2=CHCl \rightarrow \sim[CH_2-CHCl]_nCH_2-{}^{\bullet}CHCl$ is accompanied by radical transfer to the monomer

(12-1) $\sim\!\!\sim CH_2\text{–}\dot{C}HCl + CH_2\text{=}CHCl \longrightarrow \sim\!\!\sim CH_2\text{–}CHCl_2 + CH_2\text{=}\dot{C}H$

The resulting monomer radical starts a new polymer chain. The rate of this transfer reaction is much higher than the rate of termination by deactivation of two polymer radicals. The number-average degree of polymerization thus becomes practically independent of the initiator concentration; it is industrially adjusted by changes in polymerization temperatures.

Chains $CH_2\text{=}CH\text{–}[CH_2\text{–}CHCl)]_n\sim$ so formed contain vinyl endgroups. Transfer reactions also generate tertiary Cl, unsaturated groups such as $\text{–}CH\text{=}CH\text{–}CH_2Cl$ and $\text{–}CH\text{=}CH\text{–}CHCl\text{–}$, head-to-head structures $\text{–}CH_2\text{–}CHCl\text{–}CHCl\text{–}CH_2\text{–}$, etc. These groups initiate unzipping reactions, which lead to the evolution of HCl and the formation of conjugated structures $\text{–}CH_2\text{–}CHCl\text{–}(CH\text{=}CH)_i\text{–}CH_2\text{–}CHCl\text{–}$ during processing (heat) and application (effects of light and oxygen on weathering). PVCs are thus stabilized in order to avoid decolorization and the loss of mechanical properties.

About 50 % of PVC is used as "hard-PVC" (PVC-U) and the other 50 % as plasticized PVC (PVC-P). Hard-PVCs are most commonly suspension polymers; they are compounded as powder blends for the extrusion of rigid pipes. Plasticized PVCs are mainly extruded (wire and cable insulation, packaging films) and calendered (film, sheeting). Shoe soles, pipe fittings, automotive bumpers, etc., are injection molded.

Table 12-7 Six of the 163 unplasticized PVC grades from Solvay (Benvic® PVC-U). All grades contain slip agents and flame retardants. In addition, grades may contain reinforcing agents R and impact modifiers I; they may be stabilized against heat H, light L, and weather W. Recommended processing is by injection molding IM, blow extrusion BE, or extrusion EX.

Physical property	Unit	PVC-U types and grades					
PVC grade →		IR 346	PEB 918	EB 950	EB 928	ER 999	S-ER 820
Additives →		W	-	I	I	IRHLW	IRHLW
Delivery →		pellets	powder	pellets	pellets	pellets	pellets
Processing →		IM	BE	BE	BE	EX	EX
Application →		electro	food	food	food	pipes	profiles
Density	g/cm^3	1.32	1.41	1.31	1.39	1.23	1.51
Stress at yield (50 mm/min)	MPa	49	58	46	53	45	46
Strain at yield (50 mm/min)	%		3.5	4	4	2.5	3.7
Strain at break (50 mm/min)	%	12	15			28	35
Young's modulus (1 mm/min)	MPa	3300	3400	2300	2800	2700	3300
Impact strength (Izod, 23°C)	kJ/m^2		NB				NB
Impact strength (Izod, -30°C)	kJ/m^2		75				228
Notched impact strgth. (23°C)	kJ/m^2		3.5				16
Notched impact strgth. (-30°C)	kJ/m^2		3.1				5.5
Heat distortion temp. B	°C		68				73
Vicat temperature B	°C	78	77	75	76	96	82
Thermal exp.coeff. ‖ (23-80°C)	K^{-1}	$7\cdot10^5$	$7\cdot10^5$	$7\cdot10^5$	$7\cdot10^5$	$8\cdot10^5$	$3\cdot10^5$
Volume resistivity	Ω cm		$>10^{15}$				$>10^{15}$
Surface resistivity	Ω		10^{13}				$>10^{15}$
Water absorption (23°C, saturation)	%	0.3	0.3	0.3	0.3	0.3	0.2
Moist. absorpt. (23°C, 50 % RH)	%	0.03	0.03	0.03	0.03	0.03	0.04

Poly(vinylidene chloride) PVDC is never used commercially as homopolymer ~$[CH_2–CCl_2]_n$~ but always as copolymer with 15-20 % vinyl chloride or 13 % vinyl chloride + 2 % acrylonitrile (Saran®). The high chlorine content leads to high densities of 1.6-1.7 g/cm^3. PVDC has the lowest gas permeability of all commercial plastics and is thus used for packaging (shrink films, paper coatings).

12.2.4. Acrylics

"Acryl" is the name for chemical compounds containing the group CH_2=CH–CO– or its derivatives such as acrylonitrile CH_2=CHCN. The plastics term "acrylics" refers to polymers and copolymers of methyl methacrylate $CH_2=C(CH_3)(COOCH_3)$. Some textbooks use "acrylic" to denote all polymers derived from acryl compounds: poly(acrylonitrile) ~$[CH_2–CHCN]_n$~ as acrylic fibers; poly(acrylate)s ~$[CH_2–CHCOOR]_n$~ for fiber modification, paints, and adhesives; poly(acryl amide) ~$[CH_2–CONH_2]_n$~ as flocculants and paper additives; cyanoacrylate polymers ~$[CH_2–CCN(COOR)]_n$~ as acrylic adhesives; poly(methacrylimide)s by intramolecular imidization from poly(methacrylic acid-*co*-methacrylnitrile)s for hard foams; and poly(ethylene glycol methacrylate) ~$[CH_2–CCH_3(COO(CH_2)_2OH)]_n$~ for soft lenses.

Poly(methyl methacrylate) PMMA is a crystal clear thermoplastic with excellent weatherability; it was first produced in 1933 by Röhm and Haas in Germany. Its high Young's modulus (ca. 3200 MPa), moderate tensile strength at break (ca. 75 MPa), and reasonable thermal stability (static glass temperature 105°C) makes it the choice for outdoor signs, lamps, airplane windows (cross-linked polymer), dentures, etc. Its mechanical properties can be improved by orientation of cast sheets; through copolymerization of methyl methacrylate with acrylonitrile, styrene, butadiene, and acrylates; and by blending with SBR.

Vinyl ester resins of commerce are not vinyl esters despite the name but macromonomers $CH_2=C(CH_3)–CO–Z–OC–C(CH_3)=CH_2$ with two (meth)acryl end groups that are dissolved in styrene or vinyltoluene. Z is usually based on bisphenol A; for example, it may be $OCH_2CH_2O(p\text{-}C_6H_4)–C(CH_3)_2–OCH_2CH_2O$. They are cross-linked by free radical polymerization, similar to unsaturated polyesters UP. The polymers have better mechanical properties than UPs but at a higher cost.

12.2.5. Fluoroplastics

Fluoroplastics FP are based on polymers from the free radical polymerization and copolymerization of various fluorinated olefins. They are also known as teflons, the former trade name of poly(tetrafluoroethylene) PTFE. Young's moduli and fracture strengths of most fluoroplastics do not exceed those of commodity plastics; fluoroplastics are thus not engineering plastics. They do however possess good impact strengths, relatively high heat distortion temperatures, excellent weatherabilities, outstanding flame retardancies, and very unusual surface properties (Table 12-8). Due to

Table 12-8 Properties of unfilled fluoroplastics (various non-CAMPUS sources). PTFE test specimen are molded from fine powder; molded specimen from pellets have (on average) lower tensile strengths and elongations but slightly higher flexural moduli.

Property	Unit	PTFE	FEP	PFA	ETFE	PCTFE	ECTFE	CM-1	PVDF
Density	g/cm^3	2.18	2.15	2.15	1.70	<2.18	1.7	1.88	1.76
Water absorption (24 h)	%	0	0.4	0.03	0.1	0	0.01		0.03
Thermal exp.coeff. ($\cdot 10^5$)	K^{-1}	12	9	7	9	<7	5	4	9
Melting temperature	°C	327	275	305	270	216	<246	327	175
Glass temperature	°C				110	<100	-76		
Heat distortion temp. A	°C	49	51	48	74	66	78	220	90
Continuous service temp.	°C	260	205	260	150	175	150		150
Brittleness temperature	°C	-200	-100	-200	-100	-40	-100		-60
Young's modulus	MPa	420	360	700	850	2200	6700	3900	1500
Flexural modulus	MPa	630	660	660	1400	<2000	<2500	4700	1400
Flexural strength	MPa	NB	18	15	28	58	50		55
Tensile strength at break	MPa	28	21	28	45	<40	56	39	36
Elongation at break	%	200	300	300	200	<80	<250	2	37
Notched impact strength	J/m	160	NB	NB	NB	150	NB	21	200
Rel. permittivity (1 MHz)	1	2.1	2.0	2.1	2.6	<2.8	2.6	2.3	8
Dissipation factor ($\cdot 10^4$)	1	3	8	20	5	230	24	20	1800
Volume resistivity	Ω cm	$>10^{18}$	$>10^{17}$				$>10^{15}$		$2 \cdot 10^{14}$
Surface resistivity	Ω	$>10^{16}$	$>10^{15}$						
Critical surface tension	mN/m	19				31			33
Limiting oxygen index	%	95							44

its low critical surface tension γ_{crit}, poly(tetrafluoroethylene) PTFE cannot be wetted by fats and oils ($\gamma_{lv} \approx$ 20-30 mN/m) and water (γ_{lv} = 72 mN/m). PTFE also shows very low friction coefficients.

Poly(tetrafluoroethylene) PTFE with the monomer unit ~$[CF_2{-}CF_2]_n$~ crystallizes in tight helix conformations; melts have a liquid crystalline structure. The resulting high melting temperature (237°C) and very high melt viscosity (ca. 10^{10} Pa s at 380°C) as well as its insolubility in all solvents rule out conventional processing methods. PTFE powders can only be press-sintered; parts can be machined.

TFE copolymers. The difficult processing of such excellent end-use material has stimulated the search for easier-to-process plastics with similar properties. Several copolymers of tetrafluoroethylene have been developed. Ethylene can be copolymerized with tetrafluoroethylene $CF_2{=}CF_2$ (TFE) to strictly alternating copolymers E/TFE (ETFE) ~$[CF_2CF_2{-}CH_2CH_2]_n$~ or to copolymers with less than 50 mol-% ethylene units. TFE forms also copolymers with hexafluoropropylene $CF_2{=}CF(CF_3)$ (FEP) and with perfluoromethylvinyl ether $CF_2{=}CF(OC_3F_7)$ (PFA).

Poly(chlorotrifluoroethylene) (PCTFE) with the constitutional repeating unit ~$[CFCl{-}CF_2]_n$~ has a much lower melting temperature than PTFE and can be processed with the usual machines if they are stable against corrosion (decomposition of PCTFE). The alternating copolymer ECTFE of chlorotrifluoroethylene CTFE and ethylene E has mechanical properties similar to ETFE and poly(vinylidene fluoride).

Poly(vinylidene fluoride) (PVDF) with the monomer unit ~[CH_2–CF_2]$_n$~ resembles more poly(ethylene) than poly(vinylidene chloride). The alternating copolymer of vinylidene fluoride and hexafluoroisobutene CH_2=C(CF_3)$_2$ (CM-1) has the same melting temperature and similar mechanical properties as PTFE (except fracture and impact strength) but can be processed from the melt.

Poly(vinyl fluoride) PVF with the monomeric unit ~[CH_2–CHF)$_n$~ is a glass clear thermoplastic with a glass temperature of -20°C and a melting temperature of ca. 200°C. It is mainly extruded to laminating films for metal and plastics sheets.

12.3. Carbon-Oxygen Chain Thermoplastics

12.3.1. Thermoplastic Polyesters

Many polyesters ~[O–Z–CO]$_n$~ and ~[O–Z'–O–OC–Z"–CO]$_n$~ are linear chains with thermoplastic properties but the industry terms "thermoplastic polyester" and "linear polyester" refer only to poly(ethylene terephthalate) PET, poly(butylene terephthalate) PBT, and the polyester PEN from 2,6-naphthalenedicarbonic acid and ethylene glycol:

PET ~C(=O)– –C(=O)–O–CH_2–CH_2–O~

PBT ~C(=O)– –C(=O)–O–CH_2–CH_2–CH_2–CH_2–O~

PEN ~C(=O)– –C(=O)–O–CH_2–CH_2–O~

All these thermoplastic esters are obtained by melt polycondensation of diacids (or their esters) and glycols. PET is also known by the acronyms PETE (U.S. recycling), PES (polyester fibers), and No. 1 (U.S. recycling). For PBT, the abbreviation PTMT (poly(tetramethylene terephthalate)) is often used.

Poly(ethylene terephthalate). Most of PET is spun into fibers. PET engineering plastics (T_M = 265°C; T_G = 70°C) for injection molding require exact processing conditions. Nucleated resins are processed at ca. 140°C. They deliver small crystallites and clear plastics; absence of nucleating agents leads to large crystallites and opaque materials (Table 12-9). Oriented and biaxially oriented PET films have outstanding strengths. Bottle grade PET resins for extrusion blowing require high viscosity melts

Table 12-9 Properties of various poly(ethylene terephthalate)s. w_c = Degree of crystallization; HDT = heat distortion temperature (method A).

Application	Physical state	w_c/%	Clarity	HDT/°C
Blister packaging	amorphous	0 - 5	clear	< 67
Bottles	amorphous, oriented	5 - 20	clear	< 73
Food trays	crystalline	25 - 35	opaque	< 127
Films, hot-fill containers	crystalline, oriented	35 - 45	clear	< 160

(entanglements of high molar mass molecules). They must be void of nucleating agents; their crystallization is in addition suppressed by incorporation of comonomers such as isophthalic acid, cyclohexane-1,4-dimethylol $HOCH_2(1,4\text{-}C_6H_{10})CH_2OH$, neopentyl glycol $(CH_3)_2C(CH_2OH)_2$, etc.

Poly(ethylene naphthalate) PEN (T_M = 260°C, T_G = 113°C) surpasses PET in tensile strength and in temperature, dimensional, and chemical stability.

Poly(butylene terephthalate) (T_M = 232°C; T_G = 22°C) crystallizes much more rapidly than PET and can be processed at 50-80°C; its glass temperature is however much lower. PBT grades may be glass fiber-reinforced, white pigment-filled, impact-modified, etc. (Table 12-10).

Polyarylates PAR are copolyesters from diphenols such as bisphenol A and aromatic diacids such as terephthalic and isophthalic acids, probably by interfacial polycondensation. The polymers are transparent to opaque, amber-colored engineering plastics with good toughness, excellent elastic recovery, and high heat stability.

12.3.2. Polycarbonates

The industry term "polycarbonate" refers to polycarbonates $\sim[O\text{–}Z\text{–}O\text{–}CO]_n\sim$ with bisphenol A units $\text{–}O\text{–}(p\text{-}C_6H_4)\text{–}C(CH_3)_2\text{–}(p\text{-}C_6H_4)\text{–}O\text{–}$ or its derivatives as exclusive or predominant Z groups. Polycarbonates PC are synthesized by interfacial polycondensation of the disodium salt of bisphenol A and phosgene $COCl_2$ or by ester exchange polycondensation of bisphenol A and diphenyl carbonate $(C_6H_5O)_2CO$. Some polycarbonates contain additional diphenols; the distribution of the comonomeric units may be statistical or block-like. Copolymers with terephthalic acid are known as *polyphthalate carbonates*. Multifunctional comonomers lead to branched polycarbonates. Polycarbonate I uses bisphenol I, a condensation production of phenol and 3,3,5-trimethylcyclohexanone, instead of bisphenol A, the condensation product of phenol and acetone. Other polycarbonate grades contain halogenated bisphenol A or phosphonate units for the improvement of flame retardancy.

Polycarbonates are also blended with poly(butylene terephthalate) PBT, ABS resins, poly(olefin)s, and styrene-maleic anhydride copolymers. The blending of PC and PBT results in ester-ester exchange reactions and four types of species: PC, PBT, PC-*block*-PBT, and copolyesters of the two acids, the phenol and the glycol.

Table 12-10 Properties of some polymers with carbon-oxygen chains. HI = impact modified; br = branched; MVR = melt volume rate; NB = no break.
PBT = poly(butylene terephthalate)s (Pocan® grades of Bayer);
PC = polycarbonates (Calibre® grades of Dow Chemical).

Physical property Name	Unit	Polymer types and grades PBT B 1300 -	PBT B 1600 -	PBT S 1506 HI	PC 300-22 -	PC 200-4 -	PC 600-3 br
Density	g/cm^3	1.3	1.3	1.22	1.2	1.2	1.19
Viscosity coefficient	mL/g	102	138	145	53	72	65
MVR (260°C; 2.16 kg)	mL/10 min	59	18				
MVR (300°C;1.2 kg)	ml/10 min				18.3	3.33	2.96
Young's modulus (1 mm/min)	MPa	2700	2600	1700	2200	2090	2220
Creep modulus (1 h)	MPa	2500	2200	1400			
Creep modulus (1000 h)	MPa	1400	1400	700			
Stress at yield (50 mm/min)	MPa	57	55	35	61	60	62
Stress, ε = 50 % (50 mm/min)	MPa				47	46	50
Tensile strength (5 mm/min)	MPa				58	58	60
Strain at yield (50 mm/min)	%	3.7	3.8	5	11	11	5.8
Strain at break (50 mm/min)	%	30	> 50	> 50	145	183	84
Strain at break (5 mm/min)	%				135	190	93
Impact strength (Izod, 23°C)	kJ/m^2	110	200	NB	NB	NB	NB
Impact strength (Izod, -30°C)	kJ/m^2	90	150	NB	NB	NB	NB
Notched impact strgth. (23°C)	kJ/m^2	5	6	NB			86
Notched impact strgth. (-30°C)	kJ/m^2	4	5	40			15
Heat distortion temp. B	°C	150	160	100	137	141	146
Heat distortion temp. A	°C	60	65	55	124	132	143
Heat distortion temp. C	°C				113	114	140
Vicat temperature A	°C	215	210	218			152
Vicat temperature B	°C	173	170	130	146	150	146
Thermal exp.coeff. ∥ (23-80°C)	K^{-1}	$14 \cdot 10^{-5}$	$10 \cdot 10^{-5}$	$10 \cdot 10^{-5}$	$7 \cdot 10^{-5}$	$7 \cdot 10^{-5}$	$5.1 \cdot 10^{-5}$
Thermal exp.coeff. ⊥ (23-80°C)	K^{-1}	$13 \cdot 10^{-5}$	$12 \cdot 10^{-5}$	$11 \cdot 10^{-5}$			$4.8 \cdot 10^{-5}$
Relative permittivity (50 Hz)	1	3.4	3.4	3.2	3	3	3.1
Relative permittivity (1 MHz)	1	3.2	3.2	3.1	2.9	2.9	2.9
Dissipation factor (50 Hz)	1	30	20	30	9	9	10
Dissipation factor (1 MHz)	1	190	210	170	100	100	100
Dielectric strength	kV/mm	29	27	28			
Comp.tracking index (CTI)	steps	600	600	600	250	250	
Comp.tracking index (CTI M)	steps	350	425	600			
CTI M 100 drops value	steps	275	375				
Volume resistivity	Ω cm	> 10^{15}	> 10^{15}	> 10^{15}	> 10^{16}	> 10^{16}	> 10^{16}
Surface resistivity	Ω	> 10^{15}	> 10^{15}	> 10^{15}	> 10^{15}	> 10^{15}	> 10^{15}
Electric corrosion	steps	A1	A1	A1	A1	A1	A1
Flammability UL 94 (1.6 mm)	steps	HB	HB	HB	V-2	V-2	V-2
Flammability UL 94 (2nd)	steps	HB	HB	HB			
at thickness 2	mm	0.8	0.8	1.57			
Water absorption (23°C, sat.)	%	0.3	0.3	0.3	0.35	0.3	
Moist. absorpt. (23°C,50 % RH)	%	0.1	0.1	0.1	0.15	0.1	
Degree of light transmission	%				90	89	86

12.3.3. Liquid Crystalline Polyesters

Commercial LCP plastics are based on thermotropic liquid crystalline polyesters, whereas commercial LCP fibers are spun from lyotropic liquid crystalline polyamides. LCP plastics contain exclusively or predominantly rigid monomeric units such as those from *p*-hydroxybenzoic acid HB, 6-hydroxy-2-naphthoic acid NA, terephthalic acid TA, isophthalic acid IA, 4,4'-dihydroxydiphenyl DP, and bisphenol A; ethylene glycol EG may be present as minority component (Table 12-11):

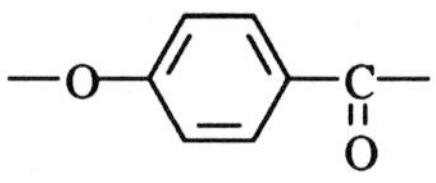

HB 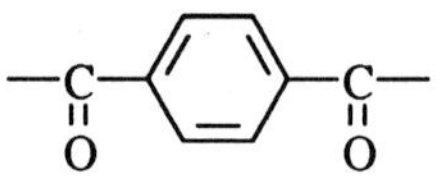 TA $-O-CH_2-CH_2-O-$ EG

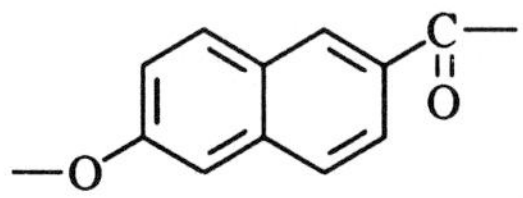

NA 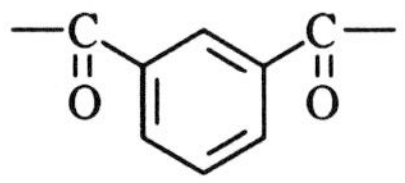 IA 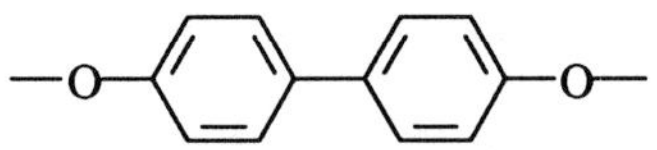 DP

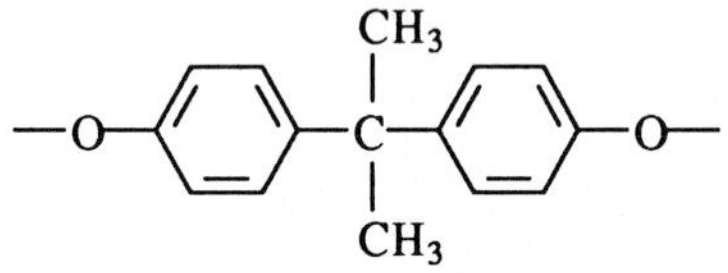

CH_3 C CH_3 BA

These polymers have very high moduli and good tensile strengths in lateral direction due to the orientation of chain axes but far lesser moduli and strengths in vertical direction because of weak interactions between chains. The anisotropy of parts can be reduced by addition of short glass fibers which however also lowers moduli and strengths in longitudinal direction. Because the reinforcement of engineering thermoplastics by glass fibers leads to similar moduli and strengths in both lateral and vertical direction as shown by very expensive unfilled LCPs, the liquid crystalline polymer X7G was withdrawn from the market.

LCPs are good for applications where complex molds require low melt viscosities. They have low thermal expansion and good dimensional stabilities; approximately 50 % of the world LCP plastics (ca. 5000 t/a) is used for cookware.

12.3.4. Acetal Polymers

Acetal polymers (acetal resins) are polymers with units $-O-CH_2-$. The repetition of these groups leads to acetal structures $-O-CH_2-O-$ in the chain and not to polyethers with the structure –Z–O–Z–, where Z is a group with at least two carbon atoms

Tab. 12-11 Properties of liquid crystalline polyesters (various non-CAMPUS sources). Eastm. = Eastman Chemical Co., Carbo. = Carborundum. BA = bisphenol A; DD = 4,4'-dihydroxydiphenyl; EG = ethylene glycol; HB = *p*-hydroxybenzoic acid; IP = isophthalic acid; NB = 2-naphthol-6-carbonic acid; TP = terephthalic acid. Numbers behind symbols: proportion of monomeric units.

Property	Unit	Eastm. X7G	Unitika Rodrun	Amoco Xydar 300	Hoechst Vectra A950	Carbo. Ekonol P3000	Dart Ekkcel I2000	Amoco Ardel D100
Composition:								
hydroxyacids		HB-60	HB	HB-68	HB-73 NA-27	HB-100	HB-?	HB-?
diacids		TA-20	TA	TA-16	-	-	TA-?	IA-?
diols		EG-20	EG	DP-16	-	-	DD-?	BA-?
Density	g/cm³	1.39		1.35	1.40	1.45	1.40	1.21
Thermal exp. coeff. (·10⁵)	K⁻¹	0		0.1	0.1	1.5	2.9	6.1
Melting temperature	°C					>550	413	
Heat distortion temp. A	°C	145	170	337	180	295	293	175
Young's modulus ∥	MPa	54 000		8300	11 000	7300	8600	2000
Young's modulus ⊥	MPa	1400			2600			
Flexural modulus	MPa	6000	10 000	13 000	8800	4800	4900	2000
Flexural strength	MPa	115	145	126		75	120	81
Tensile strength at break ∥	MPa	150	220	116	206	100	100	70
Tensile strength at break ⊥	MPa	10			50			
Strain at break	%	100	4.5	4.9	3.0	8.0	8.0	50
Notched impact strength	J/m	18	40	454	520	85	55	210
Rel. permittivity (1 MHz)	-			3.9	3.6	3.3	3.2	3.0
Dissipation factor	-			0.039	0.014	0.003		0.015
Water absorption (24 h)	%					0.02	0.025	

such as $-(CH_2)_2-$, $-(p\text{-}C_6H_4)-$, etc. Acetal polymers are also called polyacetals but this name should be restricted to the reaction products of polyols and aldehydes. Acetal polymers are also not to be confused with poly(vinyl acetal)s, which are acetalized poly(vinyl alcohol)s. Although acetal polymers are synthesized from formaldehyde HCHO or its cyclic trimer, they are not called polyaldehydes since this name is reserved for polymers from higher aldehydes such as acetaldehyde CH_3CHO.

Industry distinguishes between acetal homopolymers and acetal copolymers. Acetal homopolymers $CH_3CO[O{-}CH_2]_nOOCCH_3$ are poly(oxymethylene)s POM; they are polymers of formaldehyde, whose endgroups have been capped by acetate units. Acetal copolymers are copolymers of trioxane (cyclic trimer of formaldehyde) with small proportions of cyclic ethers as comonomers such as ethylene oxide (oxirane), dioxolane (1,3-dioxacyclohexane), and dioxepane (1,3-dioxacycloheptane); they are also called poly(oxymethylene)s but not poly(formaldehyde)s.

The acetate endgroups of acetal homopolymers protect POM from unzipping reactions; such depolymerizations to gaseous formaldehyde start at unprotected (active) chain ends because the ceiling temperature of POM (127°C; crystal to gas) is lower than then the conventional processing temperatures. Active chain ends are formed during formaldehyde polymerization (before capping) and by random thermal chain splitting during processing. Ether groups also stop the chains from further unzipping because of the much higher ceiling temperature of their polymers. They also stabilize POM chains against attack by alkali. Acetal polymers are also stabilized against heat degradation by urea, hydrazine, or polyamides and against oxidation by secondary and tertiary amines.

Poly(oxymethylene) chains crystallize in all-gauche conformation as compact 9_5 helix, whereas the trans-trans-gauche conformations of poly(oxyethylene) chains lead to a more open 7_2 helix (Fig. 12-3). POM thus has a much higher melting temperature than PEOX (184°C vs. 69°C); it is insoluble in any liquid (except under degradation) whereas PEOX is readily soluble in water. Poly(oxymethylene)s are rigid engineering thermoplastics that find use in small appliances, pump housings, etc.

Table 12-12 Properties of poly(oxymethylene) POM homopolymers and copolymers and poly-(oxy-2,6-dimethyl-1,4-phenylene) PPE. * Generic data from various sources; NB = no break.

Physical property		Polymer types and grades				
Name	Unit	POM unfilled Du Pont Delrin *	POM unfilled Hoechst Hostaform C 9021	PPE unfilled Hüls Vestoran 1300 nf	PPE 30 % GF Hüls Vestoran 1300 GF30	PPE + PA Hüls Vestoblend 1500 nf
Density	g/cm^3	1.42	1.41	1.04	1.27	1.03
Heat distortion temp. B	°C			132	140	
Heat distortion temp. A	°C	136	104	113	130	160
Heat distortion temp. C	°C			96	125	
Vicat temperature A	°C			144	151	179
Vicat temperature B	°C		151	132	138	165
Thermal exp.coeff. (23-80°C)	K^{-1}	$9 \cdot 10^{-5}$	$11 \cdot 10^{-5}$	$9 \cdot 10^{-5}$	$3.2 \cdot 10^{-5}$	$10 \cdot 10^{-5}$
Young's modulus (1 mm/min)	MPa	3100	2800	2100	8900	1550
Creep modulus (1 h)	MPa		2250	1700		
Creep modulus (1000 h)	MPa		1200	1200		
Stress at yield (50 mm/min)	MPa	69	64	50		47
Tensile strength (5 mm/min)	MPa	70			80	
Strain at yield (50 mm/min)	%		9	5		10.5
Strain at break (50 mm/min)	%	> 23	35	44	2	> 50
Impact strength (Izod, 23°C)	kJ/m^2		130	96	14	NB
Impact strength (Izod, -30°C)	kJ/m^2		100	78	16	NB
Notched impact strgth. (23°C)	kJ/m^2		5.5	15	6	47
Notched impact strgth. (-30°C)	kJ/m^2		5.5	12	6.6	13
Relative permittivity (50 Hz)	1	3.7	4.3	2.6	3	3.7
Relative permittivity (1 MHz)	1		4.3	3	3.6	3.2
Dissipation factor (50 Hz)	1		0.0027	0.0010	0.0015	0.0020
Dissipation factor (1 MHz)	1	0.0050	0.0060	0.0010	0.0027	0.0230
Dielectric strength	kV/mm			38	46	27
Comp.tracking index (CTI)	steps		600	300	200	600
CTI 100 drops values	steps		600	225	175	575
Comp.tracking index (CTI M)	steps		600	75	150	100
CTI M 100 drops value	steps			50	125	75
Volume resistivity	Ω cm		10^{15}	$> 10^{15}$	$> 10^{15}$	$7 \cdot 10^{14}$
Surface resistivity	Ω		10^{13}	$4 \cdot 10^{13}$	$3 \cdot 10^{13}$	$2 \cdot 10^{13}$

12.3.5. Polyethers

Poly(oxyethylene) ~$[OCH_2CH_2]_n$~ or poly(ethylene oxide) PEOX, has in the crystalline state a fairly open helical structure as compared to poly(oxymethylene) (Fig. 12-3); this structure is responsible for the low melting temperature of 69°C and the solubility in water.

High molar mass PEOX ($M_r \approx 10^6$) is used as thickener and size but not as plastic. Low molar mass ($M_r < 40\ 000$) compounds $HOCH_2CH_2[OCH_2CH_2]_{n-1}OH$ with hydroxyl endgroups are called poly(ethylene glycol)s; they are used for soft blocks in polyetherester thermoplastic elastomers and in cosmetics and pharmaceuticals.

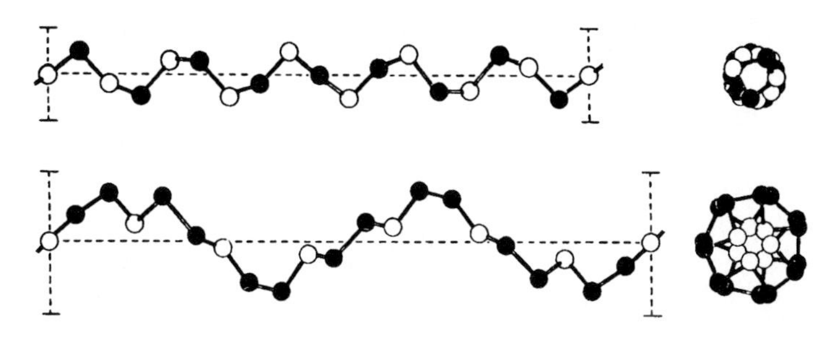

Fig. 12-3 Helix conformations of chains in crystalline poly(oxymethylene) (top) and poly(oxyethylene) (bottom). O Oxygen goups -O-; ● methylene groups $-CH_2-$.
Left: side-view; right: end-view.
Distance between vertical lines: length of chain identity period (chain repeating distance).
Horizontal lines: chain axes.
Reprinted with permission by Association for Science Documents Information, Tokyo [1]

Poly(propylene oxide) ~$[O\text{-}CH(CH_3)\text{-}CH_2]_n$~ is not used as a plastic. Segmented copolymers of D,L-propylene oxide with ethylene oxide are macromolecular detergents, and copolymers with either nonconjugated dienes or allyl glycidyl ether are elastomers for low temperature applications.

Poly(tetrahydrofuran) PTHF (poly(tetramethylene oxide) is obtained by polymerization of tetrahydrofuran (oxacyclopentane). The poly(tetramethylene ether) glycol $HO(CH_2)_4[O(CH_2)_4]_{n-1}OH$ is used as a soft segment in elastic polyurethane fibers or polyetherester elastomers.

Poly(phenylene ether) PPE or poly(phenylene oxide) PPO® is in fact the dimethyl substituted chemical compound poly(oxy-2,6-dimethyl-1,4-phenylene) = poly(2,6-xylenol) (see next page). It is produced by oxidative coupling of 2,6-dimethylphenol. The homopolymer (T_G = 209°C) is rigid and tough but very difficult to process. Most PPEs are blends, mainly with poly(styrene), and also with polyamides or poly(butylene terephthalate) (Table 12-12).

Polyetherketones contain ether and keto groups in the main chain. These engineering plastics are obtained by the reaction of the dipotassium salt of hydroquinone, $KO(p\text{-}C_6H_4)OK$ with aromatic difluoroketones such as $F(p\text{-}C_6H_4)CO(p\text{-}C_6H_4)F$. Polyether ether ketone PEEK, is a high melting polymer (Table 12-13), that can be processed at 370-400°C with conventional methods because of its not too high visco-

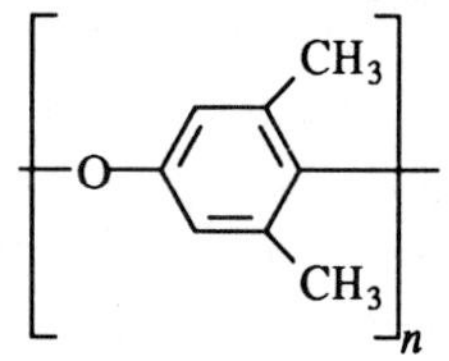

PPE 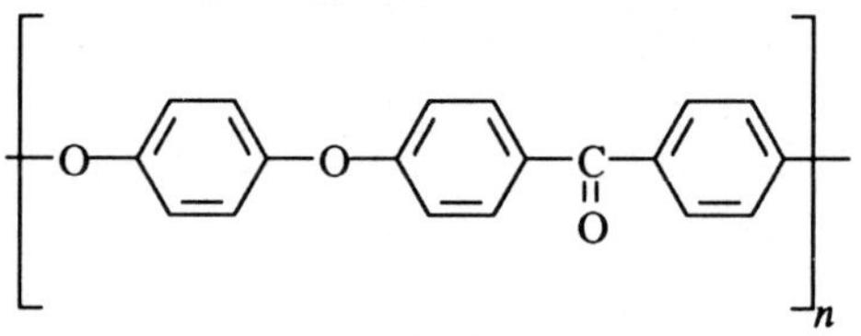 PEEK

sity. Many other polyether ketones are produced in small amounts, for example, polyether ketone PEK ~[O–(*p*-C_6H_4)–CO–(*p*-C_6H_4)]$_n$~ itself, polyether ether ketone ketone PEEKK ~[O–(*p*-C_6H_4)–O–(*p*-C_6H_4)–CO–(*p*-C_6H_4)–CO–(*p*-C_6H_4)]$_n$~, etc.

12.4. Carbon-Sulfur Chain Thermoplastics

12.4.1. Poly(phenylene sulfide)

Poly(*p*-phenylene sulfide) ~[S–(*p*-C_6H_4)]$_n$~ (PPS) results from the polycondensation of 1,4-dichlorobenzene Cl(*p*-C_6H_4)Cl with disodium sulfide Na_2S. It is a white, highly crystalline powder (T_M = 288°C) that becomes brown on heating. The high rigidity (flexural modulus 4.2 GPa) can be further increased by glass fiber-reinforcement. PPS is flame retardant and chemically resistant. It is injection molded into electrical, electronic and industrial parts, or used for powder coatings.

12.4.2. Polysulfones

All commercial polysulfones contain sulfone groups -SO_2-, ether groups, and aromatic residues in their chains. They are synthesized by polysulfonation of monomers with preformed ether groups, e.g., from C_6H_5–O–(*p*-C_6H_4)–SO_2Cl with elimination of HCl, or by polyether formation of monomers with preformed sulfone groups, e.g., from Cl–(*p*-C_6H_4)–SO_2–(*p*-C_6H_4)–OMt with elimination of metal chlorides MtCl. Both types can be performed as AB or AABB reactions. Copolycondensation of Na_2S, Cl(*p*-C_6H_4)Cl and Cl(*p*-C_6H_4)-SO_2-(*p*-C_6H_4)Cl yields polyarylenesulfidesulfones.

The various polysulfones are known under different names (Ph = *para*-phenylene), most commonly (but not always) as:

PSU	~[SO_2–Ph–O–Ph]$_n$~	polysulfone
PES	~[(SO_2–Ph–O–Ph)$_i$–*co*–(SO_2–Ph–Ph)$_j$]$_n$~	polyethersulfone ($i > j$)
	~[(SO_2–Ph–O–Ph)$_i$–*co*–(SO_2–Ph–Ph)$_j$]$_n$~	polyethersulfone ($i < j$)
PAS	~[SO_2–Ph–O–Ph–SO_2–Ph–Ph]$_n$~	polyaryl(ene)sulfone
PPSU	~[SO_2–Ph–O–Ph–Ph–O–Ph]$_n$~	polyphenyl(ene)sulfone
PSF	~[SO_2–Ph–O–Ph–C(CH_3)$_2$–Ph–O–Ph]$_n$~	bisphenol A polysulfone

All polysulfones are transparent, light yellow to amber polymers. Their high heat distortion temperatures are only surpassed by aramides, polyimides PI, and polyamideimides PAI (Table 12-13); in contrast to these, polysulfones can be injection molded. Polysulfones do not break as unnotched specimens (Izod); their notched impact strengths can be improved by blending with ABS, SAN, or EPs.

Aliphatic polysulfones such as $\sim[(CH_2)_6SO_2]_n\sim$ have very low ceiling temperatures and are thus used as resists in the manufacture of semiconductors.

Table 12-13 Properties of some high-performance polymers at 23°C unless noted otherwise (data from various sources). PI = polyimide; PEI = polyether imide; PAI = polyamide imide; PEEK = polyetheretherketone; PSU = polysulfone; PPS = poly(phenylene sulfide). * 150°C.

Property	Unit	PI Vespel SPI	PEI Ultem 1000	PAI Torlon 4203	PEEK Victrex	PSU Victrex 200 P	PPSU Radel R 5000	PSF Udel P1700	PPS Ryton 100
Density	g/cm^3	1.43	1.27	1.40	1.30	1.37	1.29	1.24	1.35
Thermal exp.coeff. $(\cdot 10^5)$	K^{-1}	5.6	6.1	4.0	4.7	5.5	5.5	5.6	4.9
Glass temperature	°C		220		144	185	220	224	185
Heat distortion temp. A	°C	260	170	274	156	210	204	174	138
Young's modulus	GPa		3.3		1.1*	2.4	2.1	2.5	3.8
Flexural modulus at 23°C	GPa	3.1	3.3	4.7	3.8	2.6	2.3	2.7	3.8
Flexural modulus at 270°C	GPa	1.8		3.0					
Flexural strength at 23°C	MPa	115		210		130	86	106	140
Flexural strength at 270°C	MPa	62		80					
Compression strength	MPa	> 280	140						113
Yield strength	MPa				91	85	72	70	
Tensile strength at break	MPa	86	105	190	70	83		76	70
Strain at break	%	8	60	12	150	60	60	75	1.6
Notched impact strength	J/m	53	153	135	48	90	640	70	22
Rel. permittivity (1 MHz)	-	3.4	3.2	4.0	3.3	3.5	3.4	3.2	3.1
Dissipation factor	-	0.0018	0.002	0.009	0.003	0.0022		0.002	0.0004
Water absorption (24 h)	%	0.32	0.25	0.28	0.14	0.43	1.1		< 0.02

12.5. Carbon-Nitrogen Chain Thermoplastics

12.5.1. Polyamides

Nomenclature. Polyamides PA contain amide groups -NH-CO- in the main chain. They are usually subdivided into AB polymers with the repeating unit -NH-X-CO- and AABB polymers with the repeating unit -NH-X^1-NH-CO-X^2-CO-, where A indicates the NH group and B the CO group.

Aliphatic polyamides contain aliphatic residues X, X^1, and X^2. These polymers are also known as "nylons" (exceptions: proteins with X = CHR and polyamide resins used in inks, adhesives, and coatings). Aliphatic polyamides are usually characterized by the number of carbon atoms per monomeric unit. An exception to this rule is the naming of the mixture of 2,2,4- and 2,4,4-trimethyl-1,6-hexamethylenediamine as 6-3; this monomer is therefore also called TMD.

AB polymers are customarily named after their monomers:

PA 6	$\sim[NH(CH_2)_5CO]_n\sim$	poly(ε-caprolactam)
PA 11	$\sim[NH(CH_2)_{10}CO]_n\sim$	poly(11-aminoundecanoic acid)
PA 12	$\sim[NH(CH_2)_{11}CO]_n\sim$	poly(laurolactam)

Other names are: PA 6 = poly(caproamide); PA 11 = poly(11-aminoundecanoamide); PA 12 = poly(dodecanolactam) or poly(dodecanoamide).

AABB polyamides are generally named after their monomeric units:

PA 4.6	$\sim[NH(CH_2)_4NH\text{-}CO(CH_2)_4CO]_n\sim$	poly(tetramethylene adipamide)
PA 6.6	$\sim[NH(CH_2)_6NH\text{-}CO(CH_2)_4CO]_n\sim$	poly(hexamethylene adipamide)
PA 6.9	$\sim[NH(CH_2)_6NH\text{-}CO(CH_2)_7CO]_n\sim$	poly(hexamethylene azelaamide)
PA 6.10	$\sim[NH(CH_2)_6NH\text{-}CO(CH_2)_8CO]_n\sim$	poly(hexamethylene sebacamide)
PA 6.12	$\sim[NH(CH_2)_6NH\text{-}CO(CH_2)_{10}CO]_n\sim$	poly(hexamethylene dodecanoamide)

The dot separating the numbers of carbon atoms in AABB polyamides is omitted in ISO symbols. An example is PA 612 = PA 6.12, the abbreviation is still pronounced "polyamide six-twelve" and not "polyamide six hundred twelve", however.

IUPAC names are too awkward for everyday use. For example, IUPAC seniority rules require the chemical structure of poly(hexamethylene adipamide) (PA 6.6 = PA 66) to be written as $\text{-}(NHCO(CH_2)_4CONH(CH_2)_6)_n\text{-}$. The IUPAC structural names of PA 6.6 are therefore poly[imino(1,6-dioxohexamethylene)iminohexamethylene] or poly[iminoadipoyliminohexamethylene].

Both copolymers and polymer blends are often symbolized by slashes between numbers. PA 6/12 may be therefore a copolymer of ε-caprolactam and laurolactam or a blend of poly(ε-caprolactam) and poly(laurolactam). Certain blending procedures may lead to transamidation reactions and the intended blend may thus be an unintended (partial) copolymer instead.

Aromatic polyamides contain at least one aromatic residue X, X^1, or X^2 in their units -NH-X-CO- or $\text{-NH-}X^1\text{-NH-CO-}X^2\text{-CO-}$. They are generally characterized by letters: I = isophthalic acid, T = terephthalic acid, MXD = *m*-xylylene diamine, etc:

PPD-T	$\sim[NH(p\text{-}C_6H_4)NH\text{-}CO(p\text{-}C_6H_4)CO]_n\sim$	poly(*p*-phenylene terephthalamide)
MXD 6	$\sim[NHCH_2(m\text{-}C_6H_4)CH_2NH\text{-}CO(CH_2)_4CO]_n\sim$	poly(*m*-xylylene adipamide)

Aromatic polyamides are also called polyaramides or just aramides. The U.S. Federal Trade Commission defines "aramide" as "a manufactured fiber in which the fiber-forming substance is a long-chain synthetic polyamide in which at least 85 % of the amide linkages are attached directly to the aromatic rings"; according to ISO, up to 50 % of the amide groups may be replaced by imide groups in aramide polymers. Aromatic polyamides are sometimes called "superpolyamides"; this name was given originally to aliphatic polyamides with molar masses > 10 000 g/mol.

Syntheses. The first commercially useful polyamide (PA 6.6) was patented in 1937 by W.H. Carothers at Du Pont de Nemours Co. Fibers from PA 6.6 were excellent substitutes for silk hosieries, hence the name "No-run", which then mutated to "Nuron", "Niron", and finally "Nylon".

One year later, the polymerization of ε-caprolactam to PA 6 was patented by W. Schlack at I.G.Farbenindustrie (Perlon® fibers). High-modulus fibers from aromatic polyamides were introduced in 1961. Polyamides are still mainly used as fibers (1990 world production: $3.7 \cdot 10^6$ t/a) but the application as engineering plastics is steadily increasing (1990 world production: $0.8 \cdot 10^6$ t/a).

Polyamides are synthesized by polycondensations of α,ω-amino acids (PA 11), internal salts of diamines and diacids (PA 4.6, 6.6, 6.10, 6.12), diesters + diamines (PA 4.2), and dicarbonic acid dichlorides + diamine dihydrochlorides (PPD-T) as well as by chain polymerizations ("addition polymerizations") of lactams (PA 6, PA 12).

Properties. Amide groups of polyamides form hydrogen bonds –NH- - -O=C– between polyamide chains. Depending on the number of carbon atoms between nitrogen atoms, chains run parallel (PA 6.6) or antiparallel (PA 6) with respect to the sequence of NH and CO groups (Fig. 12-4). The zigzag structure of the chains results in the macroconformation of a pleated sheet. A great proportion of the hydrogen bonds survives the melting temperature; the resulting physical cross-linking causes the high viscosities of polyamide melts.

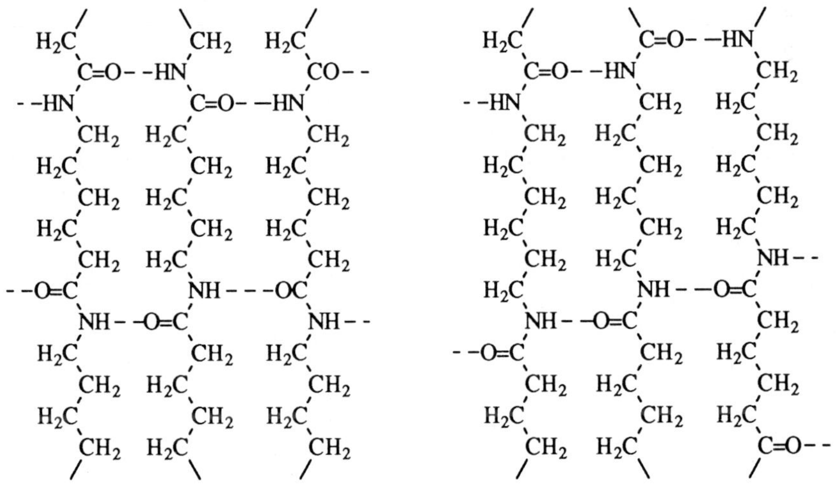

```
   /           \          /               /           \          /
H2C          CH2    H2C                H2C          C=O--HN
   C=O--HN          CO--                  C=O--HN          CH2
--HN         C=O--HN                   --HN         CH2   H2C
   CH2  H2C          CH2                  CH2  H2C          CH2
H2C          CH2    H2C                H2C          CH2   H2C
   CH2  H2C          CH2                  CH2  H2C          CH2
H2C          CH2    H2C                H2C          CH2   H2C
   CH2  H2C          CH2                  CH2  H2C          NH--
--O=C        NH---OC                   H2C          NH--O=C
   NH--O=C           NH--                 NH--O=C           CH2
H2C          CH2    H2C                --O=C        CH2   H2C
   CH2  H2C          CH2                  CH2  H2C          CH2
H2C          CH2    H2C                H2C          CH2   H2C
   CH2  H2C          CH2                  CH2  H2C          C=O--
   /          \          /                /          \          /

            PA 6                                     PA 6.6
```

Fig. 12-4 Pleated sheet structures of polyamides 6 (anticlinal) and 6.6 (isoclinal).

The hydrophilic amide groups pick up moisture. Polyamides are therefore "conditioned" at 65 % relative humidity (CAMPUS data; USA: 50 % RH) and are usually sold in this equilibrated state. Reported polyamide properties sometimes refer to the dry state, the as-molded condition, or to 100 % relative humidity instead.

The small amounts of absorbed water do not affect the densities of conditioned PAs (Table 12-14). Water molecules form hydrogen bonds with amide groups, which decreases the proportion of amide-amide interactions. The reduced interchain links make the conditioned polyamide less rigid. Young's moduli E and yield stresses σ_y decrease whereas strains at yield and at break increase. Water is not a plasticizer in the scientific sense of the word: segmental mobilities are not affected and heat distortion and Vicat temperatures remain the same in the conditioned and the dry states.

Properties of polyamides vary widely with the chemical structure, i.e., the length of the methylene sequences and the presence of substituted chain atoms or aromatic chain groups (Table 12-15). Straight-chain aliphatic AB and AABB polyamides are semicrystalline materials. Their melting temperatures decrease with decreasing proportion of amide groups; examples are 365°C (PA 2), 260°C (PA 4), 225°C (PA 6), and 190°C (PA 12). Glass temperatures are almost independent of the length $i \geq 6$ of methylene segments, indicating the same length of cooperatively moving segments. They vary only between $T_G = 53°C$ for PA 6 and $T_G = 49°C$ for PA 8.22.

Table 12-14 Effect of conditioning on properties at 25°C (unless mentioned otherwise) of three PA 6 grades from Ems Chemie. T 300 GM is an unfilled extrusion grade, PV 3 H an injection grade with 30 wt-% short glass fibers, and PM 3 H an injection grade with 30 wt-% mineral. D = dry state; C = conditioned state; NB = no break.

Physical properties Names	Test conditions	Units	Polyamide grades (Grilon®) T 300 GM		PM 3 H		PV 3 H	
			D	C	D	C	D	C
Density	-	g/cm³	1.14	1.14	1.36	1.36	1.35	1.35
Heat distortion temp. B	0.45 MPa	°C	224	224				
Heat distortion temp. A	1.8 MPa	°C	73	73	65	65	203	203
Heat distortion temp. C	5.0 MPa	°C			53	53	181	181
Vicat temperature A	10 N	°C	254	254	215	215	216	216
Vicat temperature B	50 N	°C	239	239	206	206	212	212
Young's modulus	1 mm/min	MPa	3800	1600	5000	1700	9000	5500
Yield strength	50 mm/min	MPa	95	60	90	90	210	130
Strain at yield	50 mm/min	%	5	20	5	5	3	6
Strain at break	50 mm/min	%	15	> 50	7	15	4	8
Impact strength	+23°C	kJ/m²	65	NB	NB	NB	80	95
Impact strength	- 30°C	kJ/m²	2	3	4	7	12	18
Notched impact strength	+23°C	kJ/m²	4	12	4	18	13	25
Notched impact strength	- 30°C	kJ/m²	2	3	4	7	12	18
Relative permittivity	50 Hz	1	4	12	7	18	18	25
Relative permittivity	1 MHz	1	2	3	4	4	12	13
Dissipation factor ($\cdot 10^4$)	50 Hz	1	35	700	30	750	40	1200
Dissipation factor ($\cdot 10^4$)	1 MHz	1	200	750	190	900	190	1000
Volume resistivity	-	Ω cm	10^{14}	10^{12}	10^{14}	10^{12}	10^{14}	10^{12}

Tab. 12-15 Properties of conditioned polyamides 6 (Grilon® A 28 GM; Ems), 6.6 (Grilon® T 300 FC; Ems), 6.12 (Grilon® CR 9; Ems), 12 (Grilamid® L 20 GM; Ems), 6-3 T (Trogamid T® nf; Hüls), and 6I/XT (Grivory® G 355 NZ; Ems). *Generic data. - Not applicable; NB = no break; no entry: datum not measured.

Physical properties			Polyamides					
Names	Test conditions	Units	6	6.6	6.12	12	6-3T	6I/XT
Density	-	g/cm³	1.14	1.14	1.1	1.01	1.12	1.08
Viscosity coefficient	?	mL/g	155	135	145	175		85
Melt volume index	275°C	mL/10 min	260	435	195			45
Melting temperature*	-	°C	230	265	212	180	-	
Glass temperature*, dry	-	°C	52	50	47	42	140	
Heat distortion temp. B	0.45 MPa	°C	182	226	49	111	138	124
Heat distortion temp. A	1.8 MPa	°C	58	75	39	44	119	108
Heat distortion temp. C	5.0 MPa	°C					106	
Vicat temperature A	10 N	°C	215	255	193	170		135
Vicat temperature B	50 N	°C	202	240	160	133		129
Young's modulus	1 mm/min	MPa	1100	1700	550	1000		1800
Yield strength	50 mm/min	MPa	45	60	30	15	87	55
Strain at yield	50 mm/min	%	25	20	20	15	9	6
Strain at break	50 mm/min	%	> 50	> 50	> 50	> 50	> 50	35
Impact strength	+23°C	kJ/m²	NB	NB	NB	NB	9	NB
Impact strength	- 30°C	kJ/m²	NB	50	40	NB	5	NB
Notched impact strength	+23°C	kJ/m²	25	8	NB	9	NB	NB
Notched impact strength	- 30°C	kJ/m²	5	2	3	4		20
Relative permittivity	50 Hz	1	8	6		4	4.2	3
Dissipation factor ($\cdot 10^4$)	50 Hz	1	750	600		950	205	55
Volume resistivity	-	Ω cm	10^{13}	10^{13}		10^{13}	10^{15}	10^{14}
Water absorption, saturation	23°C	%	9	10	9	1.5	1.5	7

A wide variety of polyamide grades is available for many applications. The basic semicrystalline polyamide types (6, 10, 11, 12, 6.6, 6.10, 6.12) are supplemented by nucleated, plasticized, or flame-retardant grades, copolymers (e.g., 6.6-6), amorphous, transparent polyamides (e.g., 6-3 T), amorphous, heat-resistant PAs, high-impact resistant PAs, blends of two polyamides (e.g., 12/12-XX) or blends of a polyamide and another polymer (e.g., with ABS or PPE). Properties can also be enhanced by addition of glass fibers, glass beads, carbon fibers, minerals (normally calcium carbonate or silicates), etc. (see Table 12-14). These polyamide engineering plastics are mainly used in transportation, extrusion markets, electrical and electronic applications (especially in Europe), and for many other purposes.

12.5.2. Polyimides

Polyimides PI form two groups: thermoplastic polyimide are derived from monomers with preformed imide groups and thermosetting polyimides from polycondensation reactions in which imide groups are generated.

Thermoplastic polyimides result from the polyaddition of bismaleinimides (from maleic anhydride + diamines) and chemical compounds HZ-R-ZH, where HZ may be HNH (diamines), HS (disulfides), HO-N=CH (dialdoxims), etc. The carbon-carbon double bonds of bismaleinimides are bifunctional; the polyaddition thus does not lead to branching and cross-linking, hence to thermoplastics, for example:

(12-2) $\xrightarrow{+\text{HZ-R-ZH}}$

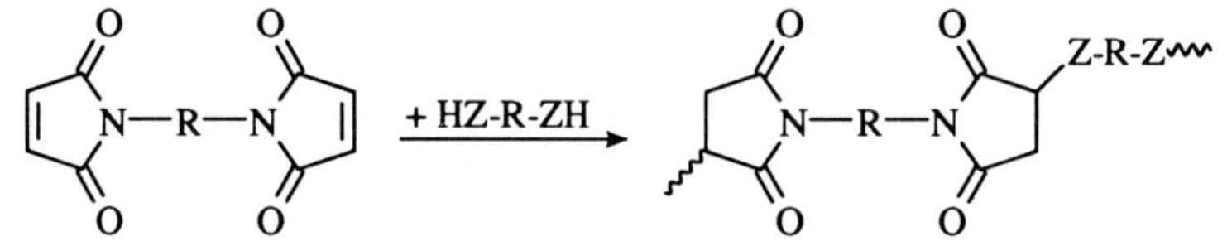

Another thermoplastic polyimide is based on isocyanate chemistry:

(12-3) + OCN–C$_6$H$_4$–NCO

$\xrightarrow{-2\,CO_2}$

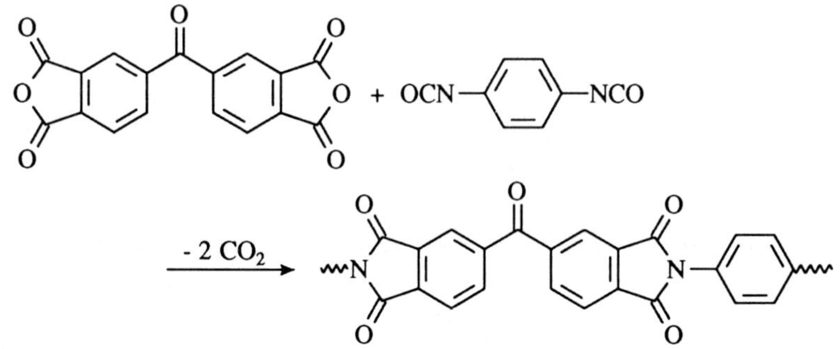

Isocyanate and anhydride groups can be joined in one monomer; AB polymers are thus available in addition to the AABB polymers from the above reaction (Eq.(12-3)).

Thermosetting polyimides are formed by polyaddition followed by polycondensation, for example, from pyromellitic dianhydride and 4,4'-diaminodiphenylether:

(12-4) + H_2N–C$_6$H$_4$–O–C$_6$H$_4$–NH_2

$\longrightarrow$

$\xrightarrow{-2\,H_2O}$

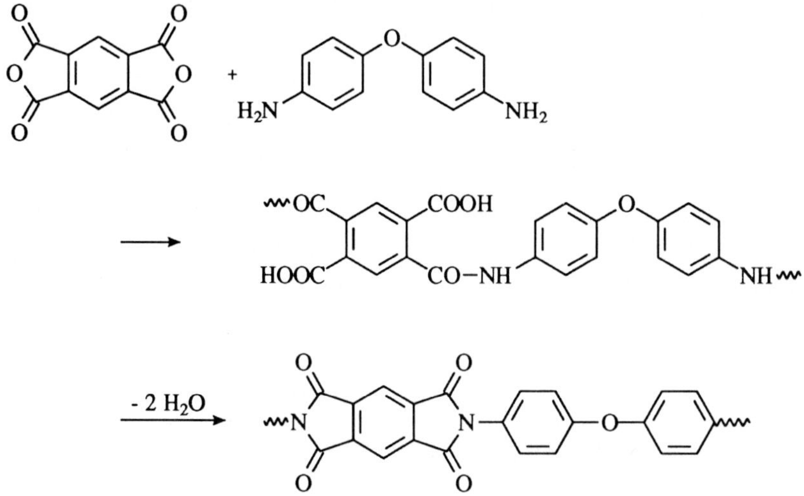

The polyamic acid from the first step (polyaddition) has a tetrafunctional repeating unit, which generates a small proportion of cross-links besides the main intramolecular ring-closure reaction. The difficult processing is avoided by polyesterimides and polyetherimides PEI that carry ester and ether groups in the monomeric dianhydrides. Polyamideimides PAI result from the polycondensation of trimellitic acid (anhydride of 1,2,4-benzenetricarboxylic acid) with diamines.

Polyimides have good thermal, oxidative, and hydrolytic stabilities and excellent electrical properties. Their mechanical properties vary widely because of the many different monomers that may be used for their manufacture; some examples are given in Table 12-13. Polyimides are used in construction and transportation.

12.6. Thermosets

Thermosets can be prepared from multifunctional monomers or prepolymers by polycondensation, polyaddition, and chain polymerization. The polymerization proceeds with cross-linking ("curing"); shaping must therefore occur simultaneously with polymerization. The resulting thermosets have high cross-linking densities; the short segments between cross-links cause low segment mobilities and thus good resistances against creep and enhanced heat distortion temperatures.

12.6.1. Phenolics

Phenolic resins PF (phenolics) are condensation products of phenols with formaldehydes. Phenol C_6H_5OH is sometimes replaced by cresols $C_6H_4(CH_3)OH$ or resorcinol $(i\text{-}C_6H_4)(OH)_2$.

Acid catalysis of a mixture with an excess of phenol leads to soluble novolacs; base catalysis with an excess of formaldehyde to soluble resols (A-stage resins), then to resitols (B-stage resins), and finally to cross-linked resits (C-stage resins). Novolacs were first used as substitutes for shellac, hence the name (Lat.: novo = new; Indian: laksha = hundred thousand, since the secretions of hundred thousand larvae of the insect *Kerria lacca* are needed to produce one ounce of lac).

Acid catalysis leads first to instable *o*-methylolphenols or *p*-methylolphenols, then to methylene compounds

$$C_6H_5OH \xrightarrow{+\ CH_2O} \left[o\text{-}HOC_6H_4CH_2OH \right] \longrightarrow o\text{-}HOC_6H_4\text{-}CH_2\text{-}C_6H_4OH\text{-}p \qquad (12\text{-}5)$$

and finally to oligomers with relative molecular weights of ca. 1000 that may also contain some open-chain formal groups instead of the desired *o,o*'-methylene bridges (with respect to OH groups) and the less desirable *p,p*'-methylene bridges (unsubstituted para-positions cure faster). Resols and resitols also possess ether bridges. Novolacs and resitols are then cured (cross-linked) with paraformaldehyde $H(OCH_2)_{6-10}OH$ by acid catalysis (mainly methylene bridges) or with hexamethylenetetramine (hexa; urotropin) by base catalysis (imine bridges). Bridging structures include:

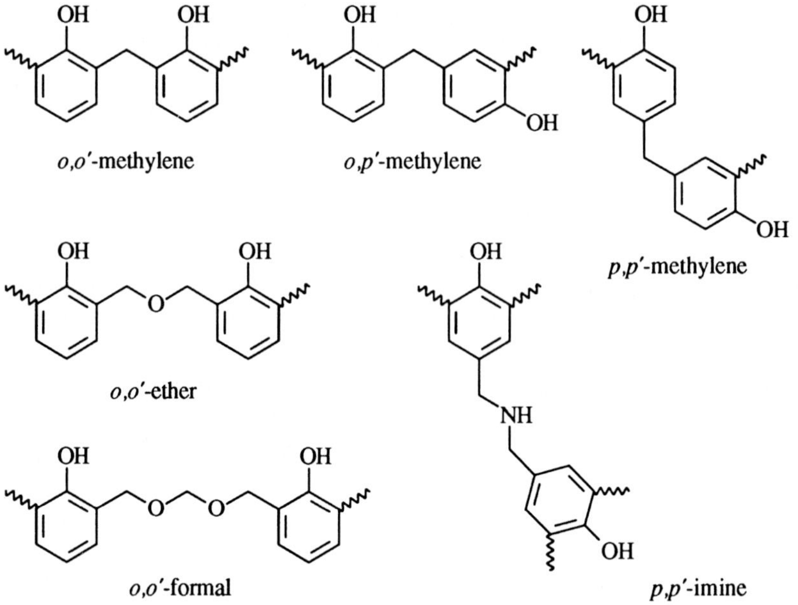

OH OH OH OH OH OH

o,o'-methylene *o,p*'-methylene *p,p*'-methylene

o,o'-ether NH *p,p*'-imine

o,o'-formal

Reaction of phenol with formaldehyde and subsequent acid curing leads also to quinonemethide groups that cause cross-linking by cycloaddition and a discoloration of phenolics with time. Quinonemethide groups can be removed by esterification; such phenolics are white products. A side reaction of the base curing with hexa generates azomethine groups that lead to the yellow-brownish color of most phenolics.

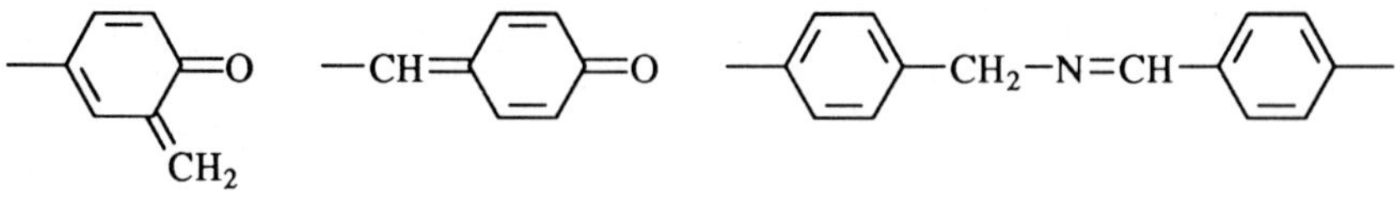

=O CH_2 —CH= =O —CH_2—N=CH—

o-quinonemethide *p*-quinonemethide azomethine

PFs have very low manufacturing costs. The processing is difficult, however: the elimination of water during hardening requires application of pressure to prevent void formation by water vapor. The high cross-linking density of phenolics leads to very low mobilities of chain segments between cross-linking points and thus to an excellent dimensional stability (heat distortion temperature A ca. 120°C). Mechanical properties are poor; low notched impact strengths (Izod; Table 12-16) are probably caused by void formation during curing.

Cured resins are mainly used in the electrical industry (fuse boxes, distributor caps, etc.) and for ion exchange resins; resitols and novolacs are utilized as adhesives, lacquers, tanning agents, binders, and for fibers.

Tab. 12-16 Properties of some unfilled thermosets (generic data from many sources). Thermoset formation by chain polymerization CPM, polycondensation PCD, or polyaddition PAD. DAP = Diallyl phthalate; EP = epoxy resin; PUR = polyurethane; PF = phenol, MF = melamine, UF = urea resins with formaldehyde. Unsaturated polyesters UP were hardened with styrene.

Property	Physical unit	CPM DAP	CPM UP	PCD PF	PCD MF	PCD UF	PAD EP	PAD PUR
Density	g/cm^3	1.27	1.3	1.25	1.48		1.2	1.05
Thermal expansion coeff. ($\cdot 10^5$)	K^{-1}	11	8	8			5	
Shrinkage	%	1.0	0.6	1.1	0.7	1.3	0.5	1.0
Heat distortion temperature A	°C	155	130	121	148		170	91
Young's modulus	GPa	2.2	3.4	2.8			2.5	
Flexural modulus	GPa	2.1						4.4
Tensile strength at break	MPa	28	70	65			70	
Elongation at break	%		2	1.8			6	
Notched impact strength	J/m	17	16	16			35	21
Hardness (Rockwell M)	-	98	90	126			95	
Relative permittivity (1 MHz)	-	3.4	3.5	4.7			3.7	3.5
Volume resistivity	Ω cm	10^{15}		10^{11}	10^{10}	10^{10}	10^{15}	
Dissipation factor ($\cdot 10^4$)	-	500	200	200			400	30
Water absorption (24 h)	%	0.2	0.4	0.15			0.13	0.2

12.6.2. Amino Resins

Amino resins are condensation products of formaldehyde (sometimes higher aldehydes and ketones) with amino group-containing chemical compounds such as urea $H_2N\text{-}CO\text{-}NH_2$ and melamine (2,4,6-triamino-1,3,5-triazine). Formaldehyde addition to amino groups results in methylol compounds of urea (left) and melamine (right):

$HOCH_2\text{-}NH\text{-}CO\text{-}NH_2$

$HOCH_2\text{-}NH\text{-}CO\text{-}NH\text{-}CH_2OH$

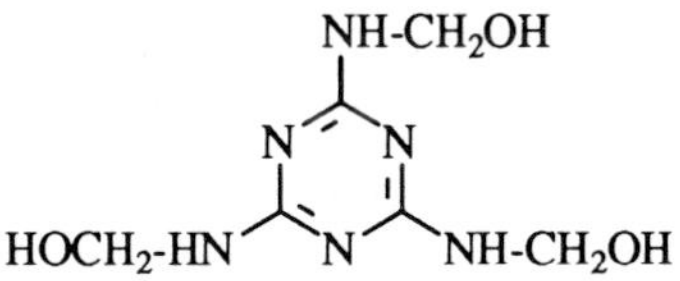

$NH\text{-}CH_2OH$

N N

$HOCH_2\text{-}HN$ N $NH\text{-}CH_2OH$

Secondary amine groups -NH- either add additional formaldehyde CH_2O or cure by reaction with CH_2O and elimination of water to branched and finally cross-linked structures such as the ones shown for urea resins (melamine reactions are similar):

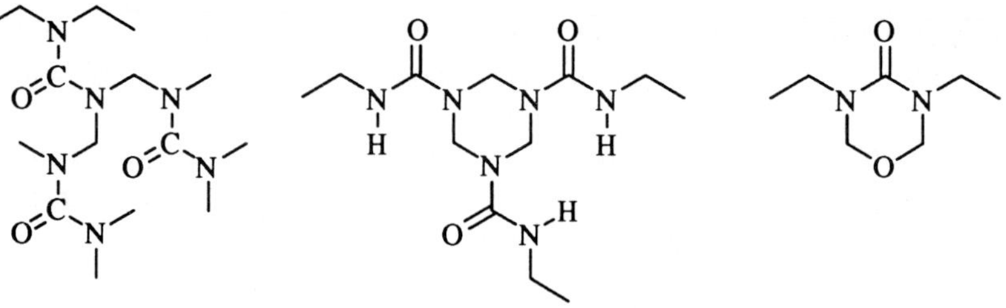

Cured amino resins are colorless products. Urea-formaldehyde plastics UF are used mainly as particle-board binders and as foams for damming purposes and to a smaller extent as coatings and in paper and textile treatment. Ca. 10 % serve as molding compounds. The more expensive melamine-formaldehyde resins are laminating resins (Formica®) or are used in mechanical parts.

12.6.3. Alkyd Resins

Alkyd resins are condensation products of multifunctional alcohols (glycerol, trimethylolpropane, pentaerythrol, etc.) with diacids (phthalic acid, succinic acid, maleic acid, etc.), or acid anhydrides, usually also as "modified" resins with fatty acids as comonomers. The name indicates their origin from *al*cohols and a*cids*. Alkyd resins from glycerol and phthalic acid (anhydride) are called glyptal resins. The polycondensation is conducted short of the gelation point (onset of chemical cross-linking); the reaction products are applied as paint resin and subsequently hardened.

12.6.4. Unsaturated Polyesters

Unsaturated polyesters UP are chemically polyesters $\sim$[O-Z-O-OC-CH=CH-CO]$_n$$\sim$ of maleic anhydride with diols; part of maleic anhydride may be replaced by other diacids such as phthalic acid (anhydride), isophthalic acid, terephthalic acid, or hexachloroendomethylene tetrahydrophthalic acid (HET acid). Diols may be ethylene glycol, 1,2-propylene glycol, neopentyl glycol, or oxethylated bisphenols. During polycondensation, most of the *cis*-carbon double bonds of maleic acid are transformed into *trans*-carbon double bonds (fumaric acid units). Up to 15 % of the carbon double bonds form ether groups by adding ethylene glycol.

The syrupy condensation products are "cured" (cross-linked) by free radical copolymerization with monomers such as styrene (less often: methyl methacrylate, α-methyl styrene, vinyl toluene, diallyl phthalates, etc.). The UP-monomer mixture is technically also called "unsaturated polyester". Industry refers to the added monomers as cross-linkers; the true cross-linking agents are the unsaturated polyester molecules.

UP resins are usually reinforced with short glass fibers or are used in sheet molding compounds or bulk molding compounds. Upon cross-linking, the resin shrinks unevenly around the immobile glass-fiber bundles of SMCs and BMCs, causing depressions and surface roughness. Surfaces become smoother if (30-40) wt-% of an incompatible thermoplastic (*low-profile polymer*) is added, for example, poly(vinyl acetate), poly(ε-caprolactone), or poly(styrene). The mixture phase-separates during cross-linking. The heat of reaction raises the temperature and the polymer molecules in the domains of the low-profile polymer expand. The expanded volume is retained upon ccoling; no hollows form at the surface. The shrinkage is reduced to 0.04 % from 0.4 %.

UPs tolerate erroneous mixing of components. The low-price mixtures have low viscosities and low curing times. Articles can be rapidly unmolded. Unreinforced UPs do have high shrinkage and low service temperatures, however.

12.6.5. Polyallyls

Polyallyls are free-radically polymerized diallyl and triallyl monomers such as diallyl diglycol carbonate $(CH_2{=}CH{-}CH_2{-}O{-}CO{-}O{-}CH_2{-}CH_2)_2O$ (DADC), diallyl phthalate DAP, and triallylcyanurate TAC:

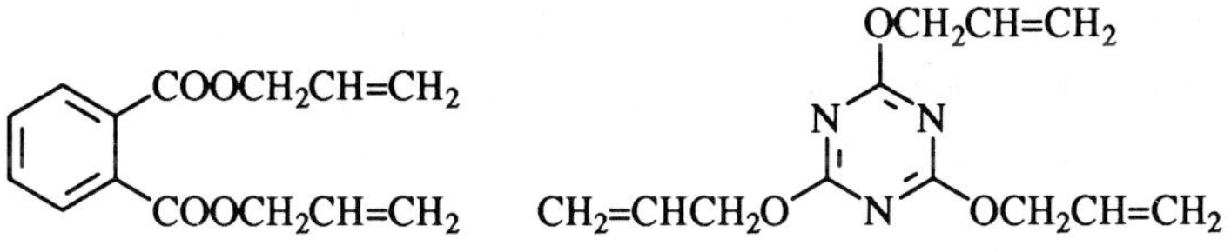

$COOCH_2CH{=}CH_2$ $COOCH_2CH{=}CH_2$ $OCH_2CH{=}CH_2$ N N $CH_2{=}CHCH_2O$ N $OCH_2CH{=}CH_2$

DAP TAC

Polymers of diallyl and triallyl esters carry ester groups as substituents; such polymers are sometimes called polyesters (which they are not). Monoallyl compounds do not lead to high molar mass polymers because of a strong chain termination by radical transfer to the monomer, which delivers a resonance-stabilized monomer radical.

DADC has a similar light transmission as PMMA but is much more scratch resistant (sun glasses). DAP is used for electrical insulators and for prepregs.

12.6.6. Epoxy Resins

Epoxy resins (epoxies; EP) are oligomeric compounds with oxirane groups (epoxide groups) (cycl-C_2H_3O)-CH_2~. Reaction products of bisphenol A and epichlorohydrin with the idealized structure comprise more than 90 % of the world production:

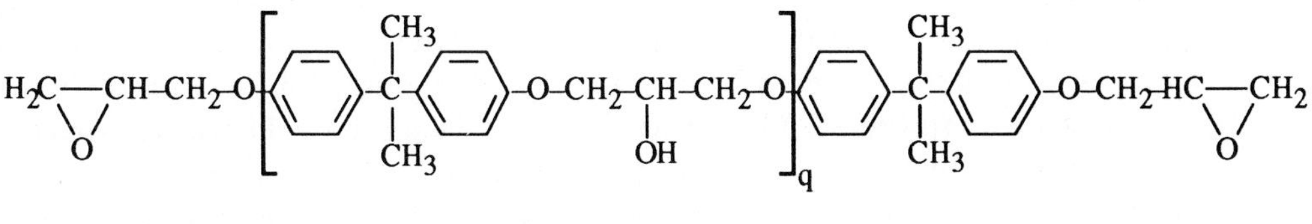

$H_2C{-}CH{-}CH_2{-}O$ O [CH_3 C CH_3 $O{-}CH_2{-}CH{-}CH_2{-}O$ OH $]_q$ CH_3 C CH_3 $O{-}CH_2{-}HC{-}CH_2$ O

Epoxies with this standard structure and $0.1 < q < 0.6$ are liquid; those with $2 < q < 25$, solid. Special epoxies contain cycloaliphatic or heterocyclic residues (higher heat stabilities). Commercial epoxies are always formulated with plasticizers, fillers, pigments, etc.

Epoxies are hardened (cured) by cross-linking reactions of their oxirane and hydroxyl groups with acid anhydrides (warm curing) or polyfunctional amines (cold curing). Warm curing leads to ester structures from oxirane and hydroxyl groups and to ether structures from oxirane groups. Cold curing generates β-hydroxypropylamine structures from oxirane groups, for example, R'R"N–CH_2–CHOH–CH_2~ from R'R"NH and (cycl-C_2H_3O)–CH_2~ and correspondingly from diamines, triamines, etc. Curing is not complete (slow diffusion of chain segments in highly viscous environment), leading to a time-dependent after-cure.

The many types of resins and hardeners and the tolerance of high filler levels allow taylor-made applications of EPs. Other advantages are cures by air-drying, low shrinkages, and high service temperatures. In contrast to unsaturated polyesters, EPs are expensive, need careful metering of resin-hardener mixtures, have high viscosities (difficult degassing of resin and impregnation of fabrics, etc.), require long curing times, and are difficult to unmold.

Epoxy resins are used as two-component adhesives, for protective coatings, in the electrical industry, and, reinforced with glass fibers, for large containers.

12.6.7. Polyurethanes

Polyurethanes PUR with the characteristic urethane group –NH–CO–O– are almost exclusively synthesized by polyaddition of isocyanate group –N=C=O containing compounds (di and tri) to hydroxyl group HO- containing compounds (di and poly). More than 95 % of all PUR are based on toluene diisocyanate TDI (an 80/20 mixture of 2,4- and 2,6-isomers) and diphenylmethane diisocyanate MDI and its derivatives:

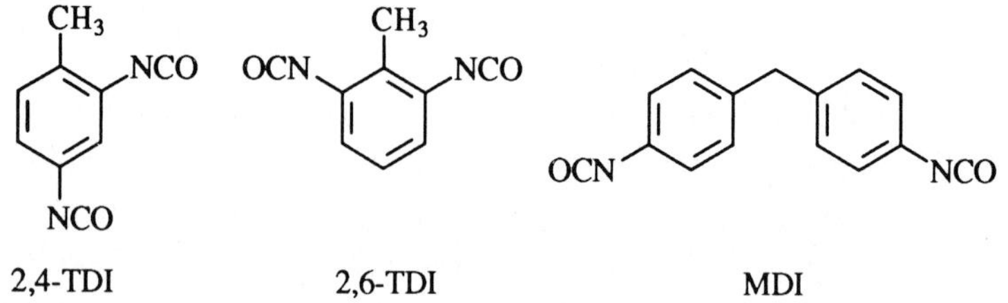

CH_3, NCO, NCO — 2,4-TDI

CH_3, OCN, NCO — 2,6-TDI

OCN, NCO — MDI

The functionality of low molar mass hydroxy compounds ranges from 3 (trimethylol propane) to 8 (saccharose), that of polyols from 2 (poly(ethylene glycol)) to very large values (modified polyether-polyols). More than 90 % of the polyols are polyether-polyols (polymers from ethylene oxide, propylene oxide, or tetrahydrofuran with hydroxyl end groups). Modified polyether-polyols are obtained by grafting of hydroxyl group-containing compounds on polymer particles (SAN, polyurea,

PUR). Polyurethanes with isocyanate end groups can be further reacted with chain extenders (low molar mass diols and diamines).

The multitude of raw materials and the resulting different properties have led to a world production of more than 5 million tons PUR per year. Polyurethanes are used as flexible foams (ca. 45 %; furniture, car seats, mattresses, packaging), rigid foams (ca. 30 %; thermal insulation), coatings, adhesives, elastomers, elastic fibers, and molding materials.

Selected PUR raw materials with carefully selected catalysts and activators can be *reaction injection molded* (RIM) to high modulus polyurethanes (Section 11.3.3.). RIM has also been developed for ε-caprolactam to PA 6, dicyclopentadiene to PDCPD, and high-modulus epoxies EP (Table 12-17).

Table 12-17 Properties of RIM plastics (data of [2]).

Property	Unit	PUR	PA 6	PDCPD	EP
Density	g/cm³	1.00	1.14	1.03	
Heat distortion temperature (0.46 MPa)	°C	80	175	115	118
Flexural modulus	MPa	830	1400	2000	2900
Tensile strength at break	MPa	28	44	46	69
Notched impact strength	J/m	240	300	430	27

12.6.8. Silicone Resins

Silicones are based on poly(dimethylsiloxane)s $\sim[O–Si(CH_3)_2]_n\sim$ and their derivatives. They are mostly used as elastomers, oils, or surfactants. A small amount is prepared as silicone resins, which are highly cross-linked siloxane systems. The cross-linking function is provided by adding trifunctional silanes such as CH_3SiCl_3 or $C_6H_5SiCl_3$ and/or the tetrafunctional silanetetrachloride $SiCl_4$ to a solution of bifunctional silanes such as $(CH_3)_2SiCl_2$or $CH_3(C_6H_5)SiCl_2$. The mixture is hydrolyzed and the liberated HCl removed. Addition of mild bases results in polycondensation and ring-opening polymerization of primarily formed linear and cyclic siloxanes such as $HO\text{-}Si(CH_3)_2\text{-}[O\text{-}Si(CH_3)_2]_i OH$ + *cyclic*-$[O\text{-}Si(CH_3)_2]_4$. The viscosity of the mixture is adjusted by the addition or removal of solvents.

The properties of the cured resins depend on the type and proportion of the cross-linking agents, the degree of cure, and the processing conditions. Greater proportions of multifunctional silanes lead to polymers with greater hardness, brittleness, stiffness, and toughness that cure faster; the tack decreases, however.

12.7. Biopolymer-Based Plastics

Plastics based on biopolymers have recently received renewed interest because of their availability from renewable resources and/or their biodegradability.

12.7.1. Cellulosics

Pure cellulose is produced by bacteria. Celluloses of plants are always accompanied by polyoses, pectin, fats and waxes, and in wood also by lignin (Section 3.1.1.). Cellulose is not used directly as plastics because of its high glass temperature (estimated as 220°C), an even higher melting temperature, and the onset of degradation above these temperatures. Cellulose derivates were however the first plastics.

Pure cellulose is a poly[β-(1→4)-anhydro-D-glucopyranose], an unbranched (linear) polysaccharide with glucose units as mers; most plant celluloses contain small proportions of other sugar units. The glucose units are interconnected in 1,4-position of the carbons; the constitutional repating units are β-cellobiose units:

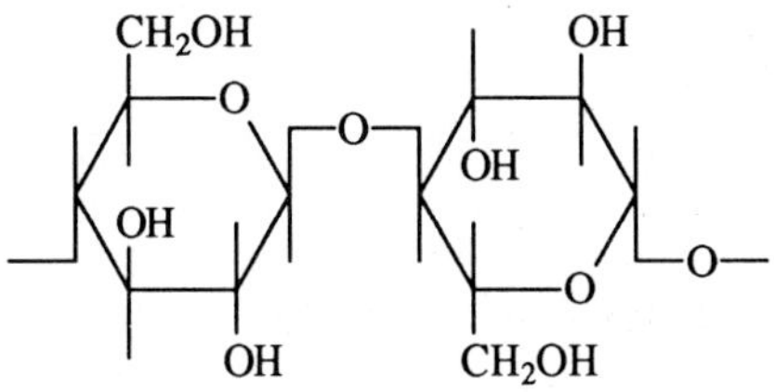

Constitutional repeating unit of cellulose (cellobiose unit)

The hydroxyl groups of cellulose can be transformed into commercially useful cellulose derivatives by replacing HO- with other substituents RO-:

Cellulose triacetate CTA with three acetate groups CH_3COO- per glucose unit is the primary product of the esterification of cellulose by acetic acid (primary acetate). It is used as triacetate fiber.

Cellulose 2 1/2-acetate CA with an average of 2/1 acetate groups per glucose residue is obtained by partial saponification of cellulose triacetate; it is thus called secondary acetate. This "cellulose acetate" of commerce is now mainly used for cigarette filters (tow); other applications include fibers, extruded tape, blister packs, signs, etc.

Cellulose acetobutyrate CAB (cellulose acetate butyrate) with 29-6 mol-% acetate CH_3COO– and 17-48 mol-% butyrate groups C_3H_7COO– is a thermoplastic for sheeting and molding (packaging, tubes, etc.).

Cellulose acetate propionate CAP with acetate groups CH_3COO– and propionate groups C_2H_5COO– is also a molding compound.

Cellulose nitrate CN ("nitrocellulose") with various degrees of substitution DS is used as gun cotton (DS = 2.7-2.9), films (DS = 2.5-2.6), lacquers (DS = 2.25-2.6), and, plasticized with camphor, as Celluloid® (DS = 2.25-2.4).

Very many cellulose ethers are produced. Most of them are used in water or organic solvents as thickeners, suspending aids, protective colloids, etc.; a smaller amount also for sheeting and molding. Cellulose ethers include:

Methyl cellulose MC (CH_3O- as substituent; various degrees of substitution DS);
Hydroxyalkylmethylcelluloses HPMC (MCs modified with ethylene, propylene, and butylene oxides; various degrees of substitution);
Ethylcellulose EC (C_2H_5O- as substituent; DS of thermoplastic grades: 2.3-2.5).
Ethylhydroxyethylcellulose EHEC (C_2H_5O- and $HO(CH_2CH_2O)_i$- as substituents);
Hydroxyethylcellulose HEC [$HO(CH_2CH_2O)_i$- as substituent];
Hydroxypropylcellulose HPC (CH_3–CHOH–CH_2–O- as substituent; OH may be further substituted by hydroxypropyl groups; DS unknown; degree of reaction usually 4);
Sodium carboxymethylcellulose CMC with the substituent NaOOC-CH_2-O- is an anionic polyelectrolyte. The food grade is known as cellulose gum.

Tab. 12-18 Properties of cellulose plastics (molding compounds); data from various sources.

Physical property Name	Phys. Unit	CN	Polymer types and grades CA	CAP	CAB	EC
Density	g/cm^3	1.35-1.40	1.22-1.34	1.17-1.24	1.15-1.22	1.14
Heat distortion temp. A	°C	60-71	44-118	44-110	45-94	
Thermal exp.coeff. ($\cdot 10^5$)	K^{-1}	8-12	8-18	11-17	11-17	
Young's modulus	GPa	1.3-1.8	0.4-2.7	0.4-1.5	0.3-1.4	
Flexural modulus	GPa		0.62-1.8	0.70-1.9	0.48-1.4	
Stress at yield	MPa		14-48	10-48	10-48	
Tensile strength	MPa	40-55	12-62	14-54	18-52	46-72
Flexural strength	MPa	62-76	14-110	20-79	12-74	
Strain at break	%	40-50	6-70	29-100	44-88	7-30
Notched impact strength	J/m	270-370	7-280	10-610	13-340	
Relative permittivity (50 Hz)	1	7.0-7.5				2.5-4.0
Volume resistivity	Ω cm	10^{11}-10^{12}	10^{10}-10^{12}	10^{12}-10^{15}	10^{10}-10^{12}	10^{12}-10^{14}
Refractive index	1	1.49-1.51	1.46-1.50	1.46-1.49	1.46-1.49	1.47

12.7.2. Starch-Based Plastics

Starch is a mixture of the polysaccharides amylose and amylopectin that is produced by plants. These polysaccharides are poly[α(1→4)glucose]s = poly[α(1→4)-anhydro-D-glucopyranose]s. Amylose is practically linear, whereas amylopectin is highly branched with approximately 1 branching point (1→6) per 18-27 glucose units.

The ratio amylose/amylopectin varies from plant to plant: corn starch contains ca. 26 % amylose and potato starch ca. 22 %, whereas hybrid amylocorn starch has 85 %

amylose and waxy rice starch none at all. Starch is mostly used as food; smaller amounts are converted into derivatives for sizes, etc., or serve as polyols in PUR foams.

Thermoplastic starch is obtained by kneading starch (presumbly potato starch) with 5-14 % water or plasticizers such as glycerol. The partially crystalline material resembles plasticized PVC. It shows rubber elasticity at temperatures between -45°C and + 65°C and glass temperatures between -60°C and 150°C, depending on the solvent content. It can be injection molded at 170°C to specimen with ultimate elongations of ca. 25 % and tensile strengths at break that are greater than those of poly-(styrene). Articles are biodegradable. Biodegradability is also claimed for blends of other plastics with 60-70 % starch.

12.7.3. Poly(hydroxybutyrate)

Poly(β-D-hydroxybutyrate) PHB $\sim[O\text{-}CH(CH_3)\text{-}CH_2\text{-}CO]_n\sim$ is stored as hydrophobic granules in the cytoplasma of certain bacteria where is serves as a reserve food. It can also be produced by bioengineered plants. Industry obtains PHB from the polymerization of glucose by the bacterium Alcaligenes eutrophus. The polymer has similar properties as isotactic poly(propylene); it is however biodegradable. Addition of propionic acid or valeric acid to the glucose delivers copolymers; the industrial product has 80 % β-hydroxybutyrate and 20 % β-hydroxyvalerate units.

Literature

12.1.A. General

J.A.Brydson, Plastic Materials, Iliffe, London, 2nd ed. 1969
W.J.Roff, J.R.Scott, Handbook of Common Polymers, CRC Press, Cleveland (OH) 1971
H.-G.Elias, New Commercial Polymers 1969-1975, Gordon and Breach, New York 1977
D.C.Miles, J.H.Briston, Polymer Technology, Chem.Publ.Co., New York 1979
C.Hall, Polymer Materials, Wiley, New York 1981
H.Saechtling, International Plastics Handbook, Hanser, Munich 1983
W.W.Flick, Industrial Synthetic Resins Handbook, Noyes Data, Park Ridge (NJ) 1985
J.M.Margolis, ed., Engineering Thermoplastics, Dekker, New York 1985
H.-G.Elias, F.Vohwinkel, New Commercial Polymers 2, Gordon and Breach, New York 1986
J.B.Rose, High Performance Polymers. Their Origin and Development, Elsevier, Amsterdam 1986
K.J.Saunders, Organic Polymer Chemistry, Chapman and Hall, New York, 2nd ed. 1988
J.A.Brydson, Plastics Materials, Butterworths, Stoneham (MA), 5th ed. 1989
J.H.Bittence, ed., Engineering Plastics and Composites, ASM International, Materials Park (OH) 1990 (trade names, manufacturers, applications)
M.Ash, I.Ash, Handbook of Plastic Compounds, Elastomers and Resins, VCH, New York 1991 (ca. 15 000 commercial products)
H.Domininghaus, Plastics for Engineers, Hanser, Munich 1993

12.1.B. Property Predictions

D.W. van Krevelen, Properties of Polymers - Correlation with Chemical Structure, Elsevier, Amsterdam, 3rd edition 1990
J.Bicerano, Prediction of Polymer Properties, Dekker, New York 1993

12.2.1. Olefin Polymers

H.P.Frank, Polypropylene, Gordon and Breach, New York 1968
I.D.Rubin, Poly(1-butene), Gordon and Breach, New York 1968
S. van der Ven, Polypropylene and Other Polyolefins. Polymerization and Characterization, Elsevier, Amsterdam 1990

12.2.2. Styrenics

R.H.Boundy, R.F.Boyer, eds., Styrene. Its Polymers, Copolymers, and Derivatives, Reinhold, New York 1952
C.A.Brighton, G.Pritchard, G.A.Skinner, Styrene Polymers: Technology and Environmental Aspects, Appl.Sci., Barking (Essex) 1979

12.2.3. Vinyls

W.S.Penn, PVC Technology, MacLaren, London, 3rd ed. 1972
R.A.Wessling, Polyvinylidene Chloride, Gordon and Breach, New York 1975
G.Butters, ed. Particulate Nature of PVC: Formation, Structure, and Processing, Appl.Sci.Publ., Barking, Essex 1982
R.H.Burgess, ed., Manufacture and Processing of PVC, Hanser International, Munich 1981; MacMillan, New York 1982
W.V.Titow, ed., PVC Technology, Elsevier Appl.Sci., New York, 4th ed. 1984
E.D.Owen, ed., Degradation and Stabilisation of PVC, Elsevier Appl.Sci., New York 1984
J.Wypych, Polyvinyl Chloride Stabilization, Elsevier, Amsterdam 1986
L.I.Nass, C.A.Heiberger, eds., Encyclopedia of PVC, Dekker, New York, 2nd ed. 1988 (4 vols.)
W.V.Titow, PVC Plastics. Properties, Processing, and Applications, Elsevier, London 1990
J.T.Lutz, Jr., D.L.Dunkelberger, Impact Modifiers for PVC, Wiley, New York 1992
E.J.Wickson, Handbook of PVC Formulation, Wiley, New York 1993

12.2.4. Acrylics

M.B.Horn, Acrylic Resins, Reinhold, New York 1960

12.2.5. Fluoroplastics

L.A.Wall, ed., Fluoropolymers, Wiley, New York 1972
M.G.Broadhurst, F.Micheron, Y.Wade, PVF_2 (Poly(Vinylidene Fluoride)), Gordon and Breach, New York 1981

12.3.1. Thermoplastic Polyesters

I.Goodman, J.A.Rhys, Polyesters, Vol.1, Saturated Polyesters, Illiffe, London 1965
R.Burns, Polyester Molding Compounds, Dekker, New York 1982
R.Meyer, Polyester Molding Compounds and Molding Technology, Chapman and Hall, New York 1987

12.3.2. Polycarbonates

W.F.Christopher, Polycarbonates, Reinhold, New York 1962
H.Schnell, Chemistry and Physics of Polycarbonates, Interscience, New York 1964

12.3.3. LC Polyesters

G.W.Gray, Ed., Thermotropic Liquid Crystals, Wiley, New York 1987
A.Ciferro, Ed., Liquid Crystallinity in Polymers, VCH, New York 1991

12.3.4. Acetal Polymers

O.Vogl, Polyaldehydes, Dekker, New York 1967

12.3.5. Polyethers

F.E.Bailey, Jr., J.V.Koleske, Poly(ethylene oxide), Academic Press, New York 1976
P.Dreyfuss, Poly(tetrahydrofuran), Gordon and Breach, New York 1981
F.E.Bailey, Jr., J.V.Koleske, Alkylene Oxides and Their Polymers, Dekker, New York 1991

12.5.1. Polyamides

M.Kohan, Ed., Nylon Plastics, Wiley, New York 1973
W.E.Nelson, Nylon Plastics Technology, Newnes-Butterworth, London 1976
H.K.Reimschuessel, Nylon 6, Chemistry and Mechanisms, Macromol.Revs. **12** (1977) 65
R.S.Lenk, Post-Nylon Polyamides, Macromol.Revs. **13** (1978) 355
E.H.Pryde, Unsaturated Polyamides, J.Macromol.Sci.-Revs.Macromol.Chem. **C 17** (1979) 1
R.Puffr, V.Kubánek, Lactam-Based Polyamides, CRC Press, Boca Raton, 1990, 1991 (2 vols.)

12.5.2. Polyimides

M.W.Ranney, Polyimide Manufacture, Noyes Data Corp., Park Ridge (NJ) 1971
C.E.Sroog, Polyimides, Macromol.Revs. **11** (1976) 161
K.L.Mittel, Ed., Polyimides, Plenum, New York 1984 (2 vols.)
M.I.Bessonov, M.M.Koton, V.V.Kudryavtsev, L.A.Laius, Polyimides - Thermally Stable Polymers, Consultants Bureau, New York 1987
C.Feger, M.M.Khojastel, J.E.McGrath, eds., Polyimides: Materials, Chemistry, and Characterization, Elsevier, Amsterdam 1990

D.Wilson, H.D.Stenzenberger, P.M.Hergenrother, eds., Polyimides, Routledge, Chapman and Hall, New York 1990
M.I.Bessonov, V.A.Zubkov, Polyamic Acids and Polyimides, CRC Press, Boca Raton (FL) 1993

12.6. Thermosets (PUR, EP, UP, PF, RIM, special monomers, additives)

W.F.Gum, W.Riese, H.Ulrich, eds., Reaction Polymers, Hanser, Munich 1992

12.6.1. Phenolics

T.S.Carswell, Phenoplasts, Interscience, New York 1947
R.W.Martin, The Chemistry of Phenolic Resins, Wiley, New York 1956
N.J.L.Megson, Phenolic Resin Chemistry, Butterworths, London 1958
D.F.Gould, Phenolic Resins, Reinhold, New York 1959
A.A.K.Whitehouse, E.G.K.Pritchett, G.Barnet, Phenolic Resins, Iliffe, London 1967
A.Knop, W.Scheib, Chemistry and Applications of Phenolic Resins, Springer, New York 1979
A.Knop, L.A.Pilato, Phenolic Resins, Springer, Berlin 1985

12.6.2. Amino Resins

J.F.Blais, Amino Resins, Reinhold, New York 1959
C.P.Vale, W.H.G.K.Taylor, Aminoplastics, Iliffe, London 1964
B.Meyer, Urea-Formaldehyde Resins, Addison-Wesley, Reading (MA) 1979

12.6.3. Alkyd Resins

K.Holmberg, High Solids Alkyd Resins, Dekker, New York 1987

12.6.4. Unsaturated Polyesters

H.B.Boenig, Unsaturated Polyesters, Elsevier, Amsterdam 1964
B.Parkyn, F.Lamb, B.V.Clifton, Polyesters, Vol. II, Unsaturated Polyesters, Iliffe, London 1967
P.F.Bruins, Unsaturated Polyester Technology, Gordon and Breach, New York 1976

12.6.5. Polyallyls

H.Raech, Allylic Resins and Monomers, Reinhold, New York 1965
H.Schildknecht, Allyl Compounds and Their Polymers, Wiley-Interscience, New York 1973

12.6.6. Epoxy Resins

I.Skeist, Epoxy Resins, Reinhold, New York 1958
H.Lee, K.Neville, Handbook of Epoxy Resins, McGraw-Hill, New York 1967
P.F.Bruins, Ed., Epoxy Resins Technology, Interscience, New York 1968

W.G.Potter, Epoxide Resins, Iliffe, London 1970
J.I.DiStasio, ed., Epoxy Resin Technology, Noyes, Park Ridge (NJ) 1982
C.A.May, Ed., Epoxy Resins. Chemistry and Technology, Dekker, New York, 2nd ed. 1988
B.Ellis, ed., Chemistry and Technology of Epoxy Resins, Blackie and Son, Glasgow 1993
E.W.Flick, Epoxy Resins, Curing Agents, Compounds, and Modifiers. An Industrial Guide, Noyes Data, Park Ridge (NJ), 2nd ed. 1993 (products of 71 manufacturers)

12.6.7. Polyurethanes

J.H.Saunders, K.C.Frisch, Polyurethanes, Chemistry and Technology, Interscience, New York, 2 vols. 1961 and 1962
B.A.Dombrow, Polyurethanes, Reinhold, New York, 2nd ed. 1965
J.M.Buist, H.Gudgeon, eds., Advances in Polyurethane Technology, Maclaren, London 1968
P.Wright, A.P.C.Cummings, Solid Polyurethane Elastomers, Maclaren, London 1969
E.N.Doyle, The Development and Use of Polyurethane Products, McGraw-Hill, New York, 2nd ed. 1971
K.C.Frisch, S.L.Reegen, eds., Advances in Urethane Science and Technology, Technomic, Westport, CN, 7 vols. 1971-1979
Z.W.Wicks, jr., New Developments in the Field of Blocked Isocyanates, Progr.Org.Coatings **9** (1981) 3
D.Dieterich, Aqueous Emulsions, Dispersions and Solutions of Polyurethanes; Synthesis and Properties, Progr.Org.Coatings **9** (1981) 281
G.Woods, Flexible Polyurethane Foams, Chemistry and Technology, Applied Science Publ., London 1982
C.Hepburn, Polyurethane Elastomers, Appl.Sci.Publ., Barking, Essex 1982
G.Oertel, ed., Polyurethane Handbook, Hanser, Munich 1985
G.Woods, The ICI Polyurethanes Book, Wiley, New York 1987

12.6.8. Silicone Resins

W.Noll, Chemistry and Technology of Silicones, Academic Press, New York 1968
G.Koerner, M.Schulze, J.Weis, Silicones: Chemistry and Technology, CRC Press, Boca Raton (FL) 1992

12.7.1. Cellulosics

F.D.Miles, Cellulose Nitrate, Interscience, London 1953
E.Ott, H.M.Spurlin, eds., Cellulose and Cellulose Derivatives, Interscience, New York, 2nd ed., vols 1-3 (1956); N.M.Bikales, L.Segal, Eds., ditto, vols. 4-5 (1971)
W.D.Paist, Cellulosics, Reinhold, New York 1958
V.E.Yarsley, W.Flavell, P.S.Adamson, N.G.Perkins, Cellulosic Plastics, Iliffe, London 1964
R.M.Rowell, R.A.Young, Eds., Modified Cellulosics, Academic Press, New York 1978
A.Hebeish, J.T.Guthrie, The Chemistry and Technology of Cellulosic Copolymers, Springer, Berlin 1981 (graft copolymers)
T.P.Nevell, S.H.Zeronian, eds., Cellulose Chemistry and Its Applications, Wiley-Horwood, Chichester 1985
J.F.Kennedy, G.O.Phillips, D.J.Wedlock, P.A.Williams, eds., Cellulose and Its Derivatives - Chemistry, Biochemistry and Its Applications, Wiley, New York 1985

R.A.Young, R.M.Rowell, Cellulose. Structure, Modification and Hydrolysis, Wiley, New York 1986
J.F.Kennedy, G.O.Phillips, P.A.Williams, Cellulose: Structural and Functional Aspects, Prentice Hall, Englewood Cliffs (NJ) 1989
L.V.Backinowski, M.A.Chlenov, Cellulose. Biosynthesis and Structure, Springer, Berlin 1990
D.N.-S.Hon, S.Shiraishi, Eds., Wood and Cellulosic Chemistry, Dekker, New York 1991

12.7.2. Starch-Based Polymers

J.A.Radley, Starch and Its Derivatives, Chapman and Hall, London, 4th ed. 1968
W.Banks, C.T.Greenwood, Starch and Its Components, Edinburgh University Press, Edinburgh 1975
J.A.Radley, Industrial Starch Technology, Noyes, Park Ridge (NJ) 1979
J.C.Johnson, Industrial Starch Technology, Noyes, Park Ridge (NJ) 1979
R.L.Whistler, J.N.BeMiller, E.F.Paschall, Starch: Chemistry and Technology, Academic Press, New York, 2nd ed. 1984

12.7.3. Poly(hydroxybutyrate)

Y.Doi, Microbial Polyesters, VCH, New York 1990
E.A.Dawes, Ed., Novel Biodegradable Microbial Polymers, Kluwer, Dordrecht 1990

References

[1] H.Tadokoro, Y.Chatani, M.Kobayashi, T.Yoshihara, S.Murahashi, K.Imada, Rep.Progr. Polym.Phys.Japan **6** (1963) 503, Fig. 1
[2] H.L.Mei, ACS Polymer Preprints **32**/3 (191) 353

13. Modified Plastics

13.1. Introduction

13.1.1. Overview

Physical structures and properties of plastics can be modified by the addition of fibers, particulates, rubbers, plasticizers, and gases. No accepted general term exists for this group of admixtures; they will be therefore called modified plastics.

"Modified plastic" fits the dictionary description of a *composite* as "a complex material ... in which two or more distinct, structurally complementary substances ... combine to produce some structural or functional properties not present in any individual component" (The American Heritage Dictionary). Examples are foamed plastics, fiber glass (glass fibers in a thermoset), and wood (a water-plasticized, air-foamed composite of cellulose fibers in a cross-linked lignin matrix). The dictionary definition is not restricted to admixtures: in biomaterials, "composite" refers accordingly to any two joined materials (e.g., polymer-coated titanium parts).

In engineering, composites are more narrowly defined as physical admixtures of various materials that are present as distinct phases; usually a continuous phase ("matrix") and a discontinuous phase. The term "phase" is used here in the descriptive and not in the thermodynamic sense. The engineering term "composite" thus includes only heterogeneous composites. Heterocomposites can be further subdivided into microcomposites (dispersed phase in the micrometer to millimeter range) and macrocomposites (dispersed phase greater than millimeters). Typical nonpolymer heteromicrocomposites are metal alloys; a typical hetero-macrocomposite is steel-reinforced concrete. In polymer engineering, the term "composite" is usually even more restricted to include only fiber-filled materials but not particulate-filled ones.

Polymer composites are defined as composites in which at least the matrix is of a polymeric nature. In contrast to the engineering use of the word, the term "polymer composite" is occasionally used to cover not only heterogeneous mixtures of a resinous matrix and another material (particulates, fibers, other plastics, elastomers, etc.), but also homogeneous (single-phase) materials of two resins (homogeneous polymer blends). Such polymer composites are in turn a subgroup of "multicomponent polymer systems". The latter term includes copolymers (two or more types of monomeric units chemically bound together).

In this book, the term "polymer composite" is used in its broadest meaning; it thus includes homogeneous and heterogeneous admixtures of polymers with various other materials. Polymer composites occur in nature (wood, bamboo, lobster shells, bone, muscle tissue, etc.) or are manufactured synthetically. Synthetic polymer composites can be subdivided into single-phase (homogeneous) and multiphase (heterogeneous) composites. Further subdivisions may be according to the size of the second component (microcomposites, macrocomposites), chemical nature (air, low molar mass plasticizer, elastomer, plastic, mineral), geometry (particulate, fiber, platelet, fabric), and macroconformation (if polymer chains: coil, rod), etc. Most of these subclasses have different technical names depending on whether the glass temperature of the matrix is above (plastics) or below (rubbers) the use temperature. The same name is often given to various subtypes.

A mixture of two (or more) different polymers is often called a *blend* if both polymers are either above (rubbers) or below (plastics) their glass temperatures. Blends are distinguished from composites of two polymers such as fiber-reinforced or rubber-toughened plastics. The term "blend" is sometimes restricted to mean only "incompatible polymer mixtures", whereas compatible polymer mixtures are called *polymer alloys*. The latter term is, however, occasionally more narrowly used for mixtures of two crystallizable polymers.

The term *miscibility* refers to admixtures on the molecular level (nanometer range); that is, true thermodynamic solubility. It is sometimes used more loosely for any polymer mixture that exhibits only one glass temperature. *Compatibility* denotes the ability of admixtures to be blended to heterogeneous microcomposites that do not separate into macroscopic phases under experimental conditions.

13.1.2. Mixture Rules

The composition dependence of properties Q of admixtures can often be described by mixture rules. The *generalized simple mixing law* relates Q to the fractions $f_A \equiv 1 - f_B$ of the components A and B with properties Q_A and Q_B, respectively:

$$(13\text{-}1) \quad Q^n = Q_A^n f_A + Q_B^n f_B = Q_A^n - (Q_B^n - Q_A^n)\phi_B$$

The generalized simple mixing law is usually applicable to single-phase admixtures without specific interactions of its components and to two-phase systems with regularly dispersed discrete phases of "infinite" dimensions. It includes three special cases that are known by different names in different fields (Table 13-1); these cases are illustrated in Figure 13-1.

Table 13-1 Names of simple mixing rules.

Field	Property exponent in generalized mixing law		
	$n = 1$	($n = 0$)	$n = -1$
Mathematics	Arithmetic mean	Harmonic mean	Geometric mean
Chemical Engineering	Rule of mixtures	Logarithmic mixture rule	Inverse rule of mixtures
Mechanical Engineering	Voigt model	-	Reuss model
Electrical Engineering	Parallel	-	Series
Materials Science	Upper bound	-	Lower bound

The *rule of mixtures* ($n = 1$) applies, for example, to various moduli if both components behave elastically. An example is Young's modulus parallel to the fiber direction for cured unsaturated polyester resins reinforced with long glass fibers (see below). The same composites follow the *inverse rule of mixtures* ($n = -1$) if the modulus is measured perpendicular to the fiber direction and no slippage occurs.

The *logarithmic mixture rule*

$$\text{(13-2)} \quad \lg Q = f_A \log Q_A + f_B \log Q_B$$

is obtained from Equation (13-1) for $n = 0$ after same mathematical manipulation. The only known case concerns the permeation of oxygen through ABS resins.

The proper fraction f_i in Equations (13-1) and (13-2) is always the volume fraction ϕ_i. If weight fractions w_i are used instead, curves depart from the lines in Figure 13-1, which may be confused with deviations from additivity or specific interactions.

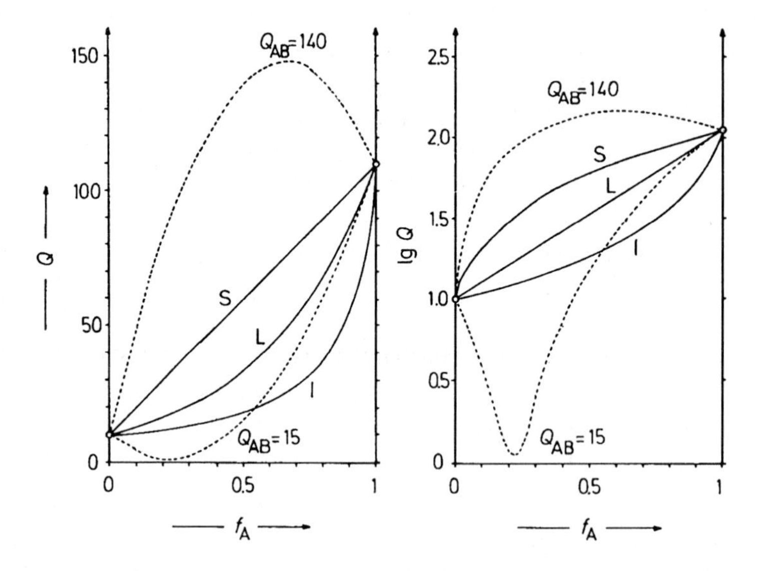

Fig. 13-1 Simple mixing law S, logarithmic mixing law L, and inverse mixing law I for composites. Left: $Q = f(f_A)$; right: $\lg Q = f(f_A)$.
Solid lines: $Q_A = 110$, $Q_B = 10$, $Q_{AB} = 0$; calculated with Eq.(13-1).
Broken lines: $Q_A = 110$, $Q_B = 10$, $Q_{AB} = 140$ or $Q_{AB} = 15$; calculated with Eq. (13-6).

A composite prepared from two components A and B may actually consist of more than two "phases". Calculated densities $\rho_{calc} = [(w_A/\rho_A) + (w_B/\rho_B)]^{-1}$ may thus differ from experimental densities ρ. They are greater if air bubbles are trapped in compounded thermoplastics or if voids were formed by water vapor during hardening of certain thermosetting resins ($\rho_{calc} > \rho$). Calculated densities can also be smaller than the experimental ones ($\rho_{calc} < \rho$). Examples are conformational changes of the coils of polymer chains of the matrix or their transcrystallization near the surface of the modifier (fiber, particulate). Such interlayers are often named "interphases" even if they are not large enough in order to fulfill the thermodynamic conditions for the existence of phases; they are better called interlayers.

True deviations from the upper bound and lower bound (Fig. 13-1) may be caused by irregularities in phase distributions (discrete phases) or phase continuities (continuous phases), phase alignments (anisotropic phases), bonding (adhesion), existence of interlayers, packing problems of dispersed particulates or fibers, and/or non-uniform stresses. The resulting functions always lie between the upper and the lower bound if specific interactions between filler and matrix are absent. Many empirical, semiempirical, and theoretical expressions have been proposed for the dependence of properties on composition of such composites.

13.1.3. Models

Composite and blend properties are often modeled by the *Halpin-Tsai equation* for a material from two components with properties Q_M (matrix) and Q_F (filler, fiber). This equation is based on a simplified Kerner model, which assumes a fiber F embedded in a cylinder of the resinous matrix M that in turn resides in matter with the properties of the composite. It gives the property Q of the material as function of the volume fraction ϕ_F of the filler where K_1 is an adjustable constant:

$$(13\text{-}3) \qquad \frac{Q}{Q_M} = \frac{1 + K_1 K_2 \phi_F}{1 - K_2 \phi_F}; \quad K_2 = \frac{Q_F - Q_M}{Q_F + K_1 Q_M}$$

The *Takayanagi model* was originally derived for a semicrystalline polymer. A fraction of the crystalline (F) part is assumed to be coupled in series with the amorphous (M) part, and the remaining crystalline part is assumed to be coupled in parallel with the crystalline-amorphous series. The property enhancement Q/Q_M with respect to the matrix property Q_M is given by

$$(13\text{-}4) \qquad \frac{Q}{Q_M} = \frac{Q_M \phi_M + Q_F (K_4 + \phi_F)}{Q_F K_4 \phi_M + Q_M (1 + K_4 \phi_F)}$$

The equations from Halpin-Tsai and the Takayanagi models are identical because Equations (13-3) and (13-4) can both be written as

$$(13\text{-}5) \qquad \frac{Q}{Q_M} = \frac{Q_M + K_4 Q_F + (Q_F - Q_M)\phi_F}{Q_M + K_4 Q_F - (Q_F - Q_M)\phi_F K_4}; \quad K_4 = 1/K_1$$

Equation 13-5 gives the upper bound for $K_4 = 0$ (i.e., $K_1 \to \infty$) and the lower bound for $K_4 \to \infty$ (i.e., $K_1 = 0$).

Specific interactions between components can be considered by an additional interaction term. This term is differently defined, for example:

(13-6) $Q = Q_A f_A + Q_B f_B + K_i f_A f_B; \quad K_i = 2(2\,Q_{AB} - Q_A - Q_B)$

where the term Q_{AB} is taken at $f_A = f_B = 1/2$. Q_{AB} may be an interaction parameter between components A and B in homogeneous mixtures or the property of the interlayer between A and B phases in heterogeneous mixtures. The upper bound for absent interactions or interlayers is retrieved for the condition $Q_{AB} = (Q_A + Q_B)/2$.

The functions for $Q_{AB} \neq (Q_A + Q_B)/2$ may lie fully or partially outside the range between upper and lower bound (Fig. 13-1); that is, synergistic effects (above the upper bound) or antagonistic effects (below the lower bound) may be present. They may also show maxima and minima, depending on the relative magnitudes of Q_{AB}, Q_A, and Q_B. Maxima or minima that exceed the highest property value of the components (Q_A in Fig. 13-1) are sometimes considered the true criteria for synergistic and antagonistic effects, respectively.

13.2. Microcomposites

Many plastics grades do not consist of neat polymers. They rather contain fillers because of the many improved properties (Table 13-2). The fillers may be extenders (spheroidal particulates) or reinforcing agents (mainly fibers, sometimes platelets); the resinous matrices may be thermoplastics or thermosets. For the purpose of this book, all types of these admixtures are called microcomposites.

Table 13-2 Effect of fillers on properties of amorphous (A) and crystalline (C) thermoplastics. [a]) Tough plastics become more brittle and brittle plastics become tough.
(↑) Weak increase, ↑ increase, ↑↑ strong increase,
(↓) weak decrease, ↓ decrease, ↓↓ strong decrease.

Property	Extenders		Reinforcing agents	
	A	C	A	C
Shrinkage	↓	↓	↓	↓
Heat deflection temperature	-	(↑↓)	-	↑
Melt flow	↓	↓	↓	↓
Young's modulus	(↑)	(↑)	↑	↑
Flexural modulus	↑	↑	↑	↑↑
Fracture strength	↓	↓	↑	↑
Brittleness	↑	↑	↑↓ [a]	↑↓[a]

The matrices for microcomposites are mostly thermosets such as unsaturated polyesters reinforced with long glass fibers and epoxide, urethane, and phenol-formaldehyde resins filled with particulates (Table 10-2). Thermoplastics are preferred for fiber-filled composites, however, because of simple formulations, excellent prepreg stability, high melt viscosity, and a relatively short processing cycle. These factors and the resulting low fabrication cost outweigh the disadvantages of thermoplastics such as difficult fiber impregnation, absent prepreg tack, and high processing temperature.

Short fibers are usually added in amounts of ca. 30 wt-%, which corresponds to ca. 17 vol-% for glass fibers (ρ = 2.55 g/cm^3) in an average polymer ($\rho \approx$ 1.2 g/cm^3). Higher fiber contents often lead to packing problems since the maximum volume fraction of three-dimensional randomly packed rods with L/d = 20 is already ca. 25 % Particulate fillers can be added in far greater amounts without sacrificing properties (see Table 10-2) because maximum packing densities of spheres are, for example, 0.74 for hexagonally most dense and 0.64 for the randomly most dense packing. At higher spheroid concentrations, a "fiber effect" may occur, however, that is, an aggregation of particles in a rod-like manner during processing.

13.2.1. Young's Modulus

Unidirectional laminae of *long glass fibers* and cured unsaturated polyester resins follow very well the theoretical predictions of the simple mixing laws (Eq.(13-1)), if the elastic moduli are measured parallel (n = 1; Voigt limit) and transverse (n = -1; Reuss limit), resp., to the fiber direction (Fig. 13-2).

Young's modulus E of *short fiber-reinforced plastics* measured parallel to the fibers can be modeled with the Halpin-Tsai and Takayanagi equations (Eqs.(13-3)-(13-5)). It is often assumed to depend on the ratio $1/K_4 = K_1 = 2L_F/d_F$ of fiber length L_F to fiber diameter d_F. This assumption may be reasonably well fulfilled for injection-molded specimens (leading to orientation of fibers in flow direction) and the dimensions of fibers *after* the processing are taken into account (fibers break during processing because of stress) (Fig. 13-2).

The same assumptions overestimate slightly the influence of *glass spheres* ($K_4 = 1/K_1 = d_F/(2\,L_F) = 1/2$) on composite moduli at low filler contents but underestimate it at higher contents (Fig. 13-2). The latter effect could be due to packing problems because particulates cannot fill the space completely (hexagonal densest packing of spheres is ca. 74 % of the available space).

The shear modulus of particulate-filled plastics has been deduced theoretically to depend on a factor $K_4 = (8 - 10\mu_M)/(7 - 5\,\mu_M)$, where μ_M is the Poisson number of the matrix. Young's moduli also seem to follow this prediction for small volume fractions of the filler ($\phi_F < 0.3$) (Fig. 13-2). The assumption underestimates the effect of fillers at greater filler contents, however.

Young's moduli E of many filled materials often increase almost linearly with the volume fraction ϕ_F of the filler at not too high filler contents (Fig. 13-2). The slope S of the function $E = f(\phi_F)$ is different from the slope $S' = E_F - E_M$ predicted by the simple mixing law, $E = E_M + S'\phi_F$. The experimental findings can be described by the introduction of an efficiency factor K for the filler (fiber)

$$(13\text{-}7) \quad E = E_M\phi_M + KE_F\phi_F = E_M + (KE_F - E_M)\phi_F$$

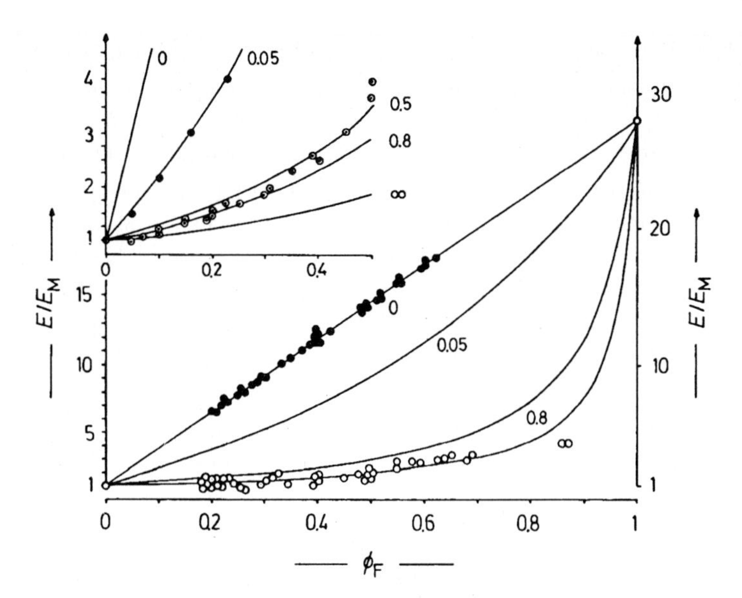

Fig. 13-2 Ratio E/E_M of elastic moduli E of composites to moduli E_M of matrices as function of the volume fraction ϕ_F of filler (fiber). Experimental data on
●, O laminae of long glass fibers in cured unsaturated polyester resins (∥ and ⊥),
⊕ short glass fibers in polyamide 6 (near line for $K_4 = 0.05$), and
O glass spheres in a cured epoxy resin; near lines with $K_4 = 0.5$ and 0.8, resp. (see text)).
Lines calculated via Equation 13-6 with
$K_4 = 0$ (upper bound),
$K_4 = 0.05$ (e.g., fibers with $L_F = 0.2$ mm (after processing) and $d_F = 20$ μm),
$K_4 = 0.5$ (e.g, spheres),
$K_4 = 0.8$ (e.g., matrix with Poisson ration of 0.4), and
$K_4 = \infty$ (lower bound).

Equation (13-7) is an approximate form of Equation (13-5) for small K_4, small E_M (Q_M) and large E_F (Q_F). It reduces to the upper bound for $K = 1$.

The efficiency factor K of a filler depends on the modulus of the matrix (Table 13-3). In general, it increases with the aspect ratio L_F/d_F of the filler particles (but see data for kaolin and talcum) and even more prominently with the specific surface a of the particles. The stiffness of composites obviously depends on the contact area between polymer segments and filler particles. Good adhesion is not necessary, however, because E is (ideally) measured for strains approaching zero.

Segments of polymer chains will straighten near filler (fiber) surfaces because the filler (fiber) represents a barrier (see Section 4.5.1). The matrix thus becomes stiffer near filler particles; the magnitude of this effect should depend also on the fractal properties of the fillers. The same effect may be caused by transcrystallization of semicrystalline polymers on surfaces of filler particles. Straightened segments have high lattice moduli in chain direction (Table 7-3), often far higher than filler moduli, but the straightened segments may be oriented at random as are their host particles. These phenomena lead to efficiency factors K for fillers which may be higher or smaller than 1, depending on the relative magnitudes of the lattice and filler moduli, the proportion and orientationof straightened chains, etc.

Table 13-3 Youngs's moduli E_F, aspect ratios L_F/d_F (after processing), and specific surface areas a_F of fillers, moduli E_M of polymer matrices, and efficiency factors K of of fillers in composites.

Filler Type	E_F/GPa	L_F/d_F	$a_F/(m^2\ g^{-1})$	Polymer Type	E_M	K
Glass spheres	72	1	0.3	PE-HD	1.5	0.10
				PA 6	3.0	0.19
Chalk	26	1-5	0.3-2.2	PP	1.3	0.37
				PA 6	3.0	0.44
Glass fibers (short)	72	20-50	0.8	PP	1.3	0.56
				POM	3.1	0.59
Kaolin	20	4-12	5-19	PBT	2.6	1.01
				PA 6	3.0	1.03
Talcum	20	5-20	6-17	PP	1.3	1.55

For spheroidal particles, the efficiency factor K is given by the interactions between particle and matrix such as adhesion, chain-straightening, transcrystallization, etc. ($K = K_i$). For nonspheroidal particles, the orientation relative to the stress direction has to be considered by an factor K_o and Equation (13-7) becomes:

$$(13\text{-}8)\quad E = E_M + (K_iK_oE_F - E_M)\phi_F = E_M + (KE_F - E_M)\phi_F$$

K_o adopts the following values for long fibers in fiber-filled composites: 1 (all fibers parallel to the stress direction), 0 (all fibers transverse to stress direction), 1/6 (fibers three-dimensionally distributed at random), and 1/3 (fibers two-dimensionally distributed at random).

13.2.2. Strengths

Elastic moduli are determined for diminishing strains but fracture strengths are obtained for maximum strains. Different criteria thus apply for moduli and strengths.

The failure mode of filled thermoplastics is matrix dominated. Electron micrographs show that at low fiber contents, planar cracks spread through the matrices and debonded fibers are pulled out of either surface of the crack. Strengths are increased if matrices adhere well to the fillers, which can be achieved by coupling agents (e.g., silanes for glass fiber-filled polymers). Coupling agents improve the bonding between fiber and matrix; their action is not well known. Incomplete bonding can be described by an adhesion factor $K_a < 1$.

Short fibers with length L_F act in plastics as stress concentrators because fiber ends do not transfer stresses from the matrix to the fibers over a distance of $L_c/2$ at each end. The average stress $\bar{\sigma}_F$ at the fibers in the composite is thus smaller than the stress at completely bonded long fibers in first approximation, $\bar{\sigma}_F L_F = \sigma_F[L_F - (L_c/2)]$. The efficiency factor for this phenomenon is thus $K_L = \bar{\sigma}_F/\sigma_F = 1 - [L_{crit}/(2\ L_F)]$.

A 95 % efficiency ($K_L = 0.95$) is thus obtained for $L_F = 10\ L_{crit}$. By analogy to Equation (13-8), the simple law of mixtures converts into:

$$\text{(13-9)} \quad \sigma_b = \sigma_M + \{K_i K_o K_a[1 - (L_{crit}/(2\ L_F))]\}\sigma_F\phi_F$$

The fiber length L_F must thus exceed a critical fiber length L_{crit} (i.e., the length of fibers that just can be pulled out of the matrix without breakage). Critical fiber lengths $L_{crit} = (\sigma_F d_F)/(2\ \tau_{F/M})$ can be calculated from the equilibrium between tensile stress σ_F and interfacial shear stress $\tau_{F/M}$. Such critical fibers lengths of glass fibers are, for example, ca. 0.3 mm in polyamide 6 and ca. 1.3 mm in poly(propylene); they are approaching (PA 6) and exceeding (PP) the lengths L_F of glass fibers in processed composites!

The experimental fracture strengths σ_b of *short fiber*-filled composites at 30 wt-% loadings are thus usually only 1.5-2 times higher than those of the amorphous and crystalline matrix polymers. These strength enhancements appear small in view of the high tensile strengths of the fillers (e.g., 3.5 GPa for E-glass fibers vs. 70 MPa for PA 6). However, they agree reasonably with estimates by Equation (13-9):

Assume a polyamide 6 (σ_M = 70 MPa) filled with w_F = 0.30 (ϕ_F = 0.17) glass fibers (σ_F = 3500 MPa) of a length L_F = (3/2) L_{crit} that are three-dimensionally oriented at random (K_o = 1/6) and are reasonably bonded (K_a = 0.7) to the matrix without segment stiffening of the latter (K_i = 1). Such a composite would have a strength enhancement of only $\sigma/\sigma_M \approx 1.8$ (experimentally found: 1.5-2).

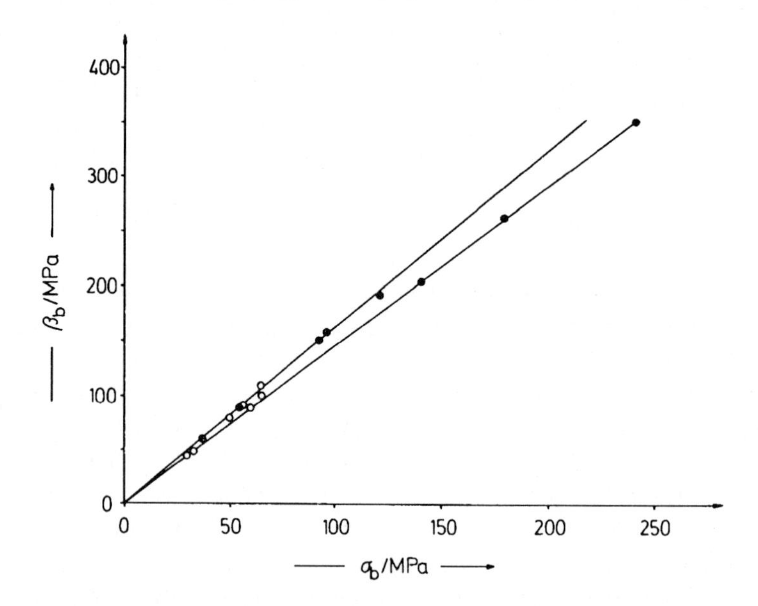

Fig. 13-3 Flexural strength β_b as function of ultimate tensile strength σ_b. Filler content: 30 wt%.
O Unfilled plastics (from left to right: PE, PP, PS, PBT, PC, PMMA, PSU);
O mineral-filled plastics (PP-talc, PA-kaolin);
⊕ glass fiber-filled plastics (two PPs, PBT, PA);
● carbon fiber-filled plastics (PBT, PA).

Strengths of composites with hard matrices may be further reduced by two additional factors. If fibers break successively, then the effects of early fiber breaks are lost for ultimate failure. Points on fibers that are adjacent to ends of neighboring fibers will furthermore experience an excessive fiber loading.

Particulate fillers usually reduce strongly the fracture strengths of composites of amorphous thermoplastics, whereas those of crystalline polymers are reduced less or not at all. The lowering of tensile strengths is probably caused by void formation at the interface filler-matrix due to the rough surface of filler particles. Voids may also be the reason why the tensile strength of fiber-filled rigid PVC is lower than that of the matrix itself.

In bending tests, the top surface is in tension and the under surface in an compression; in addition, there is always some transverse shearing. Because strengths are ideally directly proportional to moduli and the various moduli are interrelated (Chapter 7), a dependence of *flexural strengths* β_b on ultimate tensile strengths σ_b can be expected. Experimentally, a ratio $\beta_b/\sigma_b \approx 1.45\text{-}1.6$ is found (Fig. 13-3).

13.2.3. Thermal Properties

The addition of fillers (low thermal expansion) to thermoplastics (high thermal expansion) reduces the proportion of the latter and thus the *shrinkage* of plastics on processing (Table 13-2).

Heat deflection temperatures HDT (at 1.85 N/mm^2) of amorphous polymers are about the same as those for unfilled plastics (or up to 10 K higher) (Fig. 13-4, group I), probably because of the increased viscosity of the filled plastic. The HDTs of amorphous thermoplastics are almost identical with glass transition temperatures.

The HDTs of crystalline polymers depend on a number of factors. Unfilled or particulate-filled crystalline polymers show HDTs that are often considerably lower than their melting temperatures. On fiber-filling, HDT's of crystalline polymers increase strongly; they may reach values near the melting temperatures of the matrix. This behavior is thought to be due to additional polymer crystallization near, and/or polymer adsorption on, the fiber surface; epitaxial layers have been found to extend from the filler up to 150 nm into the matrix. The surface layers of fibers overlap and a physical network of fibers interconnected by interlayers is formed. The network formation reduces creep and thus increases the resistance against heat distortion.

Two groups of HDT's of glass fiber-filled crystalline polymers can be distinguished (Fig. 13-4). The HDTs of group II are raised about (50-60) K above those of their unfilled counterparts. This group comprises apolar polymers such as poly(ethylene), poly(propylene), and poly(phenylene sulfide); poly(oxymethylene) also belongs to group II since the surface of its compact helix appears apolar (see Fig. 12-3). The difference in HDTs of filled and unfilled group II polymers does not vary with melting temperatures. Group III includes polar crystalline polymers (polyesters, polyamides, etc.). For this group, the difference of HDTs of filled and unfilled polymers increases with the melting temperature of the unfilled polymer.

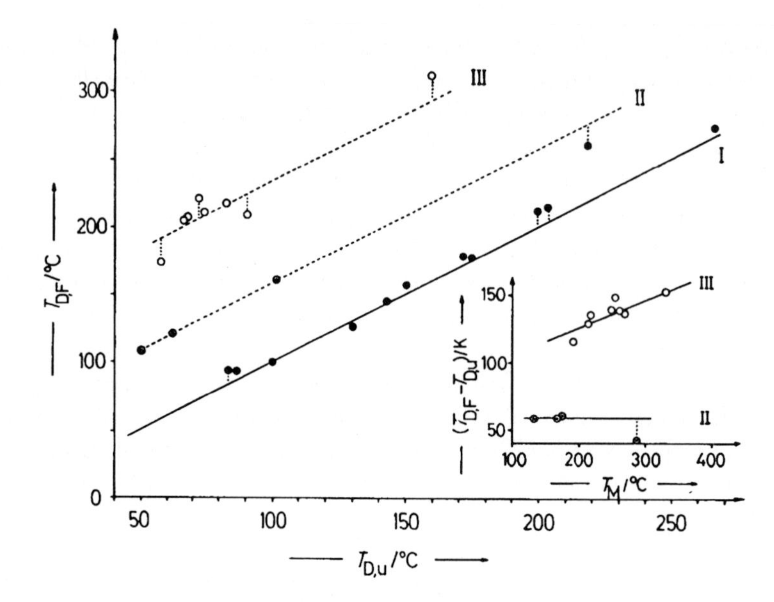

Fig. 13-4 Heat deflection temperatures $T_{D,F}$ of amorphous (●) and crystalline (O,O) thermoplastics filled with 30 wt-% short glass fibers as a function of heat deflection temperatures $T_{D,u}$ of the corresponding unfilled polymers, both at 1.85 N/mm^2 (ISO R75A).
Solid line: $T_{D,F} = T_{D,u}$; broken lines: empirical.
I = Amorphous polymers, II = apolar crystalline polymers, III = polar crystalline polymers.
Insert: Difference $T_{D,F} - T_{D,u}$ as function of melting temperatures T_M of unfilled crystalline polymers.

13.3. Macrocomposites

Polymer macrocomposites are heterogeneous composites of polymers and "macro"-sized fillers. Examples are polymer concrete, composition board, polymers reinforced by long fibers, mats, fabrics, or rovings, and plywood.

Polymer concrete is composed of polymers (usually polyester or poly(methyl methacrylate)), aggregates (sand, ground limestone, etc.), and other additives (dyes, etc.); it does not contain cement. *Polymer-concrete hybrids* include polymer-cement hybrids (cement is partially replaced by polymers) and polymer-impregnated concrete (conventional concrete saturated with a monomer that is polymerized in place). Polymer concretes have higher moduli and higher tensile and compressive strengths than ordinary concretes. They are not vulnerable to damage by freeze-thaw cycles.

Composition boards are panel products. They include particleboard, fiberboard, and hard board from flakes or shearings, etc. The fillers are mainly wood products; the resins, urea-formaldehyde (interior applications), and polyurethanes or phenolic resins (exterior).

Long fiber-filled polymer macrocomposites range in structure from polymers filled with "long" fibers of 6-12 mm length to fiber strands and mats filled with polymers. The filler content can often reach 65 wt.-%. Upon application of tensile stress, local stress concentrations are transferred by shear forces onto the polymer-fiber interface and distributed over the much greater fiber surface area. The fibers must therefore bond well to the polymer (use of coupling agents) and must have a certain length in order to avoid slipping.

Young's moduli of long glass fiber-reinforced epoxide resins follow the simple mixing law if the tensile stress is in the fiber direction (Fig. 13-2). The inverse mixing law is not observed as well for stresses perpendicular to the fiber direction because some fibers bend under transverse stress.

The simple mixing law corresponds to $K = 1$ in Eq.(13-7). For long fibers, K has the meaning of an orientation factor K_o (see Equation (13-8)). It becomes 1/2 for 90° cross-plies measured in either fiber direction and 3/8 for a uniform, planar distribution of fibers. These values are considerably higher than $K_o = 1/6$ for fibers distributed three-dimensionally at random; fiber mats reinforce considerably more than short fibers (Table 13-4).

This group of macrocomposites also includes *prepregs* (fiber mats soaked with unsaturated polyester resins) and the composites generated by filament winding (wound filaments saturated with epoxies or unsaturated polyesters), both of which are subsequently polymerized. Sheet molding compounds (SMCs) are now not only available for thermosetting resins but also for thermoplastics.

Table 13-4 Properties of cured prepregs from unsaturated polyester resins (data of [6]).

Property	Unit	Mats		Woven fabrics		Rovings
wt-% of reinforcement →		20-30	40-50	40-50	50-60	70-80
Density	g/cm^3	1.3-1.5	1.5-1.75	1.5-1.75	1.6-1.85	1.9-2.1
Thermal expansion coeff.($\cdot 10^6$)	K^{-1}	40-30	24-20	22-18	18-16	14-12
Thermal conductivity (10^{-2})	W/(K m)	14-19	22-31	19-25	25-31	37-41
Young's (or flexural) modulus	GPa	5-7	9-10	10-14	14-18	21-26
Shear modulus	GPa	2-3	4-5	2-3	3-4	-
Elongation at break	%	2	2	2	2	2
Flexural strength	MPa	115-145	180-220	220-260	260-300	400-500
Compression strength	MPa	110-135	165-200	150-180	180-200	-
Tensile strength	MPa	65-90	130-170	200-240	240-275	700-800
Impact strength	kJ/m^2	25-55	75-105	130-160	160-190	-

13.4. Molecular Composites

The problem of insufficient bonding between fiber and matrix can be dramatically reduced if so-called molecular composites are used. These materials are dispersions of semiflexible ("rod-like") polymer molecules with mesogenic units in chemically similar polymer matrices, e.g., poly(*p*-phenylenebenzobisthiazole)s in poly(benzimide)s (see Fig. 7-12). The mesogens are not truly molecular dissolved; rather, they form microdomains in the matrix. The segment axes are randomly distributed in these domains if the molecular composites are formed from polymer solutions with total polymer concentrations c below the critical isotrope phase-mesophase concentration $c_{i/n}$. Microdomains show mesophase behavior if films or fiber are formed at $c > c_{i/n}$. Such microcomposites show excellent moduli and fracture strengths (Fig. 7-12).

13.5. Homogeneous Blends

Mixtures of polymers with other polymers are called polymer blends. Such blends may be composed of two thermoplastics (plastics blends), two elastomers (rubber blends), a plastic filled with an elastomer as the dispersed phase (rubber-modified plastics), an elastomer with a plastic as the dispersed phase (polymer-filled elastomer), or a plastic filled with a polymer melt or a low-molar mass liquid (plasticized polymers). Blends may be homogeneous (true thermodynamic miscibility of components) or heterogeneous (phase-separated); in heterogeneous blends, components may also be mechanically compatible but thermodynamically immiscible.

Blends are produced because of cost advantages and/or property improvements. The homogeneous blend of inexpensive poly(styrene) PS with unfavorable mechanical properties and expensive poly(oxy-2,6-dimethyl-1,4-phenylene) PPE with excellent mechanical properties has good properties at low cost. Higher costs of blends than those of their components are acceptable if, for example, impact strengths are higher (PVC + PMMA). The market for blends of engineering plastics is presently growing with ca. 13 % per year; world production is ca. 340 000 t/a (1990).

Table 13-5 Young's moduli E, fracture strengths σ_b, impact strengths a_n, heat distortion temperatures T_{HD}, and relative costs of blends and their components. PPE/PS form a miscible blend, PVC/PMMA a compatible blend, and PC/ABS a heterogeneous blend.

Components	$\frac{E}{\text{MPa}}$	$\frac{\sigma_b}{\text{MPa}}$	$\frac{a_n}{\text{J m}^{-1}}$	$\frac{T_{HD}}{°\text{C}}$	Rel. cost
PPE	2550	72	85	192	4.10
PPE + PS	2410	66	160	129	1.31
PS	2060	33	64	88	1.00
PMMA	3100	69	25	85	2.08
PMMA + PVC	2340	45	800	75	2.44
PVC	2760	48	< 530	66	1.00
PC	2410	66	800	132	1.53
PC + ABS	2550	43	530	105	1.19
ABS	2060	35	320	86	1.00

13.5.1. Thermodynamics

If two components 1 and 2 are mixed, entropy S, enthalpy H, and volume V of the mixture become different from those of the components. In many cases, volume changes can be neglected and the Gibbs energy of mixing is given by the changes in enthalpy and entropy, i.e., $\Delta G_{mix} = \Delta H_{mix} - T\Delta S_{mix}$. These changes can be calculated by the *Flory-Huggins theory*, which assumes that the mixture can be modeled by a three-dimensional lattice on which the monomeric units (or solvent molecules) can be placed. Each monomeric unit experiences the same force field; the theory is thus a mean-field theory. The assumption of a constant force field is fairly well fulfilled for polymer blends and concentrated polymer solutions but not for dilute solutions because isolated polymer coils form "islands" in a sea of solvent molecules; thus, their segments are not homogeneously distributed throughout the solution.

The entropy and enthalpy of mixing are calculated separately. *Entropies* are calculated for ideal solutions (zero mixing enthalpy); all environment-dependent entropy changes are thus zero (translational, vibrational, and inner rotational entropies do not change). Units can however be arranged relative to each other in many ways; there is a combinatorial entropy (called "configurational entropy" in statistical mechanics).

The mixing of two components is modeled as a quasi-chemical reaction. The *mixing enthalpy* is thus determined by the change in interaction energies and the number and type of nearest neighbors. Enthalpy and entropy terms are added to give the normalized molar Gibbs energy of mixing:

$$(13\text{-}10)\quad (\Delta G_{mix})^{m}/RT = \chi\phi_1\phi_2 + \phi_1 \ln \phi_1 + (X_1/X_2)\phi_2 \ln \phi_2$$

which, on differentiation with respect to the molar amounts of the components, yields the chemical potentials, for example, of component 1,

$$(13\text{-}11)\quad \Delta\mu_1 = RT\{\chi\phi_2^2 + \ln(1 - \phi_2) + [1 - (X_1/X_2)]\phi_2^2\}$$

where χ is the parameter for the interaction between component 1 and component 2 (usually with numerical values between 0 and 2), also called the Flory-Huggins parameter. All three terms in braces must be considered if a polymer 2 is mixed with a solvent 1, for example., a plasticizer ($X_1 \ll X_2$). If both components are polymers however, then $X_1 \approx X_2$, and the entire entropy contribution comes from the relatively small logarithmic term. Little combinatorial entropy is thus gained upon mixing of two polymers and the miscibility is determined mainly by the mixing enthalpy, that is, by the term $\chi\phi_2^2$ in the expression for the chemical potential.

Phase separation occurs if the curve $(\Delta G_{mix})^{m} = f(\phi_2)$ (at T = const.) has a shape that can be touched by a tangent at two points ' and ". The compositions at these points are ϕ_2' and ϕ_2'' (with $\phi_2' < \phi_2''$). Two phases exist for the unstable range $\phi_2' < \phi_2 < \phi_2''$, which is separated from the stable ranges $\phi_2 < \phi_2'$ and $\phi_2'' > \phi_2$ by the so-called binodal. The unstable range itself is subdivided by the spinodal into two metastable ranges and one unstable one. Spinodals are characterized by inflection points in the Gibbs energy-composition functions and extremal values in the chemical potential-composition functions. Maximum, minimum, and inflection point become identical at the critical point, where the second derivative of the chemical potential

$$(13\text{-}12) \qquad \partial^2 \Delta\mu_1 / \partial\phi_2{}^2 = RT[2\,\chi - (1 - \phi_2)^{-2}] = 0$$

becomes zero. This critical point is characterized by a critical volume fraction $\phi_{2,\text{crit}} = 1/(1 + X_2{}^{1/2})$ and a critical interaction parameter $\chi_{\text{crit}} \approx (1/2) + (1/X_2)^{1/2}$.

The temperature dependence of χ is approximately given by

$$(13\text{-}13) \qquad \chi \approx A + (B/T)$$

where A and B are system-dependent constants. B is positive for endothermal mixtures, i.e., χ increases with increasing temperature and a homogeneous solution exists above an "upper critical solution temperature" UCST and a liquid 2-phase region below UCST (Fig. 13-5, I). Other liquid systems are homogeneous only below a "lower critical solution temperature" LCST (Fig. 13-5, II). The demixing is correspondingly induced either by enthalpy (UCST) or by entropy (LCST).

The terms UCST and LCST do not indicate the absolute position of demixing temperatures since some systems show LCST > UCST (Fig. 13-5, IV). Depending on the system, either an hourglass-type diagram V or a closed miscibility gap III is shown.

Phase-separation may also occur into a homogeneous liquid phase and a crystalline phase (VI) or a glassy phase (VIII). Four different regions with a eutectic point are observed if both components A and B form crystals C_A and C_B (VII). The phase diagrams get even more complicated if additional glassy phases exist (IX, X).

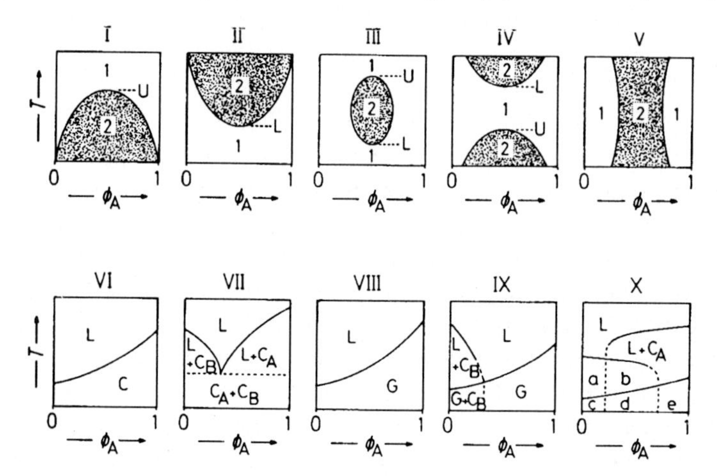

Fig. 13-5 Phase diagrams for mixtures of two components A and B: demixing temperatures T as function of volume fraction ϕ_A of component A.

Upper row: Demixing into two liquid phases; 1 = one phase, 2 = phases. U = Upper critical solution temperature UCST; L = lower critical solution temperature LCST.

Lower row: Phase separation into liquid (L), glassy (G), and crystalline (C) phases. a = L + C_B; b = L + C_A + C_B; c = G + C_B; d = G + C_A + C_B; e = G + C_A.

13.5.2. Rubber Blends

Uncross-linked, amorphous polymers behave as rubbers above their glass temperatures. Homogeneous rubber blends are usually formed if the difference in solubility parameters is smaller than 0.7 $cal^{1/2}$ $cm^{-3/2}$ = 1.44 $J^{1/2}$ $cm^{-3/2}$.

Upon vulcanization of rubbers to elastomers, chains become cross-linked but the chain segments between cross-linking sites still retain their mobilities. Miscible mixtures of rubbers do not phase-separate; they become homogeneous elastomer blends on vulcanization. Homogeneity is present if domain structures are absent under the electron microscope. The existence of only one glass temperature is no proof for a 1-phase structure, however, because the onset of glass transitions requires the cooperative movement of many chain segments, i.e., certain minimum domain sizes of more than ca. 50 nm.

Other rubber blends are heterogeneous in their uncross-linked states. Examples are the blends of SBR with natural rubber NR, *cis*-1,4-poly(butadiene) BR, butyl rubber IIR from isobutylene and 2-4 % isoprene, and EPDM at volume fractions ϕ_i between 40 and 60 %. Upon vulcanization, they retain their phase structures and become heterogeneous elastomer blends. These vulcanized blends show only one glass temperature T_G between $T_{G,A}$ and $T_{G,B}$ because their domains sizes are too small. Other heterogeneous elastomer blends with greater domain sizes exhibit two glass temperatures, both practically identical to those of the parent rubbers.

About 75 % of all elastomers are employed as blends. Heterogeneous blends of natural rubber NR + styrene-butadiene rubber SBR or cis-butadiene rubber BR + styrene-butadiene rubber SBR are used mainly for tire treads since they reduce abrasion.

13.5.3. Plastic Blends

Plastic blends are usually unmiscible because the combinatorial entropy of mixing is too small (high molar masses) and the enthalpy changes on mixing are often positive. Miscibility is observed for three types of systems:

1. Chemically similar polymers, for example, poly(styrene)poly(o-chlorostyrene), showing both LCST and UCST.
2. Systems having specific interactions between the different components, for example poly(styrene)-poly(vinyl methyl ether); these systems show only LCSTs.
3. Systems consisting of oligomers, for example, oligo(ethylene oxide)-oligo(propylene oxide); these systems possess only UCST's.

Homogeneous polymer blends show only single glass temperatures; the composition dependence of these T_G's can be described by the simple rule of mixtures. Two-phase systems always show two glass temperatures if the phase diameters are greater than ca. 3-5 nm. Transparency, on the other hand, is no indication of a single-phase blend since opacity can be observed only if the refractive indices of the two phases are sufficiently different, phase diameters are greater than ca. one-half the wavelength of incident light, and certain phase geometries are absent (Chapter 9). The presence of two phases in clear specimens can often be detected with electron microscopy by special staining techniques.

Most blends consists of engineering plastics and/or low-end performance plastics (Section 13.7.). Modified PPE, a blend of poly(1,4-oxy-2,6-dimethylphenylene) and poly(styrene), is the industrially leading *homogeneous blend.* Blends of poly(vinyl chloride) with poly(methyl methacrylate), poly(ethylene-*co*-vinyl acetate), or chlorinated poly(ethylene) are also said to be one-phase systems. Most other blends are heterogeneous.

13.6. Plasticized Polymers

Plasticization of a plastic consists of a flexibilization by either adding plasticizers (external plasticization) or by incorporating flexibilizing comonomer units into the polymer chains (internal plasticization) (see also Section 10.2.5). The plasticization of a rubber by mineral oil leads to an "oil-extended rubber".

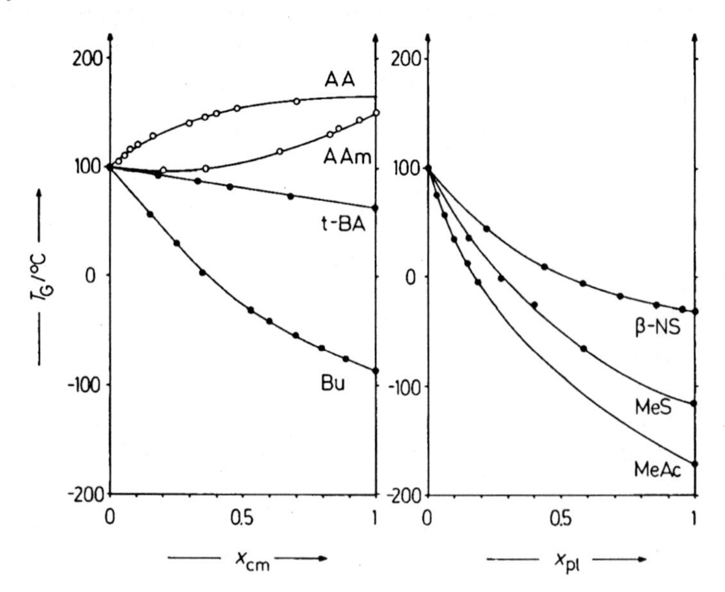

Fig. 13-6 Internal and external plasticization of poly(styrene).

Left: Internal plasticization by copolymerization. Change of glass temperatures T_G of copolymers of styrene with the mole fraction x_{cm} of the comonomers acrylic acid (AA), acrylamide AAm, t-butyl acrylate t-BA, and butadiene (Bu).

Right: External plasticization of poly(styrene) with the mole fraction x_{pl} of the plasticizers β-naphthyl salicylate β-NS, methyl salicylate MeS, or methyl acetate MeAc.

The molecular action of external plasticizers consists of a flexibilization of chain segments, which can have various causes. *Polar plasticizers* increase the proportion of gauche conformations in polar chains, thus decreasing rotational barriers. *Primary plasticizers* can dissolve helical structures and crystalline regions if the plasticizers are thermodynamically good solvents. Both *primary and secondary plasticizers* are furthermore diluents; their addition increases the distance between polymer segments and thus decreases the activation energy needed for cooperative movements of segments. Solvations do not per se increase chain flexibilities, however, because solvent shells act as substituents and may increase rotational barriers.

The molecular effect of an increase in chain flexibility by plasticization is measured as a decrease of glass temperatures with plasticizer content (Fig. 13-6). Small molecules are more efficient plasticizers than bigger ones of similar chemical constitution; their T_G's are more strongly lowered at equal plasticizer content. Plasticizer efficiency decreases with increasing thermodynamical "goodness" of the plasticizer for the polymer (increased polymer-plasticizer interaction). Good plasticizers with respect to lowering glass temperatures are thus small molecules that act as theta solvents.

These plasticizers are however bad for technological applications because they bleed (transport of plasticizer to the surface of the plastic), migrate (transport of plasticizer into another contacting material), or are extractable (e.g., transport into packaged foods). These transport phenomena can be reduced if polymer plasticizers are used. Increased molar masses lead to decreased thermodynamic compatibility, however (see above). Technical plasticizers are therefore in most cases a compromise between thermodynamic plasticizer efficiency, kinetically hindered plasticizer transport, and cost considerations.

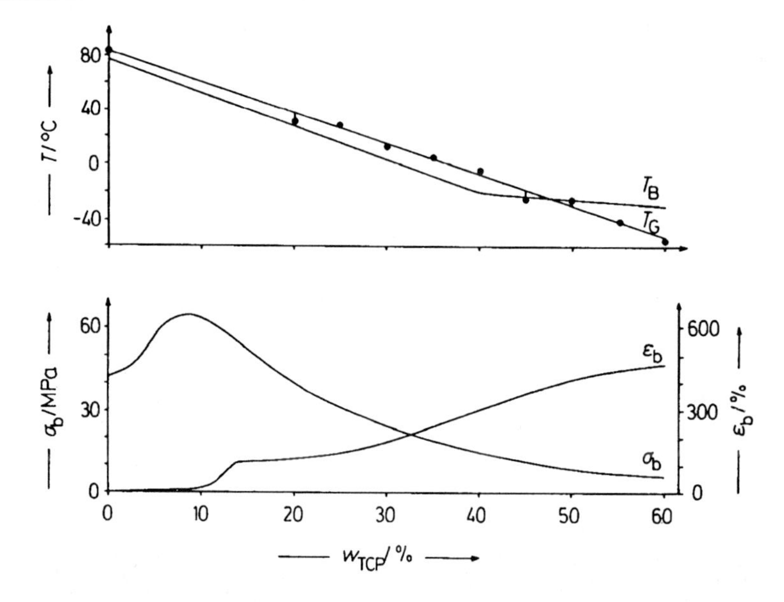

Fig. 13-7 Glass transition temperatures T_G [1], brittleness temperatures T_B [2], fracture strengths σ_b [2], and elongations ε_b at break [2] of a poly(vinyl chloride) plasticized with tricresyl phosphate.

Plasticization can also be followed indirectly by changes of certain mechanical properties such as lowered fracture strength, increased elongation, and decreased tensile modulus. Such methods sometimes show *antiplasticizations* (Fig. 13-7), for example, increases of Young's moduli, tensile and flexural strengths, and hardness and decreases of impact strengths and ultimate elongations with increasing plasticizer content at low plasticizer concentrations. Glass, brittle, and softening temperatures are not increased, however. The effect may be due to a healing of microvoids in amorphous polymers such as poly(styrene) or to an additional crystallization of slightly crystalline polymers such as poly(vinyl chloride).

13.7. Heterogeneous Blends

13.7.1. Compatible Polymers

Two polymers may be thermodynamically immiscible but still mechanically compatible. These compatible blends are two- or multiphase systems that show multiple glass temperatures (Fig. 13-8), which do not change with composition. The components do not separate, however, under mechanical stress during the life expectancy of the product.

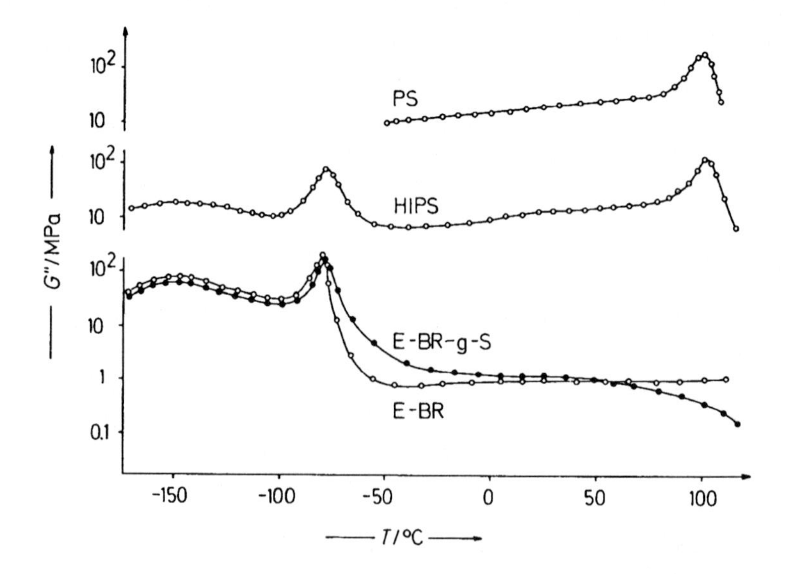

Fig. 13-8 Temperature dependence of the shear loss modul G'' of poly(styrene) PS, an emulsion-polymerized butadiene rubber E-BR, a graft copolymer of styrene on poly(butadiene) rubber (E-BR-g-S), and a high-impact poly(styrene) HIPS from the in-situ polymerization of styrenic poly(butadiene) solutions [3]. HIPS shows the loss peaks of poly(styrene) PS and the poly(butadiene) segments of E-BR and E-BR-g-S, indicating two phases.

Such mechanical compatibilities can be improved or introduced by addition of diblock copolymers as compatibilizers. One block of these diblock copolymers is compatible with one phase type of the heterogeneous blend and the other one with the other type (see Section 4.3.6). The two blocks need not necessarily be chemically identical with the two parent polymers, that is, a compatibility of poly(A) and poly(B) can be achieved not only by addition of poly(A)-*block*-poly(B) but, for example, also by poly(A)-*block*-poly(C) if the C-block is either miscible with poly(B) or can form mixed crystals with poly(B).

Diblock copolymers gather at the interface between domains; they reduce the interfacial tension between domains of amorphous polymers, regardless of the molar mass of the blocks. However, reinforcing effects are observed only if the block lengths are greater than the entanglement lengths of the domain molecules. The congregation of the diblock molecules at the interface competes furthermore with the formation of micelles of diblock copolymers in domains, which wastes diblocks.

At present, about 30 different combinations of two polymers are marketed, some with considerable number of grades. Most blends are prepared from two engineering resins or from one of these resins with a low-end performance resin (Table 13-5). Bisphenol A polycarbonate PC is blended with more polymers than any other plastic.

13.7.2. Blend Formation

Heterogeneous polymer blends can be produced by mixing together two polymers (as melts, latices, or in solution) or by in-situ polymerization of a monomer in the presence of a dissolved polymer.

The energy taken up during *melt mixing* is used for flow processes and for the generation of surfaces of new microdomains. After some time, a steady state is established and the domain size becomes constant. No macroscopic demixing occurs because of the slow diffusion resulting from the high viscosities.

Latex blending consists of the mixing of aqueous dispersions of two polymers. Far lower temperatures and lower shear fields can be employed compared to melt blending. The good mixing of the latex particles remains after coagulation. The domain size is, however, restricted to the size of the latex particles themselves; it is not altered by subsequent melting of the coagulate.

Solution blending involves the mixing of two polymer solutions. Miscible polymers can be blended to domains of molecule size. Solutions of immiscible polymers demix, however, at very low concentrations, sometimes under fractionation with respect to molar masses. Domains grow further on solvent removal by distillation or freeze-drying. The phase morphology of the blends of two immiscible polymers A and B with viscosities η_A and η_B and volume fractions ϕ_A and ϕ_B is determined in first approximation by the ratio $(\phi_A\eta_B)/(\phi_B\eta_A)$. Polymer A will form the continuous phase if $(\phi_A\eta_B)/(\phi_B\eta_A) > 1$ and polymer B the continuous phase if $(\phi_A\eta_B)/(\phi_B\eta_A) < 1$; at $(\phi_A\eta_B)/(\phi_B\eta_A) = 1$, co-continuous phases exist.

Table 13-6 Major commercial plastics blends.
For abbreviations and acronyms of polymer names see Chapter 15.

Polymer	blended with Standard plastics	Low-end performance plastics	Engineering plastics	Specialty plastics	Thermosets
Commodity plastics					
PE	PS				
PP			PA		
PS	PE		PPE		
PVC		PMMA	ABS		
Low-end performance plastics					
ASA			PBT, PC		
ABS	PVC		PA, PC, PPE, TPU		
PMMA	PVC		PC		
SAN					
Engineering plastics					
PA	PP	ABS	PC, PPE	LCP	
PBT		ASA, SMA	PC, POM		
PC		ABS, ASA, PMMA,	PA, PBT, PET, TPU	PEI	PUR
PET			PC		
POM			PBT		
PPE	PS	ABS	PA		
TPU		ABS	PC		
Specialty plastics					
LCP			PA		
PEEK				PES	
PEI			PC		
PES				PEEK	
PPS				PSU	
PSU				PPS	
Thermosets					
PUR			PC		UP
UP					PUR

In-situ polymerization involves solutions or gels of polymers A in monomers B, which are subsequently polymerized. The in-situ polymerization of styrene in a styrenic polydiene solution is industrially most important. It leads to a rubber-toughened poly(styrene) (high-impact poly(styrene), HIPS, PS-HI) (see Section 13.7.3.).

Cross-linked rubbers may be swollen by cross-linkable monomers. The subsequent polymerization of the system leads to so-called interpenetrating networks. These IPNs actually consist of interconnected domain structures and not of truly interpenetrating polymer chains because the systems phase-separate during polymerization. Industry uses various IPNs:

polyurethane + polyester (+ poly(styrene)) for sheet molding compounds
polyurethane + acrylic + poly(styrene) for sheet molding compounds
SEBS + polyester for automotive applications
poly(propylene) + EPDM for automotive applications
poly(propylene) + EPDM + poly(ethylene) for automotive and wire and cable

The various blending processes produce blends with very different morphologies (Fig. 13-9) and properties. The notched impact strengths of blends from the in-situ polymerization of styrenic *cis*-1,4-poly(butadiene) solutions are about twice as high than those from the melt blending of poly(styrene) and *cis*-1,4-poly(butadiene). Latex blends of the same polymers have very low impact strengths.

13.7.3. Toughened Plastics

Rubber-modified plastics (toughened plastics, high-impact plastics) consist of rubber domains dispersed in plastic matrices; the world production is ca. $4 \cdot 10^6$ t/a (1990).

The size of rubber domains varies with the blending process. Typical values are ca. 0.1 μm for melt-blended poly(vinyl chloride)-*blend*-acryl rubber and ca. 1 μm for polymerization-blended poly(styrene)-*blend*-poly(butadiene). The domains are often multiphased; small plastic domains are imbedded in the rubber domains (Fig. 13-9).

The morphology depends strongly on the blending process. During stirred in-situ polymerizations, phase inversions often occur if the amount of the newly formed plastic approaches that of the incipient rubber. During the phase inversion, already present plastics particles may be embedded in the newly formed rubber domains.

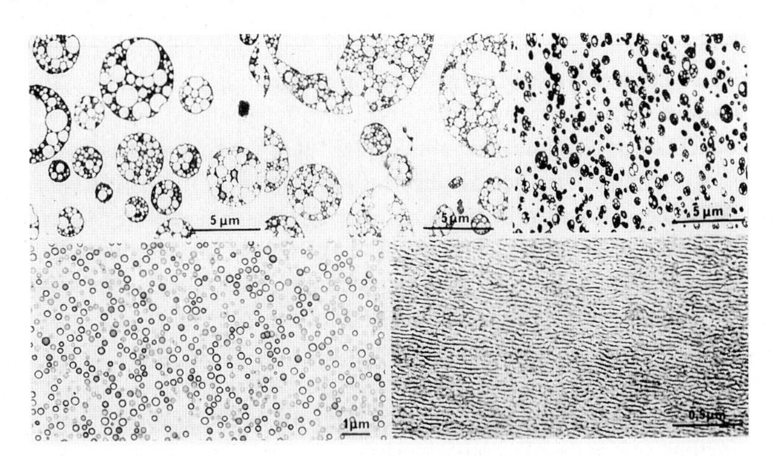

Fig. 13-9 Morphologies of various rubber-modified (impact-modified) poly(styrene)s.
Upper row; left: Conventional impact-modified poly(styrene) PS-HI;
center: PS-HI with improved resistance against stress corrosion;
right: PS-HI with high surface gloss;
Lower row; left: capsule morphology of PS-HI with improved transparency;
right: finely dispersed rubber phase in glass-clear PS-HI.
Reprinted with permission by Hanser Publ., Munich [4].

Rubber-toughened plastics are valued because of their improved impact strength. On impact, very many crazes are formed near the equators of the rubber particles. The crazes propagate until they encounter an obstacle (e.g., rubber particle or shear band) or until the stress concentration at the tip of the craze becomes very low. Many small crazes result, and the stress is evenly distributed if the rubber-phase is cross-linked and binds well to the thermoplast phase (e.g., by in-situ formed graft copolymers or deliberately added diblock copolymers as compatibilizers). In contrast, stress peaks concentrate at a few defect points in normal thermoplastics.

The high notched impact strengths of the polymerization-blended materials are caused by a strong anchoring of phases due to the formation of graft copolymers of styrene on poly(butadiene) and cross-linking within the rubber domains, both caused by free radical initiators. Such processes are less prevalent during melt blending (radical formation by high shearing at elevated temperatures) and latex blending (low shearing at ambient temperature). In-situ polymerization is thus the method of choice for unsaturated rubbers (easy cross-linking and grafting by chain transfer) and monomer-rubber pairs with favorable Q,e-values for copolymerizations. In all other cases, blends are formed by melt mixing.

13.8. Thermoplastic Elastomers

Thermoplastic elastomers (elastoplastics, thermolastics, plastomers) are processed like thermoplastics but applied like elastomers. Their unique properties follow from their molecular structures; their chains consist of "soft" and "hard" segments in block, graft, or segmented copolymers composed of monomeric units A and B. The A and B segments are mutually incompatible and form locally separated regions. With a well-designed molecular architecture, domains of the "hard" A segments (transition temperature > service temperature) act as physical cross-links in the continuous matrix of the "soft" B segments (transition temperature < service temperature) (Section 4.3.6). Such transition temperatures may be glass temperatures in amorphous polymers or melting temperatures in partially crystalline polymers. The A segments attain mobility above these transition temperatures and the elastoplasts become processible.

Thermoplastic elastomers comprise linear triblock copolymers of the poly(styrene)-*block*-poly(diene)-*block*-poly(styrene) type; radial, star, or teleblock copolymers of the same monomeric units; urethane segment or block copolymers with polyester or polyether soft segments; polyesteramides and polyesteretheramides; graft copolymers of butyl rubber on poly(ethylene), vinyl chloride on poly(ethylene-*co*-vinyl acetate), styrene-acrylonitrile on saturated acryl rubbers; and various ionomers.

Thermoplastic elastomers can also be produced by "dynamic vulcanization". The mastication of blends of conventional rubbers and crystalline poly(α-olefin)s leads to chain scissions. The resulting macroradicals cross-link the rubber domains.

The mechanical properties of thermoplastic elastomers are mainly determined by their morphologies. In styrene-butadiene-styrene triblock copolymers, for example, morphologies are ruled by the spatial requirements of the various blocks (Section 4.3.6). With a low content of hard styrene segments (e.g., 13 %), small spherical poly-(styrene) microdomains are formed in the soft poly(butadiene) matrix. These micro-domains act as physical cross-linkers. The distances between the domains are large, and the elastoplastic behaves as a weakly linked elastomer with correspondingly high extension (Fig. 13-10).

The domains are larger and their distances are shorter at 28 % styrene units and the polymer strengthens (stiffens). The stiffening becomes stronger for rod-like styrene domains (39 % S). Lamellar morphologies (53 % S) show an "unruly" behavior initially because of reorientation of lamellae. At even higher styrene contents, the polymer behaves as a plasticized tough thermoplastic (65 % S) or almost like poly-(styrene) itself (80 % S).

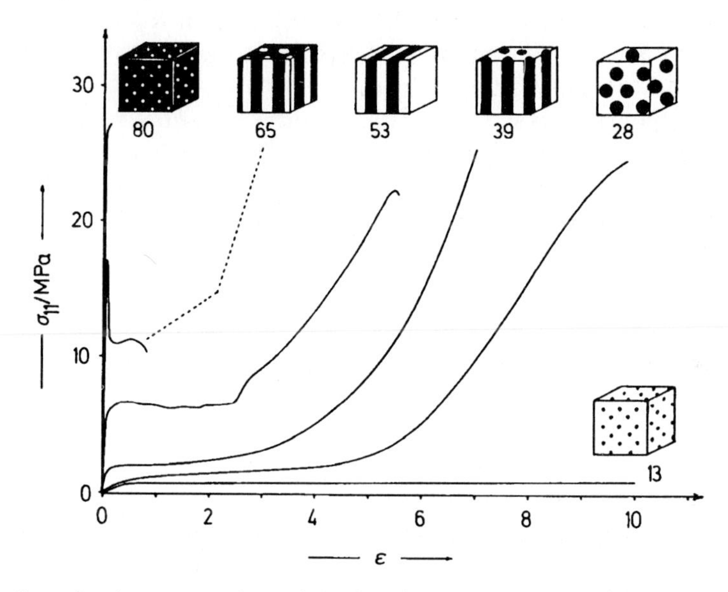

Fig. 13-10 Stress/strain curves and morphologies of styrene-butadiene-styrene triblock copolymers with various styrene contents. Stress-strain curves reprinted with permission by McLaren and Sons, London [5].

13.9. Expanded Plastics

Expanded plastics (foamed plastics, plastics foams, cellular plastics™) are blends of plastics with air. They are subdivided according to nature of the parent plastics and their density, rigidity, and cell structure.

Light plastics foams have densities of 10-100 kg/m^3 (0.01-0.10 g/cm^3); dense ones, 400-600 kg/m^3.

Rigid plastics foams are sometimes named "hard foams" and flexible ones "soft foams". Rigid plastic foams with densities greater than 320 kg/m^3 are called structural foams (but see below). Rigid expanded plastics are used mainly for thermal insulation, and flexible foams for damping and cushioning materials.

Foamed elastomers are named "foamed rubbers" or "cellular rubbers"; they are subdivided into sponge rubbers (open-celled) and expanded rubbers (close-celled). "Latex foam rubbers" are generated by frothing of a rubber latex or liquid rubber, followed by a gelling, and a vulcanization of the expanded state.

Plastic foams are defined as flexible if a test specimen (20 cm × 2.5 cm × 2.5 cm) does not rupture if it is wrapped around a 2.5 cm mandrel at (15-25)°C with a uniform rate of 1 lap per 5 seconds (ASTM test).

The rigidity ("hardness") of expanded plastics follows the properties of their parent polymers. Phenolic and urea resins thus deliver brittle-rigid; poly(styrene) and hard poly(vinyl chloride), tough-rigid; and poly(ethylene)s, polyurethanes, and plasticized PVC, semirigid to flexible foams. The elastic moduli of expanded plastics decrease approximately proportional to their polymer contents (i.e., $E/E_P = \rho/\rho_P$; simple mixing law) because the elastic moduli of gases can be neglected (Fig. 13-11). Since, however, the stiffness of an article increases with the third power of the wall thickness, the gases in expanded plastics work as enhancers that reduce material costs and weights.

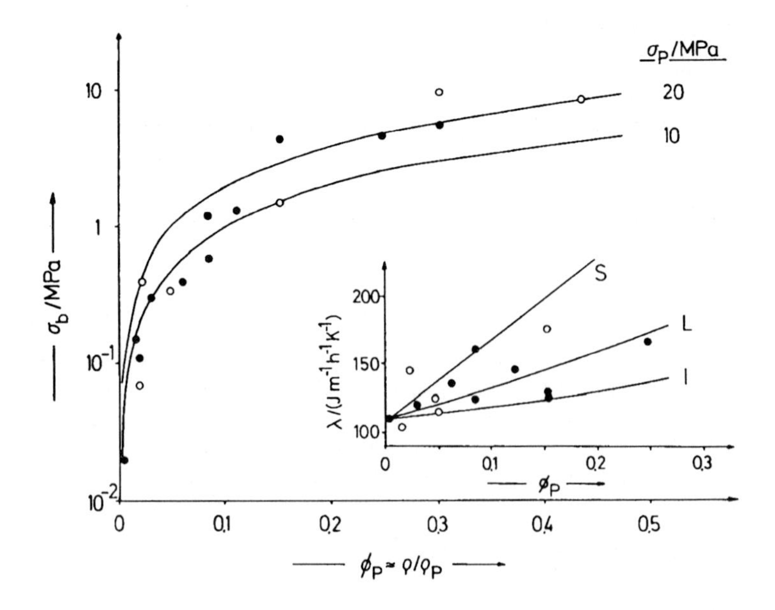

Fig. 13-11 Tensile strength σ_b of expanded plastics as function of the volume fraction ϕ_P of rigid (●) or flexible (o) parent polymers P. Solid lines: simple mixing of gases ($\sigma_G \approx 0$) and plastics with σ_P = 10 or 20 MPa. Insert: Thermal conductivities of rigid (●) and flexible (o) plastic foams. Solid lines: Simple (S), logarithmic (L), and inverse (I) mixing laws for λ_P = 700 J m^{-1} h^{-1} K^{-1} and λ_{air} = 110 J m^{-1} h^{-1} K^{-1}.

Tensile strengths, compression strengths, and thermal conductivities of foamed plastics often also follow the simple mixing law. In some cases, a power function $Q/Q_p = (\rho/\rho_p)^z$ is more useful; examples are foamed poly(ethylene) boards with $z = 4/3$ and foamed epoxide polymers with $z = 2$.

The cell structure can be open, closed, or mixed. Open-celled foams are always air-filled, regardless of the blowing or expanding gases used for foam manufacture. The trapped gases in closed-cell structures can be exchanged with the surrounding air only by slow diffusion of air through the polymer matrix. Since thermal conductivities of low-density foams ($\rho_p < 0.3$) obey, to a first approximation, a logarithmic mixing law, they are considerably affected by the thermal conductities of trapped gases [$\lambda = 93$ J m^{-1} h^{-1} K^{-1} (N_2), 56 J m^{-1} h^{-1} K^{-1} (CO_2), 26.4 J m^{-1} h^{-1} K^{-1} (CCl_3F)]. The gas exchange is considerably reduced in integral skin foams and in syntactic foams. *Integral skin foams* (structural foams, microcellular foams, self-skinning foams) have a dense skin around a low-density core. *Syntactic foams* are plastics filled with hollow spheres; the spheres may be made of glass, ceramics, or plastics and contain either gases or vacuum.

Literature

13.1. Introduction: General References

L.E.Nielsen, Mechanical Properties of Polymers and Composites, Dekker, New York 1982
M.Grayson, ed., Encyclopedia of Composite Materials and Components, Wiley, New York 1983 (reprints of articles of the 3rd edition of Encyclopedia of Chemical Technology)
R.B.Seymour, R.D.Deanin, eds., History of Polymeric Composites, Dekker, New York 1987

13.2.-13.4. Microcomposites, Macrocomposites, Molecular Composites

M.O.W.Richardson, ed., Polymer Engineering Composites, Applied Science Publ., London 1977
V.K.Tewary, Mechanics of Fibre Composites, Wiley, Chichester, Sussex 1978
R.G.Weatherhead, FRP Technology, Applied Science Publ., Barking, Essex 1980
N.L.Hancox, ed., Fibre Composite Hybrid Materials, Hanser International, Munich 1981
G.Lubin, ed., Handbook of Composites, Van Nostrand Reinhold, New York 1982
R.P.Sheldon, Composite Polymeric Materials, Applied Science Publ., London 1982
M.J.Folkes, Short Fiber Reinforced Thermoplastics, Wiley, New York 1982
M.Holmes, D.J.Just, Structural Properties of Glass Reinforced Plastics, Appl.Sci.Publ., London 1983
H.Ishida, G.Kumar, eds., Molecular Characterization of Composite Interfaces, Plenum, New York 1985
A.A.Berlin, S.A.Volfson, N.S.Enikolopian, S.S.Negmatov, Principles of Polymer Composites, Springer, Berlin 1986
S.K.Bhattacharya, Metal-Filled Polymers, Dekker, New York 1986

R.Yosomiya, K.Morimoto, A.Nakajima, Y.Ikada, T.Suzuki, Adhesion and Bonding in Composites, Dekker, New York 1989
S.Lee, ed., International Encyclopedia of Composites, VCH, New York 1990-1991 (6 vols.)
P.K.Mallick, S.Newman, eds., Composite Materials Technology, Hanser, Munich 1990
E.P.Plueddeman, Silane Coupling Agents, Plenum, New York, 2nd ed. 1990
R.B.Seymour, Reinforced Plastics: Properties and Applications, ASM, Materials Park (OH) 1991
T.L.Vigo, B.J.Kinzig, eds., Composite Applications. The Role of Matrix, Fiber, and Interface, VCH, New York 1992
H.-H.Kausch, ed., Advanced Thermoplastic Composites, Hanser, Munich 1993

13.5. and 13.7. Homogeneous and Heterogeneous Polymer Blends

J.A.Manson, L.H.Sperling, Polymer Blends and Composites, Plenum, New York 1976
D.R.Paul, S.Newman, eds., Polymer Blends, Academic Press, New York 1978 (2 vols.)
G.Olabisi, L.M.Robeson, M.T.Shaw, Polymer-Polymer Miscibility, Academic Press, New York 1979
L.H.Sperling, Interpenetrating Polymer Networks and Related Materials, Plenum, New York 1980
E.Martuscelli, R.Palumbo, M.Kryszewski, eds., Polymer Blends, Plenum, New York 1980
K.Solc, Polymer Compatibility and Incompatibility, Harwood Acad.Publ., New York 1982
D.J.Walsh, J.S.Higgins, A.Maconnachie, eds., Polymer Blends and Mixtures, Nijhoff, Dordrecht 1985 (NATO ASI Ser. E, No. 89)
L.A.Utracki, A.P.Plochocki, Industrial Polymer Blends and Alloys, Hanser, Munich 1985
L.A.Utracki, Polymer Alloys and Blends, Hanser, Munich 1989
L.A.Utracki, ed., Two Phase Polymer Systems, Hanser, Munich 1992
L.Bottenbruch, Ed., Technische Polymer-Blends (= G.W.Becker, D.Braun, Eds., Kunststoff-Handbuch **3/2**), Hander, Munich 1993

13.7.3. Toughened Plastics

J.R.Dunn, Blends of Elastomers and Plastics–A Review, Rubber Chem.Technol. **49** (1976) 978
C.B.Bucknall, Toughened Plastics, Applied Science Publ., London 1977

13.8. Thermoplastic Elastomers

N.R.Legge, G.Holden, H.Schroeder, eds., Thermoplastic Elastomers, Hanser, Munich 1987
B.M.Walker, C.P.Rader, eds., Handbook of Thermoplastic Elastomers, Van Nostrand-Reinhold, New York, 2nd ed. 1988

13.9. Expanded Plastics

K.C.Frisch, J.H.Saunders, eds., Plastic Foams, Dekker, New York 1972 and 1973 (2 vols.)
N.C.Hilyard, ed., Mechanics of Cellular Plastics, Hanser International, Munich 1982
Y.L.Meltzer, Expanded Plastics and Related Products, Noyes, Park Ridge 1983
B.C.Wendle, Structural Foam, Dekker, New York 1985
F.A.Shutov, Integral/Structural Foams, Springer, Berlin 1986
D.Klempner, K.C.Frisch, eds., Handbook of Polymeric Foams and Foam Technology, Hanser, Munich 1991

References

[1] R.S.Spencer and R.F.Boyer, J.Appl.Phys. **16** (1945) 594.
[2] G.J.van Amerongen, in R.Houwink, ed., Chemie und Technologie der Kunststoffe, Akademische Verlagsgesellschaft Geest und Portig, Leipzig, Vol. 1 (1954), p. 453 ff., data of Fig. 195
[3] H.Willersinn, Makromoleculare Chem. **101** (1967) 296, data of Figs. 3 and 9.
[4] H.Jenne, Kunststoffe **74** (1984) 551, Fig. 2.
[5] G.Holden, E.T.Bishop, N.R.Legge, Proc.Int.Rubber Conf. 1967, McLaren, London 1968, Fig. 5
[6] S.Artmeyer, K.-H.Seemann, G.Zieschank, G.Spur, W.Brockmann, P.Berns, K.Wiebusch, Ullmann's Enzyklopädie der technischen Chemie, Verlag Chemie, Weinheim, 4.Aufl., **15** (1978) 311

14. Disposal and Recycling

Four different, although frequently interconnected, aspects must be considered in the disposal of plastic waste: (1) technological demands, (2) economic considerations, (3) energy consumption, and (4) environmental concerns.

Modern technology demands certain material properties for certain applications that can be delivered only by polymers and not by other materials such as metals or ceramics. Examples range from rubber tires to high-technology applications such as polymer resists in electronic chip production. The desirability of polymers for technology-driven demands is usually undisputed; the discussion centers rather around the desirability of commodity plastics, especially for packaging purposes.

It is also not much disputed, at least amongst people that are well-versed in technology, that the polymeric materials are very economical compared to other materials in terms of production and use. However, the ecological costs are questioned: the hidden costs of the "free" use of air, water, and land and the suspected high energy consumption in the production, use and disposal of polymers.

Energy consumption has been studied by consumer agencies (e.g., NATO Science Committee), industry (e.g., BASF), and government (e.g., Bundesamt für Umweltschutz = Swiss Environmental Protection Agency). Although differences exist among the various approaches, newer data all agree that the energy consumed in plastics production agrees favorably with that of other materials if use (on a volume basis) rather than sale (on a weight basis) is considered (Table 14-1).

Table 14-1 Energy consumption for the production of various materials according to NATO Science Committee, BASF Corporation, and Swiss Environmental Protection Agency (BAFU).

Material	Consumption in kJ/cm^3		
	NATO	BASF	BAFU
Urea-formaldehyde foam	-	0.48	-
Poly(styrene), expanded	-	1.4	-
Lumber	2.0	-	-
Polyurethane, expanded	-	3.0	-
Poly(ethylene), high density	-	63	68
Poly(vinyl chloride)	11	73	84
Poly(styrene), crystal	-	84	-
Paper (Kraft, unbleached)	40	-	63
ABS plastics	-	89	87
Poly(ethylene terephthalate)	-	113	-
Glass	< 125	-	> 26
Polyamide 6	-	176	-
Tin plate	-	-	210
Aluminum	< 460	-	750

Table 14-2 Energy consumption and critical environmental data for 1000 1-L packages of milk.
*) Average Swiss disposal: 23 % landfill, 77 % incineration (55 % with use of energy).
**) Solid wastes from PE (ash from production processes), paper (ash from production and paper), glass (with 60 % recycling assumed), aluminum, and tin plate (100 %, no recycling).

	PE tube	PE bottle	Paper carton	Glass bottle
Use cycles (life)	1	1	1	40
Weight per life (in kg)	7	22	25	40
Energy consumption (in MJ)*	400	1510	1770	650
Environmental charges				
Air (in 10^6 m^3)	5.9	20.5	30.9	5.4
Water (in m^3)	3.0	17.2	154	9.3
Solids (landfill) (in L)	2.3	9.8	8.3	6.8
Solids (average Swiss disposal) (L)**	0.35	3.0	1.3	3.3

Energy considerations must however to be based on specific applications of materials. For example, these considerations must take into account the current possible use of different materials for its manufacture, its use (incl. cleaning), possible recycling, and/or disposal by landfill or incineration (with partial use of heat). This is shown for the example of 1000 1 L-packages of milk (Table 14-2).

Energy data are again favorable for plastics as are the critical amounts of air and water employed in the production, use, and disposal (i.e., volumes to which air and water can be maximally charged without exceeding Swiss environmental laws). It must be mentioned in this context that only ca. 5 % of the world petroleum production is used for the manufacture of polymers; the remainder is burned (as gasoline, heating oil, etc.), often with very low energy efficiencies (ca. 30 % in cars) and large amounts of harmful byproducts (exhaust gases).

Environmental concerns include possible harmful effects by plastics articles, plastics themselves, polymers, incineration products, and byproducts of polymer and plastics manufacture. Environmentalists estimate the amount of plastics discarded annually into the high seas as 1 million tons. Certain plastics articles are undoubtedly harmful to wildlife, for example, six-pack rings and old fishing nets to fish and other animals. The use of biodegradable plastics for such articles may be justified.

Biodegradable plastics either contain biodegradable polymers themselves or are compounds of conventional polymers with biodegradable polymers such as starch. Biodegradation may occur via hydrolysis (starch, polyesters) or via UV radiation (polymers containing carbonyl groups) or both. The degradation however is slow, taking from several months under laboratory conditions to several years in landfills. Biodegraded products may no longer be visible, but whether biodegradation proceeds to small molecules and whether these molecules are harmfull to living organisms is still not known. Harmful polymer additives may be released during biodegradation. Conventional polymer molecules, on the other hand, are not harmful to living beings, at least not to higher ones, because they can be neither resorbed nor digested. The energy content of plastics is of course wasted during biodegradation.

Landfills. Biodegradable polymers thus do not seem to be the general solution of the plastics waste problem. Contrary to popular opinion, the relative amount of plastics in household refuse is relatively small, ca. 7 wt-% vs. ca. 50 % paper, and almost constant over at least 25 years according to archeological studies of landfills. Weight fractions and not volume fractions of plastics must be used for comparison with other materials because hollow plastic containers and expanded plastics are compressed by the weight of the landfill (glass bottles are not compressed). Biodegradation processes are very slow in landfills (absence of light, air, and even microorganisms); some buried papers could be read after 25 years in a landfill. This leaves recycling and incineration as possible ways for the disposal of plastics.

Recycling. Most industrial thermoplastics wastes are already recycled or incinerated. Even cross-linked polyurethane waste and discarded parts can be recycled: preheated pellets "flow" together in molds at 180-185°C and 350 bar. The molded parts do not have the same surface smoothness as parts from virgin PUR though.

The problem is the collection, cleaning, and recycling of household refuse. Only ca. 2 % of all consumed plastics were recycled in the United States (Table 14-3) and the Federal Republic of Germany in 1988. The recycling of PET bottles now approaches 25 %. For comparison: the gross discards recovered in the U.S. were 25 % of aluminum, 23 % of paper and paper board, and 9 % of glass. In Switzerland, 95 % of the glass is recycled but none of the aluminum.

Table 14-3 Recycling of plastics in the United States of America (1988).

Polymer	Mass in 1000 t		% Recycled
	Consumed	Recycled	
Poly(olefins)	11 538	343	3.0
Poly(vinyl chloride)	3 779	34	0.9
Styrenics	3 516	39	1.1
Poly(ethylene terephthalate) (bottles)	911	54	5.9
Polyamides	253	27	10.7
Other	6 219	< 1	< 0.01
Total	*26 216*	*498*	*1.9*

Since unsorted recycled polymers represent mixtures of different grades and even different polymers, they can be used only for low-value articles unless generally applicable compatibilizers are developed. Several processes exist for the separation of polymer mixtures according to the density of the ground-up refuse or via the solubility of polymers in solvents; none of these processes seems to be economical.

The sorting of discarded plastics bottles and containers is made more easy by recently introduced codes which are stamped on the bottom of containers or other inconspicuous places. Such codes have been recommended by the International Standardization Organization (ISO), the American Society for Testing and Materials (ASTM), and the Society of the Plastics Industry (SPI).

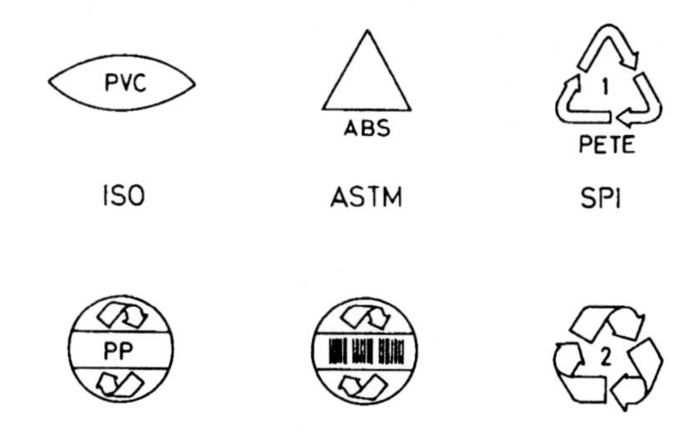

Fig. 14-1 Recommended or presently (1992) used symbols for recyclable plastics.
Upper row from left to right: Recommended symbols by ISO (International Standardization Organization), ASTM (American Society for the Testing of Materials), and SPI (Society of the Plastics Industry (USA)).
Lower row: Various other used symbols with letter, bar, or number codes.

SPI codes are now generally used for bottles and containers in the United States. These codes consist of three broken arrows, arranged in a triangle, and a number, sometimes also combined with letters and/or a bar code (Fig. 14-1). The letter combinations and numbers indicate the type of plastics; they are not always identical with the acronyms and abbreviations recommended by ISO, ASTM, DIN, IUPAC, etc., for the same polymers. The following number and letter codes are used by SPI for the recycling of plastics:

1 = PETE = poly(ethylene terephthalate),
2 = HDPE = high density poly(ethylene),
3 = V = vinyl chloride polymers,
4 = LDPE = low density poly(ethylene),
5 = PP = poly(propylene),
6 = PS = poly(styrene),
7 = = all other plastics, including multilayered materials.

Incineration of refuse is the present method of choice of many European communities: 77 % of all household refuse is incinerated in densely populated Switzerland, 30 % in West Germany, but only 10 % in the United States where 80 % of the refuse is accumulating rapidly in landfills. Furthermore, concerns exist about the emission of HCl from the burning of PVC, the generation of dioxins, and the disposal of heavy metal-containing ashes.

In the old states of the Federal Republic of Germany (West Germany), only 18 % of the HCl comes from the incineration of refuse (and not more than half of it from PVC); the primary sources of HCl are power plants (1978 data). This HCl is removed by scrubbing, etc., and may be used to bind heavy metals in compost from incineration. Dioxins are generated by the burning of all organic materials. Their amounts can be reduced significantly by high-temperature incinerators, which of course command higher investment costs.

Literature

J.E.Guillet, ed., Polymers and Ecological Problems, Plenum, London 1973
J.Leidner, Plastics Waste: Recovery of Economic Value, Dekker, New York 1981
G.Menges, W.Michaeli, eds., Recycling von Kunststoffen, Hanser, Munich 1991
R.J.Ehrig, Plastics Recycling, Hanser, Munich 1992

15. Appendix

15.1. Physical Quantities and Units

Many physical properties can be described in quantitative terms by *quantity calculus*. The value of a physical quantity (symbols in *italics*) equals the product of a numerical value (upright numbers) and a physical unit (symbols in upright letters):

$$\text{physical quantity} = \text{numerical value} \times \text{physical unit}$$

This equation can be manipulated by the ordinary rules of algebra. If, for example, a certain item has a length of 0.002 meters, then this may be written as

$$L = 0.002 \text{ m} \quad \text{or} \quad L = 2 \cdot 10^{-3} \text{ m} \quad \text{or} \quad 10^3 L/\text{m} = 2 \quad \text{or} \quad L = 2 \text{ mm} \quad \text{or} \quad L/\text{mm} = 2$$

but not as $10^{-3} L/\text{m} = 2$. A column head $10^2 F/(\text{N m}^{-2})$ for a column entry of 7.35 thus indicates a force $F = 7.35 \cdot 10^{-2}$ N/m^2. Literature data often do not follow these SI rules (Systéme International) by the International Standardization Organization (ISO), which are adopted by IUPAP (International Union of Pure and Applied Physics), IUPAC (International Union of Pure and Applied Chemistry). etc. Instead one finds various nonrational notations such as F, N m^{-2} or F [N m^{-2}] for a column figure of $7.35 \cdot 10^{-2}$, often with wrong algebraic statements such as $10^{-2} F$, N m^{-2}.

American technical literature and sometimes also American scientific literature still uses mainly Anglo-Saxon units although by law all U.S. government units were supposed to convert to the SI system by the end of 1992. Only SI units can be lawfully used for commercial purposes in many other countries. In order to provide access to older and Anglo-Saxon technical literature, the following tables list names and symbols of SI units, SI prefixes, and conversions from SI units (in American notation) to Anglo-Saxon units. In many cases, IUPAC/IUPAP recommendations for symbols for physical quantities are also given.

Table 15-1 Names and symbols for physical quantities and their SI units.

Physical quantity Symbol	Name	SI unit Name	Symbol
Base units			
l	length	meter	m
m	mass	kilogram	kg
t	time	second	s
I	electric current	ampere	A
T	thermodynamic temperature	kelvin	K
n	amount of substance	mole	mol
I_v	luminous intensity	candela	cd
Supplementary units			
α, β, γ	plane angle	radian	rad = m m^{-1}
ω, Ω	solid angle	steradian	sr = m^2 m^{-2}

Table 15-2 SI derived units with special names and symbols.

Physical quantity Symbol	Name	Physical unit Name	Symbol	Derived units	Base units
ν	frequency	hertz	Hz[a)]	$= s^{-1}$	
E	energy, work, heat	joule	J	= N m	$= m^2\ kg\ s^{-2}$
F	force	newton	N	$= J\ m^{-1}$	$= m\ kg\ s^{-2}$
p	pressure, stress	pascal	Pa	$= N\ m^{-2}$	$= m^{-1}\ kg\ s^{-2}$
Q	electric charge	coulomb	C	= A s	
U	electric potential, electromotive force	volt	V	$= J\ C^{-1}$	$= m^2\ kg\ s^{-3}\ A^{-1}$
R	electric resistance	ohm	Ω	$= V\ A^{-1}$	$= m^2\ kg\ s^{-3}\ A^{-2}$
G	electric conductance	siemens	S	$= \Omega^{-1}$	$= m^{-2}\ kg^{-1}\ s^3\ A^2$
C	electric capacitance	farad	F	$= C\ V^{-1}$	$= m^{-2}\ kg^{-1}\ s^4\ A^2$
P	power, radiant flux	watt	W = V A	$= J\ s^{-1}$	$= m^{-1}\ kg\ s^{-2}$
B	magnetic flux density	tesla	T	$= V\ s\ m^{-2}$	$= kg\ s^{-2}\ A^{-1}$
Φ	magnetic flux	weber	Wb	= V s	$= m^2\ kg\ s^{-2}\ A^{-1}$
L	inductance	henry	H	$= V\ A^{-1}\ s$	$= m^2\ kg\ s^{-2}\ A^{-2}$
A	activity (radioactive)	becquerel	Bq	$= s^{-1}$	
D	absorbed dose, radiation	gray	Gy	$= J\ kg^{-1}$	$= m^2\ s^{-2}$
-	dose equivalent	sievert	Sv	$= J\ kg^{-1}$	$= m^2\ s^{-2}$
θ	Celsius temperature	degree Celsius	°C	b)	

a) The unit hertz should be used only for frequency in the sense of cycles per second. Radial (circular) frequency and angular velocity have the units rad s^{-1} which may be simpified to s^{-1} (see Table 15-1) but *not* to Hz.

b) The Celsius temperature is defined by $(\theta/°C) \equiv (T/K) - 273.15$. The SI unit of the Celsius temperature interval is the degree Celsius (°C), which is equal to the kelvin (not "degree kelvin") K. IUPAC now recommends the symbol θ for the Celsius temperature. A capital theta (Θ) has been traditionally used in polymer science for the theta temperature.

Table 15-3 SI prefixes.

Multiple	Prefix	Symbol	Submultiple	Prefix	Symbol
10^{18}	exa	E	10^{-18}	atto	a
10^{15}	peta	P	10^{-15}	femto	f
10^{12}	tera	T	10^{-12}	pico	p
10^9	giga	G	10^{-9}	nano	n
10^6	mega	M a)	10^{-6}	micro	μ
10^3	kilo	k b)	10^{-3}	milli	m
10^2	hecto	h	10^{-2}	centi	c
10	deca	da	10^{-1}	deci	d

a) The letter "M" is frequently used in the U.S. to symbolize 10^3, especially in financial circles and in the gas industry. The same groups use "MM" for 10^6, "B" for 10^9 (American billion) and "T" for 10^{12} (American trillion).

b) Computer jargon uses "K" for $2^{10} = 1024 \approx 1000$.

Table 15-4 Older non-SI units which may be used with SI prefixes and together with SI units.

Physical quantity	Physical unit Name	Symbol	Value in SI units
time	minute	min	60 s
time	hour	h	3600 s
time	day	d	86 400 s
length	ångstrøm	Å	10^{-10} m
area	barn	b	10^{-28} m^2
volume	liter	l, L	10^{-3} m^3
mass	ton[a]	t	10^3 kg
pressure	bar	bar	10^5 Pa
energy	electronvolt	eV	$\approx 1.60218 \cdot 10^{-19}$ J
mass	unified atomic mass unit[b]	$u = m_a(^{12}C)/12$	$\approx 1.66054 \cdot 10^{-27}$ kg
plane angle	degree	°	(π/180) rad
plane angle	minute	'	(π/10 800) rad
plane angle	second	"	(π/648 000) rad

[a] The metric "ton(ne)" is not to be confused with the "long ton" (≈ 1016.047 kg) and "short ton" (≈ 907.185 kg), which are often used in the Anglo-Saxon technical and commercial literature without the adjectives "long" and "short".
[b] The unified atomic mass is also sometimes called the dalton (symbol Da); in the biosciences, "dalton" has come to mean the relative molecular mass (unit 1) or the molar mass (unit g/mol)!

Table 15-5 Conversion of out-dated and Anglo-Saxon units into SI units. Non-SI units marked with * may be used with SI prefixes and together with SI units.

Name	Old unit	= SI unit
Length		
light year	1 ly	$= 9.4605 \cdot 10^{15}$ m
nautical mile (sea mile)	1 n	= 1852 m
statute mile (land mile)	1 mile	= 1609.344 m
furlong	1 furlong	= 201.168 m
rod (perch, pole)	1 rod	= 5.029 2 m
fathom	1 fathom	= 1.828 8 m
yard	1 yd	= 0.914 4 m
foot	1 ft = 1'	= 0.304 8 m = 30.48 cm
inch	1 in = 1"	= 0.025 4 m = 2.54 cm
mil	1 mil	$= 2.54 \cdot 10^{-5}$ m = 25.4 µm
micron	1 µ	$= 10^{-6}$ m = 1 µm
millimicron	1 mµ	$= 10^{-9}$ m = 1 nm
ångstrøm*	1 Å	$= 10^{-10}$ m = 0.1 nm
Area		
square mile	1 sq. mile	$= 2.589\ 988\ 110 \cdot 10^6$ m^2
hectare	1 ha	= 10 000 m^2
acre	1 acre	= 4046.856 m^2

Table 15-5 (continued) Conversion of out-dated and Anglo-Saxon units into SI units.

Name	Old unit	= SI unit
Area (continued)		
square yard	1 sq. yd.	= 0.836 127 36 m^2
square foot	1 sq. ft.	= 9.20 304$\cdot 10^{-2}$ m^2
square inch	1 sq. in.	= 6.451 6$\cdot 10^{-4}$ m^2
Volume		
stere	1 st	= 1 m^3
cubic yard	1 cu. yd.	= 0.764 554 857 m^3
imperial barrel	1 barrel	= 0.1636 m^3
US barrel petroleum	1 bbl = 42 US gal	= 0.158 987 m^3
US barrel	1 barrel	= 0.119 m^3 = 119 L
bushel	1 bu	= 3.524$\cdot 10^{-2}$ m^3 = 35.24 L
cubic foot	1 cu. ft.	= 2.381 684 659 2$\cdot 10^{-2}$ m^3
peck (US)	1 peck	= 8.810 L
board foot	12·12·1 cu. in.	= 2.3597$\cdot 10^{-3}$ m^3
gallon (British or Imperial)	1 gal	= 4.546 09$\cdot 10^{-3}$ m^3 = 4.545 96 L
gallon (US dry)	1 gal	= 4.405$\cdot 10^{-3}$ m^3
gallon (US liquid)	1 gal = 4 US qt.	= 3.785 41$\cdot 10^{-3}$ m^3 = 3.785 412 L
liter (cgs)	1 L	= 1.000 028$\cdot 10^{-3}$ m^3
liter*	1 L	≡ 1$\cdot 10^{-3}$ m^{-3}
quart (US dry)	1 qt.	= 1.101 L
quart (US liquid)	1 qt. = 2 US pints	= 0.946 335 L
pint (US liquid)	1 pt. = 2 US cups	= 0.473 168 L
pint (U.S. dry)	1 pt.	= 0.550 6 L
cup (US)	1 cup = 8 fluid oz.	= 0.236 534 L
British liquid ounce	1 oz.	= 0.028 413 L
US fluid ounce	1 oz. = 2 table sp.	= 0.029 574 L
cubic inch	1 cu. in.	= 0.016 387 064 L
tablespoon	1 tbl.sp. = 3 tea sp.	= 0.014 79 L
teaspoon	1 tea sp.	= 0.004 93 L
dram (US fluid)	1 dram	= 0.003 697 L
Mass		
long ton (UK)	1 ton = 2240 lbs	= 1016.046 909 kg
short ton (US)	1 sh.ton = 2000 lbs	= 907.184 74 kg
quintal	1 quintal	= 100 kg
hundred weight (UK)	1 cwt.	= 50.802 3 kg
short hundred weight	1 sh.cwt.	= 45.359 2 kg
slug	1 slug	= 14.59 39 kg
stone	1 stone = 14 lbs.	= 6.350 293 18 kg
pound (avoirdupois, US)	1 lb = 16 oz.	= 453. 592 37 g
pound (apothecaries' or troy, US)	1 lb = 8 drams	= 373. 242 g
ounce (avoirdupois, US)	1 oz.	≈ 28.349 5 g
ounce (troy)	1 oz.	≈ 31.103 5 g
dram (apothecaries')	1 dram	= 3.888 g
dram (avoirdupois)	1 dram	= 1.772 g

Table 15-5 (continued) Conversion of out-dated and Anglo-Saxon units into SI units.

Name	Old unit	= SI unit
Mass (continued)		
pennyweight	1 pennyweight	= 1.555 g
carat	1 ct	= 0.2 g
grain	1 gr	= 64.798 91 mg
unified atomic mass unit*	1 mu	= $1.660\ 540\ 2\ (10)\cdot 10^{-27}$ kg
electron mass*	-	≈ $9.109\ 39\cdot 10^{-31}$ kg
Time		
year	1 a	≈ 365 days (for statistical purposes only)
month	1 mo	≈ 30 days (for statistical purposes only)
Temperature		
degree Fahrenheit	(z°F - 32°F)(5/9)	= y°C
degree Celsius	$(\theta/°C)$	≡ $(T/K) - 273.15$
Density ($1\ kg\cdot m^{-3} = 1\cdot 10^{-3}\ g\cdot cm^{-3}$)		
-	1 lb/cu.in.	= 27.679 904 71 g cm^{-3}
-	1 oz/cu.in.	= 1.729 993 853 g cm^{-3}
-	1 lb/cu.ft.	= $1.601\ 846\ 337\cdot 10^{-2}$ g cm^{-3}
-	1 lb/gal US	= $7.489\ 150\ 454\cdot 10^{-3}$ g cm^{-3}
Energy, work (1 J = 1 N·m = 1 W·s)		
quadrillion BTU (US)	1 Quad = 10^{15} BTU	= 1.055 EJ
coal unit (German)	1 ton SKE	∴ 29.31 GJ
coal unit (US)	1 ton coal	∴ 27.92 GJ
coal unit (UK)	1 ton coal	∴ 24.61 GJ
short ton bituminous coal	1 T	∴ 26.58 GJ
kilowatt-hour	1 kWh	= 3.6 MJ
horsepower-hour	1 hph	= 2.685 MJ
cubic foot-atmosphere	1 cu.ft.atm.	= 2.869 205 kJ
British thermal unit	1 BTU_{mean}	= 1.055 79 kJ
British thermal unit	1 BTU_{IT}	= 1.055 06 kJ
-	1 cu.ft.lb(wt)/sq.in.	= 195.237 8 J
liter atmosphere (cgs)	1 L atm	= 101.325 J
-	1 m kgf	= 9.806 65 J
calory, international	1 cal_{IT}	= 4.186 8 J
calory, thermochemical	1 cal_{th}	= 4.184 J
-	1 ft-lbf	= 1.355 818 J
-	1 ft-pdl	= 4.215 384 J
-	1 erg	= $1\cdot 10^{-7}$ J = 0.1 μJ = 1 g $cm^2\ s^{-2}$
electronvolt*	1 eV	≈ $1.602\ 18\cdot 10^{-19}$ J
Force		
-	1 ft-lbf/in.	= 53.378 64 N
kilogram-force	1 kgf	= 9.806 65 N
pound force	1 lbf	= 4.448 22 N

Table 15-5 (continued) Conversion of out-dated and Anglo-Saxon units into SI units.

Name	Old unit	= SI unit
Force (continued)		
ounce-force	1 oz.f.	= 0.2780 N
poundal	1 pdl	= 0.138 255 N
pond	1 p	= $9.806\ 65 \cdot 10^{-3}$ N
dyne	1 dyn	= $1 \cdot 10^{-5}$ N
Length-related force		
-	1 kp/cm	= 980.665 N m^{-1}
-	1 lbf/ft	= 14.593 898 N m^{-1}
Pressure, stress (1 MPa = 1 N mm^{-2})		
physical atmosphere	1 atm = 760 torr	= 0.101 325 MPa
bar*	1 bar	≡ 0.1 MPa
technical atmosphere	1 at	= 0.098 065 MPa
-	1 kp/cm^2	= 0.098 065 MPa
-	1 kgf/cm^2	= 0.098 065 MPa
pound-force per square inch	1 psi = 1 lb/sq.in.	= $6.894\ 757 \cdot 10^{-3}$ MPa
inch mercury (32° F)	1 in.Hg	= $3.386\ 388 \cdot 10^{-3}$ MPa
inch water (39.2°F)	1 in.H_2O	= 249.1 Pa
torr	1 torr	= (101 325/760) Pa ≈ 133.322 Pa
millimeter mercury	1 mm Hg	= 13.5951·9.806 65 Pa ≈133.322 Pa
-	1 dyn/cm^2	= $1 \cdot 10^{-5}$ MPa
millimeter water	1 mm H_2O	= $9.806\ 65 \cdot 10^{-6}$ MPa
-	1 pdl/sq.ft.	= $1.488\ 649 \cdot 10^{-6}$ MPa
Power (1 W = 1 J·s^{-1})		
horsepower (UK)	1 hp	= 745.7 W
horsepower (boiler)	1 hp	= 9810 W
horsepower (electric)	1 hp	= 746 W
horsepower (metric)	1 PS	= 735.499 W
-	1 BTU/h	= 0.293 275 W
-	1 cal/h	= $1.162\ 222 \cdot 10^{-3}$ W
Heat conductivity		
-	1 cal/(cm s °C)	= 418.6 W m^{-1} K^{-1}
-	1 BTU/(ft h °F)	= 1.731 956 W m^{-1} K^{-1}
-	1 kcal/(m h °C)	= 1.162 78 W m^{-1} K^{-1}
Heat transfer coefficient		
	1 cal/(cm^2 s °C)	= $4.186\ 8 \cdot 10^4$ W m^{-2} K^{-1}
	1 BTU/(ft^2 h °F)	= 5.682 215 W m^{-2} K^{-1}
	1 kcal/(m^2 h °C)	= 1.163 W m^{-2} K^{-1}
Fineness = titer = linear density		
tex*	1 tex	= $1 \cdot 10^{-6}$ kg m^{-1}
denier	1 den	= $0.111 \cdot 10^{-6}$ kg m^{-1}

Table 15-5 (continued) Conversion of out-dated and Anglo-Saxon units into SI units.

Name	Old unit	= SI unit
Tenacity		
-	1 gf/den	= 0.082 599 N tex^{-1} = 0.082 599 $m^2 s^{-2}$
		= 98.06 MPa × (density in g/cm^3)
Dynamic viscosity		
poise	1 P	= 0.1 Pa s
centipoise	1 cP	= 1 mPa s
Kinematic viscosity		
stokes	1 St	= $1 \cdot 10^{-4} m^2 s^{-1}$
Heat capacity		
clausius	1 Cl	= 1 cal_{th}/K = 4.184 J K^{-1}
Molar heat capacity		
entropy unit	1 e.u.	= 1 $cal_{th} K^{-1} mol^{-1}$ = 4.184 J $K^{-1} mol^{-1}$
Electrical conductivity		
reciprocal ohm	1 mho	= 1 S^{-1}
Electrical resistance		
Average international ohm	-	= 1.000 49 Ω
US international ohm	-	= 1.000 495 Ω
Electrical field strength		
-	1 V/mil	= $3.937\,008 \cdot 10^4$ V·m^{-1}
Electrical dipole moment		
Debye	1 D	= 10^{-18} Fr cm ≈ $3.335\,64 \cdot 10^{-30}$ C m
		(Fr = Franklin)
Radioactivity		
Curie	1 Ci	= 37 GBq
Röntgen	1 R	= $2.58 \cdot 10^{-4}$ C·kg^{-1}
	1 rem	= 10^{-2} Sv
	1 rad	= 0.01 J·kg^{-1} = 0.01 Gy

Table 15-6 Physical Terms

Concentrations

Concentrations measure the abundance of a substance. The following concentrations are in use:

Mass fraction ($w_A = m_A/\Sigma_i m_i$): mass of substance A per sum of masses m_i of all substances i. The value of 100 w_A is called weight percent (wt-%) and the value of $1000w_A$ = wt-‰. The Anglo-Saxon literature also uses part per million (1 ppm = 10^{-4} %), part per (American) billion (1 ppb = 10^{-7} %), and part per (American) trillion (1 ppt = 10^{-10} %).

Volume fraction ($\phi_A = V_A/\Sigma_i V_i$): volume of substance A per sum of volumes V_i of all substances i. The volumes V_A and V_i are the volumes of substances *before* the mixing process.

Mole fraction = amount fraction = number fraction ($x_A = n_A/\Sigma_i n_i$): amount of substance A per sum of amounts n_i of all substances i; it was called "mole number" in the literature.

Mass concentration = mass density (in polymer science as $c_A = m_A/V$): mass of substance A per volume V of mixture after the mixing. IUPAC recommends the symbols γ_B or ρ_B instead of c_B but ρ_B may be confused with the same symbol for the mass density of a *neat* substance.

Number concentration = number density of entities ($C_A = N_A/V$): number of entities A (molecules, atoms, ions, etc.) per volume V of mixture.

Amount-of-substance concentration = amount concentration, often only called **concentration** (in polymer science in general as $[A] = n_A/V$): amount of substance A per volume of mixture (IUPAC recommends $c_B = n_B/V$, which may be confused with the symbol c_B for the mass concentration). The amount concentration is often called molarity and given the symbol M; the latter symbol is not recommended by IUPAC and should not be used with SI prefixes.

Molality of a solute ($a = n_A/m_B$): amount n_A of substance A per mass m_B of solvent B. Molalities are often denoted by m, which should not be used as symbol for the unit mol kg^{-1}.

Ratios of physical quantities

The terms "normalized", "relative", "specific", and "reduced" are sometimes used with different meanings in the literature although they are clearly defined:

Normalized requires that the quantities in the numerator and the denominator are of the same kind. A normalized quantity is always a fraction (quantity of the subgroup divided by the quantity of the group); the sum of all normalized quantities is usually normalized to unity.

Relative also refers to quantities that are of the same kind in numerator and denominator but the quantity in the denominator may be in any defined state. Example 1: relative viscosity $\eta_i = \eta/\eta_1$ as the ratio of the viscosity η of a solution to the viscosity η_1 of the solvent 1. Example 2: relative humidity = ratio of water content of air to water content of air saturated with water (both at the same temperature and pressure).

Specific refers to a physical quantity divided by the mass. Example: specific heat capacity = heat capacity (in heat per temperature) divided by mass. The "specific viscosity" $(\eta - \eta_1)/\eta_1$ is not a specific quantity.

Reduced refers to a quantity that is divided by a specified other quantity. Example: reduced osmotic pressure Π/c = osmotic pressure Π divided by the mass concentration c.

15.2. Plastics Names and Standards

Several organizations have issued recommendations for the use of abbreviations and acronyms of thermoplastics, thermosets, fillers, fibers, elastomers, etc., for example:

ASTM = American Society for the Testing of Materials;
ANSI = American National Standards Institute;
DIN = German Industrial Standards;
ISO = International Standardization Organization;
IUPAC = International Union of Pure and Applied Chemistry ;
"Other" = abbreviations used but not recommended by ASTM, DIN, ISO, or IUPAC.

Abbreviations and acronyms used in this book are indicated by **bold letters**; an alphabetized list of these abbreviations and acronyms is in the front matter of this book.

15.2.1. Abbreviations and Acronyms

For plastics names, the following standards apply: D 1600-86a [ASTM/ANSI], 7728 (1988) [DIN], 1043-1: 1987(E) [ISO], and the list in Pure Appl.Chem. **59** (1987) 691 [IUPAC].

The recommended abbreviations and acronyms for plastics are often neither identical with those of fibers of the same chemical structures nor with those of elastomers containing the same monomeric units; some of these also deviate from those recommended for recyclable plastics (see Chapter 14). Several abbreviations have different meanings in either two or more of the ASTM, DIN, ISO, and IUPAC systems and/or the technical literature, respectively; symbols with more than one meaning are characterized by # in Tables 15-7 to 15-10.

Abbreviations and acronyms are in part based on the poly(monomer) nomenclature, i.e., on the names of the monomers used in the manufacture of polymers, sometimes however without a prefix "P" for "poly" (Table 15-7). The names of monomers for copolymers are given in alphabetical order without regard to their prevalence.

Other abbreviations and acronyms are based on characteristic groups of polymers (Table 15-8) or indicate polymers synthesized by chemical transformation of base polymers (Table 15-9). Special abbreviations apply to blends, reinforced polymers, etc. (Table 15-10).

Table 15-7 Polymers by chain polymerization (addition polymerization), copolymerization, polycondensation (condensation polymerization), and polyaddition, which are characterized by abbreviations based on the poly(monomer) nomenclature.
Registered trademarks: [a] SAN in Japan and the USA; [b] PAN in Europe.

Polymers of	ASTM	DIN	ISO	IUPAC	Other
Acrylester + acrylonitrile + butadiene	ABA	A/B/A	A/B/A		
Acrylester + acrylonitrile + styrene		**ASA**	**ASA**		
Acrylester + ethylene	EEA#		E/EA		
Acrylic acid	PAA				PAS#
Acrylonitrile	**PAN**	**PAN**		**PAN**	
Acrylonitrile + butadiene	PBAN				

Table 15-7 (continued)

Polymers of	ASTM	DIN	ISO	IUPAC	Other
Acrylonitrile + butadiene + methyl methacrylate + styrene					MABS
Acrylonitrile + butadiene + styrene	**ABS**	**ABS**	**ABS**	**ABS**	**ABS**
Acrylonitrile + chlorinated poly(ethylene) + styrene		A/PE-C/S	A/CPE/S		
Acrylonitrile + ethylene-propylene-diene + styrene		A/EPDM/S	A/EPDM?S		
Acrylonitrile + methyl methacrylate	AMMA	A/MMA	A/MMA		
Acrylonitrile + styrene	**SAN**[a)]	**SAN**	**SAN**	**SAN**	AS
Acrylonitrile + styrene + chlorinated poly(ethylene)		A/PE-C/S			
Adipic acid + hexamethylenediamine	**PA 6.6**	PA 66	PA 66	**PA 6.6**	
Adipic acid + hexamethylenediamine + sebacic acid			PA 66/610		
Adipic acid + tetramethylene glycol					PTMA
Allyl diglycol carbonate	ADC				
p-Aminobenzoic acid					PAB
Aminotriazol					PAT
11-Aminoundecanoic acid		**PA 11**	**PA 11**		
Azelaic acid anhydride					PAPA
Bisphenol A + phosgene	**PC**#	**PC**#	**PC**#	**PC**#	
Bitumen + ethylene				ECB	
Butadiene + methyl methacrylate + styrene		**MBS**	**MBS**		
Butadiene + styrene (thermoplastic)	PBS	S/B	S/B	S/B	**SB**
Butene-1	**PB**	**PB**	**PB**		BT
Butyl acrylate		PBA	PBA	PBA	
Butylene glycol + terephthalic acid	**PBT**	**PBT**	**PBT**		PTMT
Butyl methacrylate					PBMA
ε-Caprolactam	**PA 6**	**PA 6**	**PA 6**	**PA 6**	
ε-Caprolactam + ω-dodecanolactam			PA 6/12		
Chlorotrifluoroethylene	**PCTFE**	**PCTFE**	**PCTFE**	**PCTFE**	
Chlorotrifluoroethylene + ethylene					**ECTFE**
Cresol + formaldehyde	**CF**	**CF**	**CF**	**CF**	
1,4-Cyclohexanedimethylol + terephthalic acid					PCDT
Diallyl chloroendate (= diallyl ester of 4,5,6,7,7-hexachlorobicyclo-[2,2,1]-5-heptane-2,3-dicarbonic acid)	PDAC				
Diallyl fumarate	PDAF				
Diallyl isophthalate	PDAIP				
Diallyl maleate	PDAM				
Diallyl phthalate	**PDAP**	**PDAP**	**PDAP**	**PDAP**	DAP
1,4-Dichlorobenzene + disodium sulfide	**PPS**	**PPS**			
2,6-Dimethylphenol + oxygen	POP				PPO™
ω-Dodecanolactam (laurolactam)	**PA 12**	**PA 12**	**PA 12**	**PA 12**	
Ethyl acrylate					PEA

Table 15-7 (continued)

Polymers of	ASTM	DIN	ISO	IUPAC	Other
Ethyl acrylate + ethylene	**EEA#**	E/EA	E/EA		
Ethylene	**PE**	**PE**	**PE**	**PE**	PL
-, chlorinated polymer	CPE	**PE-C**	**PE-C**		
-, cross-linked polymer			**PE-X**		VPE
-, high density polymer	HDPE	**PE-HD**	**PE-HD**		
-, high-impact polymer			PE-HI		
-, linear low density polymer	LLDPE		**PE-LLD**		
-, low density polymer	LDPE	**PE-LD**	**PE-LD**		
-, medium density polymer	MDPE		PE-MD		
-, very low density linear polymer					VLDPE
-, ultrahigh molar mass polymer	UHMW-PE		**PE-UHMW**		
Ethylene + methacrylic acid	EMA				
Ethylene + methyl acrylate + vinyl chloride		VC/E/MA	VC/E/MA		
Ethylene + methyl methacrylate		E/MA	E/MA		
Ethylene + propylene		E/P	E/P		PEP
Ethylene + propylene (+ diene)		**EPDM**	**EPDM**		
Ethylene + tetrafluoroethylene	**ETFE**	E/TFE	E/TFE		
Ethylene + vinyl acetate	EVA	E/VA	E/VAC		**EVAC**
Ethylene + vinyl acetate + vinyl chloride		VC/E/VAC	VC/E/VAC		
Ethylene + vinyl chloride		VC/E	VC/E		
Ethylene glycol					PEG
Ethylene glycol + maleic anhydride	**UP**	**UP**	**UP**	**UP**	
Ethylene glycol + terephthalic acid(ester)	**PET**	**PET**	**PET**	PETP	PETE
-, ditto, with glycol comonomer	PETG				
-, fast crystallizing polymer					CPET
-, oriented polymer					OPET
Ethylene oxide	PEO	**PEOX**	**PEOX**	PEO	
Formaldehyde (or trioxane)	**POM#**	**POM#**	**POM#**	**POM#**	
Formaldehyde + furan	FF		FF		
Formaldehyde + melamine	**MF**	**MF**	**MF**	**MF**	
Formaldehyde + melamine + phenol		MPF			
Formaldehyde + phenol	**PF**	**PF**	**PF**	**PF**	
Formaldehyde + urea	**UF**	**UF**	**UF**	**UF**	
Furfural + Phenol	PFF				
Hexafluoropropylene + tetrafluoroethylene	**FEP**	**FEP**	**FEP**		
Hexamethylenediamine + sebacic acid	PA 6-10	PA 610	PA 610		**PA 6.10**
p-Hydroxybenzoic acid	POB				PHB#
3-Hydroxybutyric acid					PHB#
2-Hydroxyethylmethacrylate					PHEMA
Isobutene (isobutylene)	**PIB**	**PIB**	**PIB**	**PIB**	IM
Isocyanurates		PIR	PIR		
Laurolactam (dodecanolactam)	**PA 12**	**PA 12**	**PA 12**	**PA 12**	
Linseed oil, epoxidized		ELO	ELO		
Maleic anhydride + styrene	**SMA**	S/MA	**SMA**		
Methacrylimide		PMI	PMI		

Table 15-7 (continued)

Polymers of	ASTM	DIN	ISO	IUPAC	Other
Methyl acrylate					PMA
Methyl acrylate + vinyl chloride		VC/MA	VC/MA		
Methyl α-chloromethacrylate	PMCA				
Methyl methacrylate	**PMMA**	**PMMA**	**PMMA**		**PMMA**
Methyl methacrylate + vinyl chloride			VC/MMA		
4-Methylpentene-1	**PMP**	**PMP**	**PMP**		TPX
α-Methylstyrene		PMS	PMS		PAMS
α-Methylstyrene + styrene	SMS	S/MS	S/MS		
Octyl acrylate + vinyl chloride		VC/OA	VC/OA		
Perfluoro alkoxy alkane	**PFA**	**PFA**	**PFA**		
Phosphoric acid (polymer)					PPA[#]
Propylene	**PP**	**PP**	**PP**	**PP**	
-, oriented polymer					OPP
-, biaxially oriented polymer					BOPP
Propylene + tetrafluoroethylene (alt.)					TFE-P
Propylene oxide	PPOX	PPOX	PPOX		PPO
Soy bean oil, epoxidized		ESO	ESO		
Styrene	**PS**	**PS**	**PS**	**PS**	
-, expanded (foamed) polymer					EPS
-, high-impact polymer			**PS-HI**		HIPS
-, impact resistant polymer	SRP	IPS			
-, oriented polymer					OPS
-, biaxially oriented polymer					BPS
Tetrafluoroethylene	**PTFE**	**PTFE**	**PTFE**	**PTFE**	
Tetrahydrofuran					PTHF
-, polymer with hydroxy endgroups					PTMEG
Triallyl cyanurate	PTAC				
Trifluoroethylene				P3FE	
Trioxane (+ comonomers!)	**POM**[#]	**POM**[#]	**POM**[#]	**POM**[#]	
Vinyl acetate	**PVAC**	**PVAC**	**PVAC**	**PVAC**	PVA
Vinyl acetate + vinyl chloride	PVCA	VC/VAC	VC/VAC		
N-Vinylcarbazole	**PVK**	**PVK**	**PVK**		
Vinyl chloride	**PVC**	**PVC**	**PVC**	**PVC**	PCU; V
-, polymerized in bulk					M-PVC
-, polymerized in emulsion					E-PVC
-, polymerized in suspension					S-PVC
-, polymer as flexible film					FPVC
-, polymer as oriented film					OPVC
-, polymer as rigid film					RPVC
Vinyl chloride + vinylidene chloride		VC/VDC	VC/VDC		
Vinyl fluoride	**PVF**	**PVF**	**PVF**	**PVF**	
Vinylidene chloride	**PVDC**	**PVDC**	**PVDC**	**PVDC**	
Vinylidene fluoride	**PVDF**	**PVDF**	**PVDF**	**PVDF**	
Vinyl methyl ether					PVME
N-Vinylpyrrolidone	PVP	PVP	PVP		

Table 15-8 Abbreviations for polymers named after a characteristic polymer group.

Characteristic polymer group	ASTM	DIN	ISO	IUPAC	Other
Amide	**PA**	**PA**	**PA**		
Amide, aromatic (aryl amide)	PARA				**PAR**
-, metal coated films	PA**				
-, saran coated films	PA*				
Amide-imide	**PAI**	**PAI**	**PAI**		
Arylene sulfone					PAS#
Benzimidazole					**PBI**
Bisphenol A-sulfone					PSF
Carbodiimide					PCD
Carbonate, aromatic	**PC#**	**PC#**	**PC#**	**PC#**	
Epoxide (epoxy)	**EP**	**EP**	**EP**	**EP**	
-, glass fiber-reinforced					GEP
Ester, saturated (polyester)		SP	SP		
-, thermoplastic	TPES				
-, -, metallized polymer film					MPE
Ester, unsaturated (polyester)	**UP#**	**UP#**	**UP#**	**UP#**	
-, -, glass fiber-reinforced					FRP
Ester-alkyd	PAK				
Ester-ether (thermoplastic elastomer)	TEEE				
Ester-imide	**PEI#**		**PEI#**		
Ester-urethane			PEUR		
Ether-block-amide	**PEBA**	**PEBA**	**PEBA**		
Ether ether ketone	**PEEK**		**PEEK**		
Ether-imide		**PEI#**			
Ether-sulfone	**PES**	**PES**	**PES**		**PES**
	PESU				
Imide	**PI**	**PI**	**PI**		
Parabanic acid					PPA#
Phenylene ether	**PPE**	**PPE**	**PPE**		PPO™
Phenylene sulfide	**PPS**		**PPS**		
Phenylene sulfone	PPSU	PPSU	PPSU		PSU
Silicone (plastics)	SI	SI	SI		
Sulfone	PSU	PSU	PSU		PSU
Urethane	**PUR**	**PUR**	**PUR**	**PUR**	
-, thermoplastic	TPUR				TPU
-, thermoset	TSUR				

Table 15-9 Abbreviations for polymers produced by chemical transformation of other polymers.

Resulting polymer	ASTM	DIN	ISO	IUPAC	Other
Carboxymethyl cellulose	**CMC**	**CMC**	**CMC**	**CMC**	
Carboxymethyl hydroxyethyl cellulose					CMHEC
Casein, cross-linked with formaldehyde	**CS**	**CS**	**CS**	**CS**	

Table 15-9 (continued)

Resulting polymer	ASTM	DIN	ISO	IUPAC	Other
Cellulose, as Cellophan®	C				
-, ditto, coated with Saran	C*				
Cellulose acetate	**CA**	**CA**	**CA**	**CA**	
Cellulose acetobutyrate	**CAB**	**CAB**	**CAB**	**CAB**	
Cellulose acetopropionate	**CAP**	**CAP**	**CAP**	**CAP**	
Cellulose nitrate	**CN**	**CN**	**CN**	**CN**	NC
Cellulose plastics, unspecified	CE				
Cellulose propionate	**CP**	**CP**	**CP**	**CP**	
Cellulose triacetate	CTA	CTA	CTA		TA
Ethyl cellulose	**EC**	**EC**	**EC**	**EC**	
Hydroxyethyl cellulose					HEC
Hydroxypropyl cellulose					HPC
Hydroxypropyl methyl cellulose					HPMC
Methyl cellulose		**MC**	**MC**		
Poly(ethylene), chlorinated	CPE	**PE-C**	**PE-C**		
-, chlorosulfonated	CSM				CSR
Poly(ethylene-*co*-vinyl alcohol)		E/VAL	E/VAL		EVAL™ **EVOH**
Poly(propylene), chlorinated		PPC			
Poly(vinyl alcohol)	**PVAL**	**PVAL**	**PVAL**	**PVAL**	PVOH
Poly(vinyl butyral)	**PVB**	**PVB**	**PVB**		
Poly(vinyl chloride), chlorinated	CPVC	**PVC-C**	**PVC-C**		PC#, PeCe
Poly(vinyl formal)	PVFM	PVFM	PVFM	PVFM	

Table 15-10 Abbreviations for blends, reinforced polymers, etc.

ISO recommends for mixtures of polymers to put the symbols for the parent polymers separated by a plus sign in parentheses [for example, (PMMA + ABS) for a physical blend of poly(methyl methacrylate) and ABS].

ASTM recommends GP for "general purpose" and SS for "single stage".

Resulting polymer	ASTM	DIN	ISO	IUPAC	Other
Bulk molding compound					**BMC**
Elastomers, thermoplastic					
-, -, containing ester and ether groups	TEEE				
-, -, olefin-based	TEO				
-, -, styrene-based	TES				
Epoxide, glass fiber-reinforced			**EP-G**		GEP
Plastic, carbon fiber-reinforced		KEK			CFP
-, glass fiber-reinforced					GFK
-, man-made fiber-reinforced					CFK
-, metal fiber-reinforced					MFK

Table 15-10 (continued)

Resulting polymer	ASTM	DIN	ISO	IUPAC	Other
Poly(acrylonitrile-co-styrene) + chlorinated poly(ethylene)					ACS®
Poly(styrene), rubber-modified	SRP				
Sheet molding compound					SMC
-, with high glass fiber content					HMC

Table 15-11 ISO codes for data block 1 and corresponding DIN codes.

The code letters are arranged after the symbol for the base polymer according to ISO and DIN; in the technical literature, they are however commonly placed in front of the base symbol, following ASTM, which uses similar code letters and asterisks.

ASTM: * saran-coated polymer film;
** metallized polymer film.

Technical literature: BO = biaxially oriented and O = oriented.

Code	ISO data block 1	DIN
A	-	-
B	block copolymer	-
C	chlorinated	chlorinated
D	-	density
E	polymerization in emulsion	expanded or expandable
F	-	flexible or fluid (liquid state)
G	-	-
H	homopolymer	high
I	-	impact resistant
J	-	-
K	-	-
L	-	linear or low
M	bulk polymerization	medium or molecular
N	-	normal or novolac
O	-	-
P	plasticized	plasticized
Q	mixtures of different polymers	-
R	random copolymer	resol; formerly also: raised (enhanced)
S	polymerization in suspension	-
T	-	-
U	unplasticized	ultra or unplasticized
V	-	very
W	-	weight (mass)
X	no indication	cross-linked or cross-linkable
Y	-	-
Z	-	-

Table 15-12 ISO codes for items in data block 2.

Code	Position 1	Code	Positions 2-4
A	adhesives	A	antioxidant for processing
B	blow molding	B	antiblocking agent
C	calendering	C	colored
D	disc manufacture	D	powder (dry blend)
E	extrusion (pipes, profiles, sheets)	E	additive for expansion
F	extrusion (films, foils)	F	special burning characteristics
G	general purpose	G	granules (pellets)
H	coating	H	heat aging stabilizer
K	cable and wire insulation	K	-
L	extrusion (monofilament)	L	light and weather stabilizer
M	injection molding	M	-
N	-	N	natural (not colored)
P	paste resin	P	polymer modifier
Q	compression molding	Q	-
R	rotational molding	R	mold release agent
S	powder spray sintering	S	slip agent, lubricant
T	tape manufacture	T	improved transparency
W	-	W	stabilized against hydrolysis
X	no indication	X	-
Y	textile yarn	Y	-
Z	-	Z	antistatic agent

Table 15-13 ISO codes for extending and reinforcing fillers (data block 4).

Code	Position 1 (material)	Code	Position 2 (form or structure)
B	boron	B	balls, beads, spheres
C	carbon	C	chips, cuttings
D	-	D	powder (dry blend)
E	clay	E	-
F	-	F	fiber
G	glass	G	ground, grinding stock
H	-	H	whisker
K	calcium carbonate	K	knitted fabric
L	cellulose	L	layer
M	mineral, metal	M	mat (thick)
N	-	N	nonwoven (fabric, thin)
P	mica	P	paper
Q	silicon	Q	-
R	aramid	R	roving
S	synthetic, organic	S	scale, flake (spheres: now B)
T	talcum	T	cord
V	-	V	veneer
W	wood	W	woven fabric
X	not specified	X	not specified
Z	others	Z	others

15.2.2. ASTM Standards

Properties measured by ASTM standards are in general not comparable to those measured by ISO standards. ASTM properties are commonly given in Anglo-Saxon units.

F.T.Traceski, Specifications and Standards for Plastics and Composites, ASM International, Materials Park (OH) 1990

Table 15-14 Some ASTM standards.

CAMPUS	ASTM and other names	ASTM standard
	Properties of Plastics	
General properties		
Density	Specific gravity	D 792, D 941, D 1505
Optical properties		
	Color	D 1209
	Haze	D 1003
	Gloss	D 2457
	Transparency	D 1746
Refractive index	Refractive index	D 542
Mechanical properties		
Stress at yield	Yield strength, tensile yield strength	D 638, D 882, D 1708
Strain at yield	Yield elongation, tensile yield	D 638, D 882, D 1708
Tensile strength	Tensile strength at break, failure, or fracture; ultimate tensile strength	D 638, D 882, D 1708
Strain at break	Elongation at break, failure, fracture; ultimate elongation; extensibility	D 638, D 882, D 1708
Young's modulus	Tensile modulus, initial modulus	D 638, D 882
	Flexural modulus, bending modulus	D 747, D 790
	Flexural (bending) strength or stress	D 790
	Compressive modulus	D 695
	Compressive stress or strength	D 695
	Shear strength	D 732
	Torsional modulus	D 1043
	Deformation under load	D 621
Impact strength	Impact strength	D 256
	Izod impact strength	D 621
Notched impact strength	Notched Izod impact strength	D 256
	Tear strength; tear resistance; tear resistivity	D 470, D 624
	Fatigue resistance, flexing fatigue life, flex fatigure limit	D 761
	Split tear resistance	D 1938
	Hardness, Shore	D 695, D 2240
	Hardness, Rockwell	D 785
	Coefficient of friction	D 1894
	Abrasion resistance	D 1044

Table 15-14 (continued)

CAMPUS	ASTM and other names	ASTM standard
Thermal properties		
	Melting temperature (or point)	D 2117
	Glass (transition) temperature	D 846
	Brittleness temperature (under impact) low temperature embrittlement	D 746
Heat distortion temperature	Heat deflection temperature	D 648
Vicat temperature	Vicat softening temperature	D 1525
Thermal expansion coefficient	Coefficient of thermal expansion	D 696, E 831
	Thermal conductivity	C 177
Electrical properties		
Relative permittivity	Dielectric constant	D 150, D 1531
Dissipation factor	Dissipation factor, dielectric loss factor, loss factor	D 150, D 1531
	Power factor, loss tangent	D 150
Dielectric strength	Dielectric strength, breakdown str.	D 149
Comparative tracking index	Tracking resistance	D 2132
CTI 100 drops values		
Volume resistivity	Volume resistivity	D 150, D 257
Surface resistivity	Surface resistivity	D 257
	Surface arc resistance	D 495
	Corona endurance	D 2275
Electric corrosion		
Barrier properties		
Water absorption	Water absorption at saturation	D 570
	Water vapor transmission	D 570, E 96
	Gas permeability	D 1434
Moisture absorption		
Environmenatal properties		
Flammability UL 94		
	Properties of Melts	
Thermal properties		
Melt density	g/mL	
Melt thermal conductivity	$W\ m^{-1}\ K^{-1}$	
Melt specific heat capacity	$J\ kg^{-1}\ K^{-1}$	
Effective thermal diffusivity	m^2/s	
No-flow temperature	°C	
Freeze temperature	°C	
Rheological properties		
Melt volume rate	Melt-flow index, melt flow number	D 2116

16. Index

Entries are listed in strict alphabetical order; they may consist of a single word, abbreviations, acronyms, or combinations thereof. Qualifying letters and numbers as well as parentheses, brackets, and braces in names of chemical compounds such as 1-, 1,4-, α-, β-, *o*-, *m*-, *p*-, [L], ( ), [ ], etc., were disregarded for alphabetization.

The following abbreviations are used: def. = definition; eq. = equation; ff. = and following; M = molar mass; pm = polymerization; synth. = synthesis.